COMPARATIVE
PLANT VIROLOGY

SECOND EDITION

COMPARATIVE PLANT VIROLOGY

SECOND EDITION

ROGER HULL

Emeritus Fellow
Department of Disease and Stress Biology
John Innes Centre
Norwich, UK

ELSEVIER

AMSTERDAM • BOSTON • HEIDELBERG • LONDON
NEW YORK • OXFORD • PARIS • SAN DIEGO
SAN FRANCISCO • SINGAPORE • SYDNEY • TOKYO

Academic Press is an imprint of Elsevier

Cover Credits:

BSMV leaf — Mild stripe mosaic; Symptom of BSMV in barley. Image courtesy of A.O. Jackson.

BSMV genome: The infectious genome (BSMV) is divided between 3 species of positive sense ssRNA that are designated α, β, and γ. Image courtesy of Roger Hull.

BSMV particles. Image courtesy of Roger Hull.

Diagram showing systemic spread of silencing signal: The signal is generated in the initially infected cell (bottom, left hand) and spreads to about 10–15 adjacent cells where it is amplified. It moves out of the initially infected leaf via the phloem sieve tubes and then spreads throughout systemic leaves being amplified at various times. Image courtesy of Roger Hull.

Elsevier Academic Press

30 Corporate Drive, Suite 400, Burlington, MA 01803, USA
525 B Street, Suite 1900, San Diego, California 92101-4495, USA
84 Theobald's Road, London WC1X 8RR, UK

This book is printed on acid-free paper. ∞

Library of Congress Cataloging-in-Publication Data
Hull, Roger, 1937-
    Comparative plant virology / Roger Hull. –
2nd ed.
        p. cm.
    ISBN 978-0-12-374154-7 (hardcover : alk. paper) 1. Plant viruses. I. Title.
    QR351.H85 2009
    579.2′8–dc22

                                                                    2008040333

British Library Cataloguing in Publication Data
A catalogue record for this book is available from the British Library

ISBN 13: 978-0-12-374154-7

For all information on all Elsevier Academic Press publications
visit our Web site at www.elsevierdirect.com

Printed in China
09 10 9 8 7 6 5 4 3 2 1

Contents

Section II

WHAT IS A VIRUS MADE OF?

Chapter 5. Architecture and Assembly of Virus Particles

Section III

HOW DO PLANT VIRUSES WORK?

Section IV

PLANT VIRUSES IN AGRICULTURE AND INDUSTRY

Chapter 12. Plant-to-Plant Movement

Chapter 13. Plant Viruses in the Field: Diagnosis, Epidemiology, and Ecology

Chapter 14. Conventional Control

Chapter 15. Transgenic Plants and Viruses

Preface

This book has been developed from and is a revision to *Fundamentals of Plant Virology* written by R. E. F. Matthews in 1992. Since then major advances have been made in the understanding of the molecular biology of viruses, how they function and how they interact with their hosts. This has revealed similarities and differences between viruses infecting members of the different kingdoms of living organisms, plants, animals, fungi, and bacteria. In this changing environment of teaching virology, this book does not just deal with plant viruses alone but places them in context in relation to viruses of members of other kingdoms.

This book has been written for students of plant virology, plant pathology, virology, and microbiology who have no previous knowledge of plant viruses or of virology in general. An elementary knowledge of molecular biology is assumed, especially of the basic structures of DNAs, RNAs, and proteins, of the genetic code, and of the processes involved in protein synthesis. As some of these students may not have a grounding in the structure and function in plants including the main subcellular structures found in typical plant cells, these features

which are important to the understanding of how viruses interact with plants are illustrated. In each chapter there is a list of further reading to enable the student to explore specific topics in depth.

The fifteen chapters in this book can be divided into four major sections that form a logical progression in gaining an understanding of the subject. The first four chapters introduce plant viruses describing: what is a virus, giving an overview of plant viruses, discussing other agents that cause diseases that resemble plant virus diseases, and considering factors that are involved in virus evolution. The points raised in this latter chapter are equally relevant to viruses of other kingdoms. The next four chapters deal with what viruses are made of. The chapter on virus architecture and assembly is also very relevant to viruses of other kingdoms as are the major points raised in chapters on plant virus genome organization, genome expression, and genome replication. The next section on how do plant viruses work is more specific to plant viruses and highlights differences and similarities between virus interactions with plant, animal, and bacterial hosts.

These interactions are described at the plant level (movement of the virus within the plant and effects on plant metabolism) and at the molecular level including a chapter devoted to the newly understood host defence system of RNA silencing. The last four chapters deal with plant viruses and agriculture and industry. The description of how plant viruses move between hosts which often involves specific molecular interactions leads into discussion of the epidemiology of viruses in the field and how they are controlled. The last chapter is on the use of recombinant DNA technology in controlling viruses and also in using them commercially in, for instance, the pharmaceutical and nanotechnology industries.

A unique feature of this book is a series of "profiles" on 32 plant viruses that feature in the text. These profiles describe briefly the major properties of the viruses including their taxonomic position, their biology, their particles, and their genomes. References are given to enable students to acquire even more information on these targeted viruses.

I am very grateful to a large number of colleagues for their helpful discussion on various topics and in providing material prior to publication. I am especially indebted to John Carr, Andy Jackson, Mark Stevens, and Peter Waterhouse for their helpful comments on various sections of the book and on providing illustrative material. My eternal gratitude goes to my wife who has tolerated "piles of paper" around the house and who has given me continuous encouragement.

Roger Hull
Norwich, UK
July, 2008

List of Abbreviations

3'OH	3' hydroxyl group
Å	Angström (10^{-10} meter)
AAB	Association of Applied Biologists
DPV	Descriptions of Plant Viruses
ADP	Adenosine diphosphate
AR	Aberrante ratio
ATP	Adenosine triphosphate
cDNA	Complentary (or copy) DNA
CI	Cylindrical inclusion
DdDp	DNA-dependent DNA polymerase
D RNA/ DNA	Defective RNA or DNA
DI RNA/ DNA	Defective interfering RNA or DNA
dsDNA	Double-stranded DNA
dsRNA	Double-stranded RNA
EDTA	Ethylene diamino tetra-acetic acid
ELISA	Enzyme-linked inmuno-sorbent assay
EM	Electron Microscope
ER	Endoplasmic reticulum
GTP	Guanosine triphosphate
GM	Genetically modified
HC-Pro	Helper component protease
HEL	Helicase
HR	Hypersensitive response
HSP	Heat-shock protein
ICR	Inter-cistronic region
ICTV	International Committee on the Taxonomy of Viruses
IRES	Internal ribosome entry site
ISEM	Immunoabsorbent electron microscopy
kb	Kilobase
kDa	Kilodalton
LRR	Leucine-rich repeat
Mab	Monoclonal antibody
MP	Movement protein
mRNA	Messenger RNA
MiRNA	MicroRNA
MTR	Methyl transferase
MW	Molecular weight
NBS	Nucleotide binding site
NI	Nuclear inclusion
nm	Nanometer
ORF	Open reading frame
PCD	Programmed cell death
PCR	Polymerase chain reaction (IC-PCR, immune-capture PCR; RT-PCR, reverse transcription PCR)
PDR	Pathogen-derived resistance

Pol	Polymerase	**SDS-**	Sodium dodecyl sulphate
PR	Pathogenesis-related (protein)	**PAGE**	polyacrylamide gel
(protein)			electrophoresis
PRO	Protease	**SEL**	Size exclusion limit
PTGS	Post-transcriptional gene	**Sg RNA**	Subgenomic RNA
	silencing	**Si RNA**	Small interfering RNA
Rb	Retinoblastoma (protein)	**ssDNA**	Single-stranded DNA
(protein)		**SSEM**	Serologically-specific electron
RdRp	RNA-directed RNA Polymerase		microscopy
RF	Replicative form	**ssRNA**	Single-stranded RNA
RFLP	Restriction fragment length	**TAV**	Transactivator
	polymorphism	**TGB**	Triple gene block
RI	Replicative intermediate	**TGS**	Transcriptional gene silencing
RISC	RNA-induced silencing complex	**tRNA**	Transfer RNA
RNAi	RNA interfering	**UTR**	Untranslated region
RT	Reverse transcriptase	**VIGS**	Virus-induced gene silencing
SA	Salicylic acid	**VPg**	Virus protein genome-linked
SAR	Systemic acquired resistance		

INTRODUCTION TO PLANT VIRUSES

1

What Is a Virus?

This chapter discusses broad aspects of virology and highlights how plant viruses have led the subject of virology in many aspects.

I. INTRODUCTION

Plant viruses are widespread and economically important plant pathogens. Virtually all plants that humans grow for food, feed, and fiber are affected by at least one virus. It is the viruses of cultivated crops that have been most studied because of the financial implications of the losses they incur. However, it is also important to recognise that many "wild" plants are also hosts to viruses. Although plant viruses do not have an immediate impact on humans to the extent that human viruses do, the damage they do to food supplies has a significant indirect effect. The study of plant viruses has led the overall understanding of viruses in many aspects.

II. HISTORY

Although many early written and pictorial records of diseases caused by plant viruses are available, they are do not go back as far as records of human viruses. The earliest known written record of what was very likely a plant virus disease is a Japanese poem that was written by the Empress Koken in A.D. 752 and translated by T. Inouye:

> In this village
> It looks as if frosting continuously
> For, the plant I saw
> In the field of summer
> The colour of the leaves were yellowing

The plant, which has since been identified as *Eupatorium lindleyanum*, has been found to be susceptible to *Tobacco leaf curl virus*, which causes a yellowing disease.

In Western Europe in the period from about 1600 to 1660, many paintings and drawings were made of tulips that demonstrate flower symptoms of virus disease. These are recorded in the Herbals of the time and some of the earliest in the still-life paintings of artists such as Ambrosius Bosschaert. During this period, blooms featuring such striped patterns were prized as special varieties, leading to the phenomenon of "tulipomania" (Box 1.1).

Because of their small genomes, viruses have played a major role in elucidating many of the concepts in molecular biology, and the study of plant viruses has produced several of the major findings for virology in general. The major steps in reaching the current understanding of viruses are shown in the timeline in Figure 1.1.

Details of these "breakthroughs" can be found in Hull (2002; plant viruses), Fenner, (2008; vertebrate viruses), and Ackermann (2008; bacterial viruses). Plant viruses played a major role in determining exactly what a virus was. In the latter part of the nineteenth century, the idea that infectious disease was caused by microbes was well established, and filters were available that would not allow the known bacterial pathogens to pass through. In 1886, Mayer (see Figure 1.2A) described a disease of tobacco that he called *Mosaikkrankheit*, which is now known to be caused by the *Tobacco mosaic virus* (TMV). Mayer demonstrated that the disease could be transmitted to healthy plants by inoculation with extracts from diseased plants. A major observation was made in 1892 by Iwanowski, who showed that sap from tobacco plants displaying the disease described by Mayer was still infective after it had been passed through a bacteria-proof filter candle. However, based on previous studies, it was thought that this agent was a toxin. Iwanowski's experiment was repeated in 1898 by Beijerinck (see Figure 1.2B), who showed that the agent multiplied in infected tissue and called it *contagium vivum fluidum* (Latin for "contagious living fluid") to distinguish it from contagious corpuscular agents (Figure 1.2C).

Beijerinck and other scientists used the term *virus* to describe the causative agents of such transmissible diseases to contrast them with bacteria. The term *virus* had been used more or less synonymously with **bacteria** by earlier workers, but as more diseases of this sort were discovered, the unknown causative agents came to be called "filterable viruses." Similar properties were soon after reported for some viruses of animals (e.g., the filterable nature of

---

BOX 1.1

TULIPOMANIA

Tulips were introduced into the Netherlands in the late sixteenth century. Bulbs that produced "broken-coloured" flowers were in great demand and created a rapidly expanding market, leading to hyperinflation.

(continued)

BOX 1.1 *(continued)*

Semper Augustus tulip with flower colour break (one of the most favoured varieties)

One bulb cost 1,000 Dutch florins (guilders) in 1623, and by 1635, 6,000 florins. To understand the value of this, one Viceroy tulip bulb was exchanged for goods that were valued at almost 2,400 florins:

4 tons of wheat (448 florins)	4 barrels of beer (3 florins)
8 tons of rye (558 florins)	2 barrels of butter (192 florins)
4 fat oxen (480 florins)	1,000 lbs cheese (120 florins)
8 fat pigs (240 florins)	1 bed with accessories (100 florins)
12 fat sheep (120 florins)	1 silver goblet (60 florins)
2 hogsheads of wine (70 florins)	

By 1636 there was much speculation, and futures were being taken out on these bulbs. In early 1637 one bulb was valued at 10,000 florins, but a few weeks later, the bubble burst and many people were left bankrupt. It was not until the 1920s that the viral aetiology of tulip flower breaking was discovered and that the symptoms were caused by an aphid-transmitted potyvirus. Today, 100 florins is equivalent to about U.S. $30,000.

Plant	Animal	Bacteria
Prehistory		
752 AD Plant virus in Japanese poem	1350 BC Smallpox recorded in Egypt	
1600-1637 Tulipomania	1796 Jenner developed smallpox vaccine	
Recognition of viral entity		
1886 Meyer Transmission of TMV		
1892 Iwanowski Filterability of TMV		
1898 Beijerink Viruses as an entity	1898 Filterability of PV and FMDV	
		1915. Filterability of phage
Biological age		
1900-1935 Descriptions of many viruses	1900– Descriptions of many viruses	1915- Descriptions of many viruses
	1901 Mosquito transmission of YFV	Early 1920s Infection cycle understood
Biophysical/biochemical age		
1935 Purification of TMV		
1936 TMV contains pentose nucleic acid		
1939 EM TMV rod-shaped particles	1940 VACV contains DNA	
	1949 PV grown in cultured cells	1940-1970 Phage genetics
1951 TYMV RNA in protein shell		
1956 Virus particles made of identical protein subunit		
1955/56 Infectious nature of TMV RNA		
1962 Structure of isometric particles		
1983 Structure of TBSV to 2.9Å	1985 Structure of poliovirus to 2.9Å	
Molecular age		
1960 Sequence of TMV coat protein	1979 Sequence of PV VPg	
	1970 Recognition of reverse transcriptase	
	1981 Infectious transcript of PV	1978 Infectious transcript of Qβ
1980 Sequence of CaMV DNA genome		
1982 Sequence of TMV RNA genome	1981 Sequence of poliovirus RNA genome	
1984 Infectious transcripts of multicomponent BMV		
1986 Transgenic protection of plants against TMV		
1996 Recognition of RNA silencing		
1997 Recognition of virus suppressors of silencing		

Abbreviations: BMV, *Brome mosaic virus*; CaMV, *Cauliflower mosaic virus*; FMDV, *Foot and mouth disease virus*; PV, *Poliovirus*; TBSV, *Tomato bushy stunt virus*; TMV, *Tobacco mosaic virus*; TYMV, *Turnip yellow mosaic virus*; VACV, *Vaccinia virus*; YFV, *Yellow fever virus*.

FIGURE 1.1　Timeline of development of virology.

FIGURE 1.2 A. Adolf Eduard Mayer (1843–1942); B. Martinus Willem Beijerinck (1851–1931); C. Page from lab journal of W.M. Beijerinck from 1898 relating to TMV. A and B courtesy of the historical collection, Agricultural University, Wageningen, Netherlands; C. (© Kluyver Institute) Courtesy Curator Kluyver Laboratory Collection, Delft School of Microbiology Archive, Delft University of Technology.

A   B

C

the agent causing foot and mouth disease in 1898) and of bacteria in 1915. Over the course of time, the word *filterable* has been dropped, leaving just the term *virus*.

As shown in the timeline in Figure 1.1, in the subsequent development of virology, many of the studies ran in parallel for viruses of plants, vertebrates, invertebrates, and bacteria. In fact, when viewed overall, there is evidence of much cross-feeding between the various branches of virology. However, there were differences mainly due to the interactions that these viruses have with their hosts. For instance, vertebrates produce antibodies that counter viruses, whereas plants, invertebrates, and bacteria do not. Another factor that has contributed to advances is the simplicity of the system exemplified by studies on bacteriophage being linked to studies on bacterial genetics.

The development of plant, and other, virology can be considered to have gone through five major (overlapping) ages. The first two, Prehistory and Recognition of viral entity, were just described. After these two came the Biological age, between 1900 and 1935, when it was determined that plant viruses were transmitted by insects and that some of these viruses multiplied in, and thus were pathogens of, insects in a manner similar to some viruses of vertebrates. One of the constraints to plant virology was the lack of a quantitative assay, until Holmes in 1929 showed that local lesions produced in some hosts after mechanical inoculation could be used for the rapid quantitative assay of infective virus. This technique enabled properties of viruses to be studied much more readily and paved the way for the isolation and purification of viruses a few years later.

The Biochemical/Physical age started in the early 1930s. The high concentration at which certain viruses occur in infected plants and their relative stability was crucial in the first isolation and chemical characterisation of viruses because methods for extracting and purifying proteins were not highly developed.

In 1935, Stanley announced the isolation of this virus in an apparently crystalline state but considered that the virus was a globulin containing no phosphorus. In 1936, however, Bawden and his colleagues described the isolation from TMV-infected plants of a liquid crystalline nucleoprotein containing nucleic acid of the pentose type. Around 1950, Markham and Smith showed that the RNA of *Turnip yellow mosaic virus* was encapsidated in a protein shell and was important for biological activity. This led to the classic experiments of Gierer, Schramm, Fraenkel-Conrat, and Williams in the mid-1950s that demonstrated the infectivity of naked TMV RNA and the protective role of the protein coat.

In parallel with these biochemical studies, physical studies in the late 1930s using X-ray analysis and electron microscopy confirmed that TMV had rod-shaped particles and obtained accurate estimates of the size of the rods. Attention turned to the structure of these particles, and in 1956, Crick and Watson suggested that the protein coats of small viruses are made up of numerous identical subunits arrayed either as helical rods or as a spherical shell with cubic symmetry. This led to Caspar and Klug (1962) formulating a general theory that delimited the possible numbers and arrangements of the protein subunits forming the shells of the smaller isodiametric viruses (see Chapter 5). Our recent knowledge of the larger viruses with more complex symmetries and structures has come from electron microscopy using negative-staining and ultrathin-sectioning methods.

The current Molecular age started in about 1960 when the full sequence of 158 amino acids in the coat protein of TMV was determined. The sequence of many naturally occurring strains and artificially induced mutants was also determined at about the same time. This work made an important contribution to establishing the universal nature of the genetic code and to our understanding of the chemical basis of mutation. This age continued with the sequencing of representatives of most, if not

III. DEFINITION OF A VIRUS

all, virus genera leading to a greater understanding of how viruses function and interact with their hosts. The results from these studies are described in detail in this book and in the suggested further reading.

III. DEFINITION OF A VIRUS

A. How Viruses Differ from Other Plant Pathogens

In the size of their nucleic acids, viruses range from a monocistronic mRNA in the satellite virus of tobacco necrosis virus (STNV) to a genome larger than that of the smallest cells (Figure 1.3). A biologically more meaningful way of comparing genome sizes is to consider the information content—that is, the number of genes that they contain; some examples are given in Table 1.1. Before attempting to define what viruses are, we must consider briefly how they differ from other entities such as cellular parasites, plasmids, and transposable genetic elements. The three simplest kinds of parasitic cells are the Mycoplasmas, the *Rickettsiae*, and the *Chlamydiae*.

Mycoplasmas and related organisms are not visible by light microscopy. They are 150–300 nm in diameter with a bilayer membrane but no cell wall, and they contain RNA along with ribosomes and DNA. They replicate by binary fission, and some that infect vertebrates can be grown *in vitro*. Their growth is inhibited by certain antibiotics. Some mycoplasmas are plant pathogenic (see Chapter 3).

The **Rickettsiae**, for example, the agent of typhus fever, are small, nonmotile bacteria, usually about 300 nm in diameter. They have a cell wall, plasma membrane, and cytoplasm with ribosomes and DNA strands. They are obligate parasites and were once thought to be related to viruses, but they are definitely cells because they multiply by binary fission, and they contain enzymes for ATP production.

The **Chlamydiae**, for example, the agent that causes psittacosis, include the simplest known type of cell. They are obligate parasites that grow by infecting eukaryotic cells and lack an energy-generating system. They are as small as, or smaller than, many viruses. *Chlamydiae* have two phases to their life cycle. Inside host cells they take on an *intracellular* replicative form (termed the *reticulate body*) and rely on the host cell energy-yielding system; outside the cell they survive by forming infectious *elementary bodies* about 300 nm in diameter, which is smaller than some pox viruses. *Chlamydiae* can be grown only where their host cells grow and cannot be propagated in bacterial culture media.

Several criteria *do* and *do not* distinguish all viruses from all cells (see Table 1.2).

Plasmids are autonomous extrachromosomal genetic elements found in many kinds of bacteria. They consist of closed circular DNA. Some can become integrated into the host chromosome and replicate with it. Some viruses that infect prokaryotes have properties like those of plasmids and, in particular, the ability to integrate into the host cell chromosome. However, viruses differ from plasmids in the following ways:

1. Normal viruses have a particle with a structure designed to protect the genetic material in the extracellular environment and to facilitate entry into a new host cell.
2. Virus genomes are highly organised for specific virus functions of no known value to the host cell, whereas plasmids consist of genetic material that is often useful for the survival of the cell.
3. Viruses can kill cells or cause disease in the host organism, but plasmids cannot.

Transposons, or mobile genetic elements (sometimes called "jumping genes"), are sequences of DNA that can move around to different positions within the genome of a single cell, a process termed *transposition*. Two types of mobile genetic elements exist, based on their

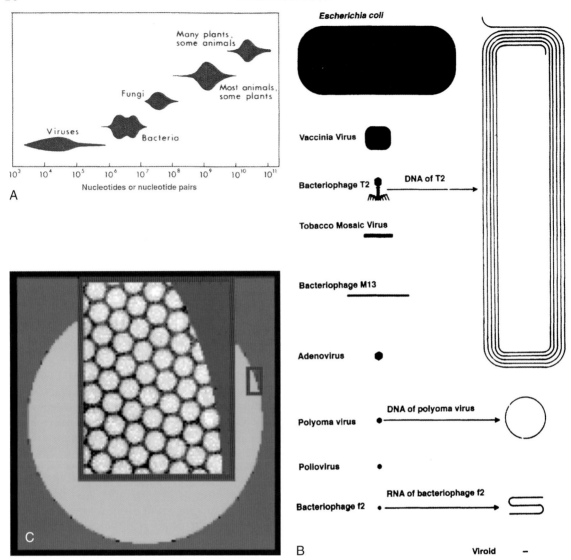

FIGURE 1.3 Size comparison of different organisms. A. Organisms classified according to genome size. The vertical axis gives an approximate indication of numbers of species within the size range of each group. B. Size comparison among a bacterium, several viruses, and a viroid. C. Comparison of size of rhinovirus and a pinhead. A. Modified from Hinegardner [1976; in *Molecular Evolution*, (F.J. Ayala, Ed.), pp. 179-199, Sinauer, Sunderland, MA]; B. With kind permission from Springer Science + Business Media: *Arch. Virol.*, Interference between proflavine treated reovirus and related and unrelated viruses, vol. 15, 1965, pp. 200–2009, E. Zalan; *Arch. Virol.*, Die Interferenz zwischen dem Polyoma-virus und dem Stomatitis-vesicularis-Virus in der Maus, vol. 15, 1965, pp. 210-219, D. Falke; *Arch. Virol.*, Properties of a new attenuated type 3 polio-virus, vol. 15, 1965, pp. 220-233, J. Šimon. C. http://web.uct.ac.za/depts/mmi/stannard/linda.html.

mechanism of transposition. Class I mobile genetic elements, or *retrotransposons*, move in the genome by being *transcribed* to RNA and then back to DNA by *reverse transcriptase*. Class II mobile genetic elements move directly from one position to another within the genome using a *transposase* to "cut and paste" them within the genome. In many properties, retrotransposons

TABLE 1.1   Information Content of Genomes of Various Organisms

Type of Organism	Example	Size of Genome	Number of Genes (Open Reading Frames)
Higher plant	Rice	3.9×10^8 kbp	>37,000
Vertebrate	Human	3.3×10^9 kbp	20,000–25,000
Invertebrate	Drosophila	1.2×10^8 kbp	~13,400
	Yeast	1.2×10^7 kbp	~5,770
Eubacteria	*Escherichia coli*	4.6×10^6 kbp	4,377
Mycoplasma	*Mycoplasma genitalium*	5.8×10^5 kbp	485
Large virus infecting vertebrates	*Vaccinia virus*	190 kbp	~250
Large virus infecting chlorella-like algae	*Paramecium bursarum Chlorella virus 1*	330 kbp	697
Large virus infecting invertebrates	*Autographa californica multiple nucleopolyhedrosis*	133.9 kbp	~150
Small virus infecting angiosperms	*Tobacco mosaic virus*	6395 nt	4
Smallest known virus	*Tobacco necrosis satellite virus*	1239 nt	1

TABLE 1.2   Distinguishing Criteria for Viruses

Criteria That Distinguish Viruses from Cells	Criteria That Do Not Distinguish Viruses from Cells
1. Lack of continuous membrane separating virus from host during replication	1. Size
2. Absence of protein-synthesising system	2. Nature and size of genome
3. Contain either DNA or RNA	3. Contain both DNA and RNA
4. Replication is by synthesis of a pool of components and not by binary fission	4. Absence of rigid cell envelope
	5. Obligate cell parasitism
	6. Absence of energy-yielding system
	7. Complete dependence on host cell for amino acids

resemble retroviruses, and they are classified as Metaviruses and Pseudoviruses. However, there is debate as to whether these are really viruses in the strictest sense. We can now define a virus, as shown in Box 1.2.

To be identified positively as a virus, an agent must normally be shown to be transmissible and to cause disease in at least one host. One of the basic tenets of pathology is that to prove that a disease is caused by a certain infectious agent, one must fulfill Koch's postulates, which were devised for bacteria; modifications of the postulates have been suggested to account for specific properties of viruses (Table 1.3). Today, however, it is not always possible to fulfill these postulates for viruses. For instance, plant cryptoviruses rarely cause detectable disease and are not transmissible by any mechanism except through seeds or pollen. Usually, it is satisfactory to show a clear association of the viral genome sequence with the disease after eliminating the possibility of joint infection with another virus.

BOX 1.2

DEFINITION OF A VIRUS

A *virus* is a set of one or more nucleic acid template molecules, normally encased in a protective coat or coats of protein or lipoprotein, that is able to organise its own replication only within suitable host cells. Within such cells, virus replication is (1) dependent on the host's protein-synthesising machinery, (2) organised from pools of the required materials rather than by binary fission, (3) located at sites that are not separated from the host cell contents by a lipoprotein bilayer membrane, and (4) continually giving rise to variants through several kinds of change in the viral nucleic acid.

TABLE 1.3   Koch's Postulates for Bacteria and Viruses

Bacteria	Viruses[a]
1. Demonstrate that the agent is regularly found in the diseased host	1. Isolation of virus from diseased host
2. Cultivate the agent on a suitable medium	2. Cultivate virus in experimental host or host cells
3. Reproduce the disease in the host by reintroducing the cultured agent	3. Prove lack of larger pathogens
4. Reisolate the agent from the artificially infected host	4. Produce comparable disease in original host species or in related ones
	5. Reisolate the virus

[a]Rivers (1937).

The structure and replication of viruses have the following features.

1. The infectious nucleic acid may be DNA or RNA (but never both) and be single- or double-stranded. If the nucleic acid is single-stranded it may be of positive or negative sense. (Positive sense has the sequence that would be used as an mRNA for translation to give a virus-coded protein.)

2. The mature virus particle may contain polynucleotides other than the genomic nucleic acid.

3. Where the genetic material consists of more than one nucleic acid molecule, each may be housed in a separate particle or all may be located in one particle.

4. The genomes of viruses vary widely in size, encoding between 1 and about 250 proteins. Plant viral genomes are at the small end of this range, mostly encoding between 1 and 12 proteins. The plant virus-coded proteins may have functions in virus replication, in virus movement from cell to cell, in virus structure, and in transmission by invertebrates or fungi. Animal and bacterial viruses may contain more genes associated with their interactions with their hosts.

5. Viruses undergo genetic change. Point mutations occur with high frequency as a result of nucleotide changes brought about by errors in the copying process during genome replication. Other kinds of genetic change may be due to recombination, reassortment of genome pieces, loss of genetic material, or acquisition of nucleotide sequences from unrelated viruses or the host genome.

6. Enzymes specified by the viral genome may be present in the virus particle. Most of these enzymes are concerned with nucleic acid synthesis.
7. Replication of many viruses takes place in distinctive virus-induced structures in the cell.
8. Some viruses share with certain nonviral nucleic acid molecules the property of integration into host-cell genomes and translocation from one integration site to another.
9. A few viruses require the presence of another virus for their replication.

B. Are Viruses Alive?

This question is asked very frequently. The definitions of a *living organism* vary widely, with the most accepted one being "A living organism has cellular structure and is manifest by growth through metabolism, reproduction, and the power of adaptation to the environment through changes that originate internally." While viruses reproduce and adapt, they are not cellular and do not metabolise; they rely on their host cell metabolism. Thus, technically they are not living organisms and the term *virus life cycle* should not be used; *virus replication cycle* describes the making of a new virus particle from an input particle.

IV. CLASSIFICATION AND NOMENCLATURE OF VIRUSES

In all studies of natural objects, humans seem to have an innate desire to name and to classify everything. It has been said that taxonomy is "the craft of making dead things look alive." Virologists are no exception. Virus classification, as with all other classifications, arranges objects with similar properties into groups, and even

though this may be a totally artificial and human-driven activity without any natural base, it does have certain properties:

- It gives a structured arrangement of the organisms so that the human mind can comprehend them more easily.
- It helps with communication among virologists and between virologists and other interested parties.
- It enables properties of new viruses to be predicted.
- It could reveal possible evolutionary relationships.

In theory, it is possible to consider the problems of naming and classifying viruses as separate issues. In practice, however, naming soon comes to involve classification.

From the 1930s to 1960s, various classification systems were proposed for plant (and other) viruses. This led to much confusion, and at the International Congress for Microbiology, held in Moscow in 1966, the first meeting of the International Committee for the Nomenclature of Viruses was held. An organisation was set up for developing an internationally accepted taxonomy and nomenclature for all viruses. Rules for the nomenclature of viruses were laid down. This committee developed into the International Committee for Taxonomy of Viruses (ICTV), which has since produced eight reports, the most recent being Fauquet *et al.* (2005). These reports give the definitive descriptions of the various taxa of viruses.

A. Virus Classification

A detailed list of the criteria used for virus classification and taxonomy is given in Murphy *et al.* (1995). The criteria come under four major headings: virion properties, such as size and shape, type of genome, properties of proteins; genome organisation and replication; antigenic properties; and biological properties, such as host range, pathogenicity, and transmission.

A problem arises as to how much weight is put onto each character. In practice, the nature and sequence of the genomic nucleic acid are the major characters that are used, but other properties, such as particle shape and composition, antigenic relationships, and biology, are also considered to be important. Any classification of viruses should be based not only on evolutionary history, as far as this can be determined from the genotype, but should also be useful in a practical sense. Most of the phenotypic characters used today in virus classification will remain important even when the nucleotide sequences of most viral genomes have been determined.

B. Families, Genera, and Species

The main building block of a biological classification is the species. In day-to-day practice, virologists use the concept of a "virus" as being a group of fairly closely related strains, variants, or pathovars. A virus defined in this way is essentially a species in the sense suggested for angiosperms and defined by the ICTV. In 1991, the ICTV accepted the concept that viruses exist as species, adopting the following definition:

> A viral species is a polythetic class of viruses that constitutes a replicating lineage and occupies a particular ecological niche. [**Polythetic** denotes a taxonomic group classified on the basis of several characters, as opposed to a monothetic group.]

The species has formed the basis of modern virus classification being established in subsequent ICTV reports, especially the seventh and eighth, in which a List of Species-Demarcating Criteria is provided for each genus. This enables viruses to be differentiated as species and tentative species, which are viruses that have not yet been sufficiently characterised to ensure that they are distinct and not strains of an existing virus or do not have the full characteristics of the genus to which they have been assigned. Of the 1,037 plant viruses listed in the eighth ICTV report, 751 are true species and 286 are tentative species. Further studies will provide enough data to classify the tentative species. A common problem is determining whether a new virus is truly a new species or a strain of an existing species. Conversely, what was considered to be a strain may, on further investigation, turn out to be a distinct species. This is due to the population structure of viruses that, because of continuous production of errors in replication, can be considered a collection of quasi-species. The concept of quasi-species is discussed in more detail following.

With the species forming the basis of the classification system, they can be grouped into other taxa on various criteria. To date, the taxonomic levels of *order, family,* and *genus* have been defined by the ICTV, and it is likely that there will be pressure for further higher and intermediate taxa. No formal definition for a *genus* exists, but it is usually considered "a population of virus species that share common characteristics and are different from other populations of species." Currently, 80 genera of plant viruses are recognised. In some cases—such as the *Rhabdoviridae*—numerous viruses are recognised that obviously belong to that family but for which there is not enough information to place them either in existing genera or for creating new genera; these viruses are listed as "unassigned." Genera are named either after the type species—for example, *Caulimovirus* after *Cauliflower mosaic virus*—or are given a descriptive name, often from a Greek or Latin word, for a major feature of the genus—for example, *Closterovirus*, from the Greek κλωστηρ (kloster), which is a spindle or thread, or that describes the virus particle shape, such as *Geminivirus*, from the Latin *geminus*, meaning "twins."

Similarly, genera are grouped together into families on common characteristics (Table 1.4). There are 17 families recognised for plant viruses; some, such as *Reoviridae* and *Rhabdoviridae*, are in common with animal virus families.

TABLE 1.4   Criteria Demarcating Different
Virus Taxa

I Order

Common properties between several families including:

Biochemical composition

Virus replication strategy

Particle structure (to some extent)

General genome organisation

II Family

Common properties between several genera including:

Biochemical composition

Virus replication strategy

Nature of particle structure

Genome organisation

III Genus

Common properties with a genus including:

Virus replication strategy

Genome size, organisation, and/or number of
segments

Sequence homologies (hybridisation properties)

Vector transmission

IV Species

Common properties within a species including:

Genome arrangement

Sequence homologies (hybridisation properties)

Serological relationships

Vector transmission

Host range

Pathogenicity

Tissue tropism

Geographical distribution

Seventeen of the genera have not yet been assigned to families and are termed "floating genera." The acquisition of further data on these floating genera, together with changing attitudes on virus classification, will no doubt lead to the designation of further plant virus families. The family is either named after the type member genus—for example, *Caulimoviridae*, named after the genus *Caulimovirus*—or given a descriptive name, as with the genus, for a major feature of the family—for example, *Geminiviridae*, which describes the virus particles.

Only three orders have been accepted thus far by the ICTV. The *Mononegavirales* contains, among other families, the *Rhabdoviridae*, which contains two plant virus families. In practice, genome nucleic acid sequence data are increasingly being used to delimit genera, species, and strains (Figure 1.4). A detailed discussion of virus classification, the currently accepted taxa, and how the ICTV operates are provided in Fauquet *et al.* (2005).

C. Naming Viruses (Species)

Questions of virus nomenclature have generated more heat over the years than the much more practically important problems of how to delineate distinct virus species. When a family or genus is approved by the ICTV, a type species is designated. Some virologists favour using the English vernacular name as the official species name. Using part of a widely known vernacular name as the official species name may frequently be a very suitable solution, but it could not always apply (e.g., with newly discovered viruses). Other virologists favour serial numbering for viruses (species). The experience of other groups of microbiologists is that, although numbering or lettering systems are easy to set up in the first instance, they lead to chaos as time passes and changes must be made in taxonomic groupings. The idea of Latinized binomial names for viruses was supported by the ICTV for many years but never implemented for any viruses.

In successive editions of the ICTV reports, virus names in the index have been listed by the vernacular name (usually English) followed by the family or genus name—for example,

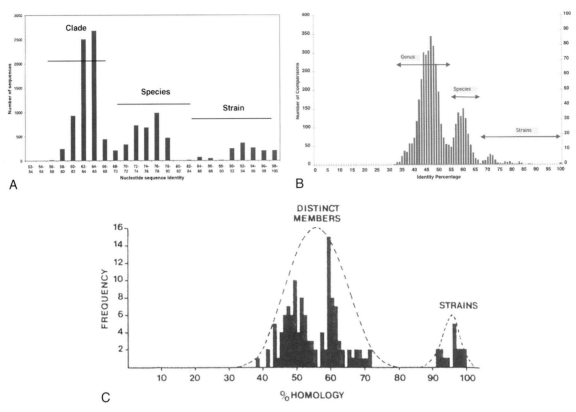

FIGURE 1.4 Differentiation of taxa by pairwise identities of sequences of variants of A. RT/RNaseH nucleotide sequences of *Banana streak virus* isolates; B. Nucleic acid sequences of the L1 gene of members of the Family *Papillomaviridae*; C. Amino acid sequences of coat proteins of potyviruses. A. With kind permission from Springer Science+ Business Media: *Arch. Virol.*, The diversity of *Banana streak virus* isolates in Uganda, vol. 150, 2005, pp. 2407-2420, G. Harper; B. From Virus Taxonomy, 8th Report of the National Committee on the Taxonomy of Viruses, Fauquet *et al.*, p. 5, Copyright Elsevier (2005); C. Reichmann *et al.* (*Journal of General Virology* 73, 1–16, 1992).

tobacco mosaic *Tobamovirus*, Fiji disease *Fijivirus*, and *Lettuce necrotic yellows rhabdovirus*. This method for naming a plant virus is becoming increasingly used in the literature.

D. Acronyms or Abbreviations

Abbreviations of virus names have been used for many years to make the literature easier to read and more succinct to present. The abbreviation is usually in the form of an acronym using the initial letters of each word in the virus name. As the designation of the acronym was by the author of the paper, it was leading to much

overlap and confusion. For instance, among plant viruses, AMV was used to designate *Alfalfa mosaic virus* and *Arabis mosaic virus* and could also justifiably be used for *Abutilon mosaic virus*, *Agropyron mosaic virus*, *Alpina mosaic virus*, *Alstromeria mosaic virus*, *Alternantha mosaic virus*, *Aneilema mosaic virus*, or *Anthoxanthum mosaic virus*. Therefore, in 1991 the Plant Virus section of the ICTV initiated a rationalisation of plant virus acronyms and has subsequently updated the list regularly in ICTV reports (Box 1.3).

There are no efforts to create a common acronym system for viruses from different kingdoms. Thus, CMV can mean *Cucumber*

BOX 1.3

RULES FOR VIRUS ABBREVIATIONS OR ACRONYMS

- Abbreviations should be as simple as possible.
- An abbreviation must not duplicate any other previously coined term or one still in use.
- The word *virus* in a name is abbreviated as *V*.
- The word *viroid* in a name is abbreviated as *Vd*.
- *M* is usually used for "mosaic" and *Mo* for "mottle."
- The word *ringspot* is abbreviated as *RS* and *symptomless* as *SL*.
- Abbreviations for single words should not normally exceed two letters.
- Where a particular combination of letters has been adopted for a particular plant, subsequent abbreviations for viruses of that host should use the same combination.
- The second (or third) letter of a host plant abbreviation is in lowercase—for example, Ab for Abutilon.
- When several viruses have the same name and are differentiated by a number, the abbreviation will have a hyphen between the letters and number—for example, *Plantain virus 6* is abbreviated as PlV-6.
- When viruses end with a letter, the letter is added to the end of the abbreviation without a hyphen—for example, *Potato virus X* is abbreviated PVX.
- When viruses are distinguished by their geographical location, a minimum number of letters (two or three) are added to the abbreviation with a hyphen—for example, *Tomato yellow leaf curl virus* from Thailand is TYLCV-Th.
- When a virus name comprises a disease name and the words *associated virus*, these are abbreviated *aV*—for example, *Grapevine leafroll associated virus 2* is abbreviated GLRaV-2.

A set of guidelines is laid out in Fauquet and Mayo (1999). Although these and the acronyms derived from them, are not officially sanctioned by the ICTV, the acronyms are used in the ICTV reports.

---

mosaic virus (of plants), *Canine minute virus* (of vertebrates), or *Clo Mor virus* (of invertebrates). Thus, acronyms have to be taken in context.

E. Plant Virus Classification

The current classification of plant viruses is shown in Figure 1.5.

F. Virus Strains

A virus species is not a uniform population because in each infected cell, a wide range of variants is present. This situation is termed a *quasi-species* (Box 1.4).

The quasi-species concept makes it difficult to strictly define a *strain*. However, one must describe variants within a species and, in reality, take a pragmatic approach. Characters have to be weighed up as to how they would contribute to making subdivisions and to communication, not only between virologists but also to plant pathologists, extension workers, farmers, and many other groups. An example is the luteovirus *Beet western yellows virus* (BWYV), which has a wide host range, including sugar beet in the United States. For many years, *Beet mild yellows virus*, which infected sugar beet in Europe, was regarded as a strain of BWYV. Confusion arose when it was discovered that the European luteovirus that was most closely related to BWYV did not infect sugar beet but was common in the oilseed rape crop. This caused many problems in explaining

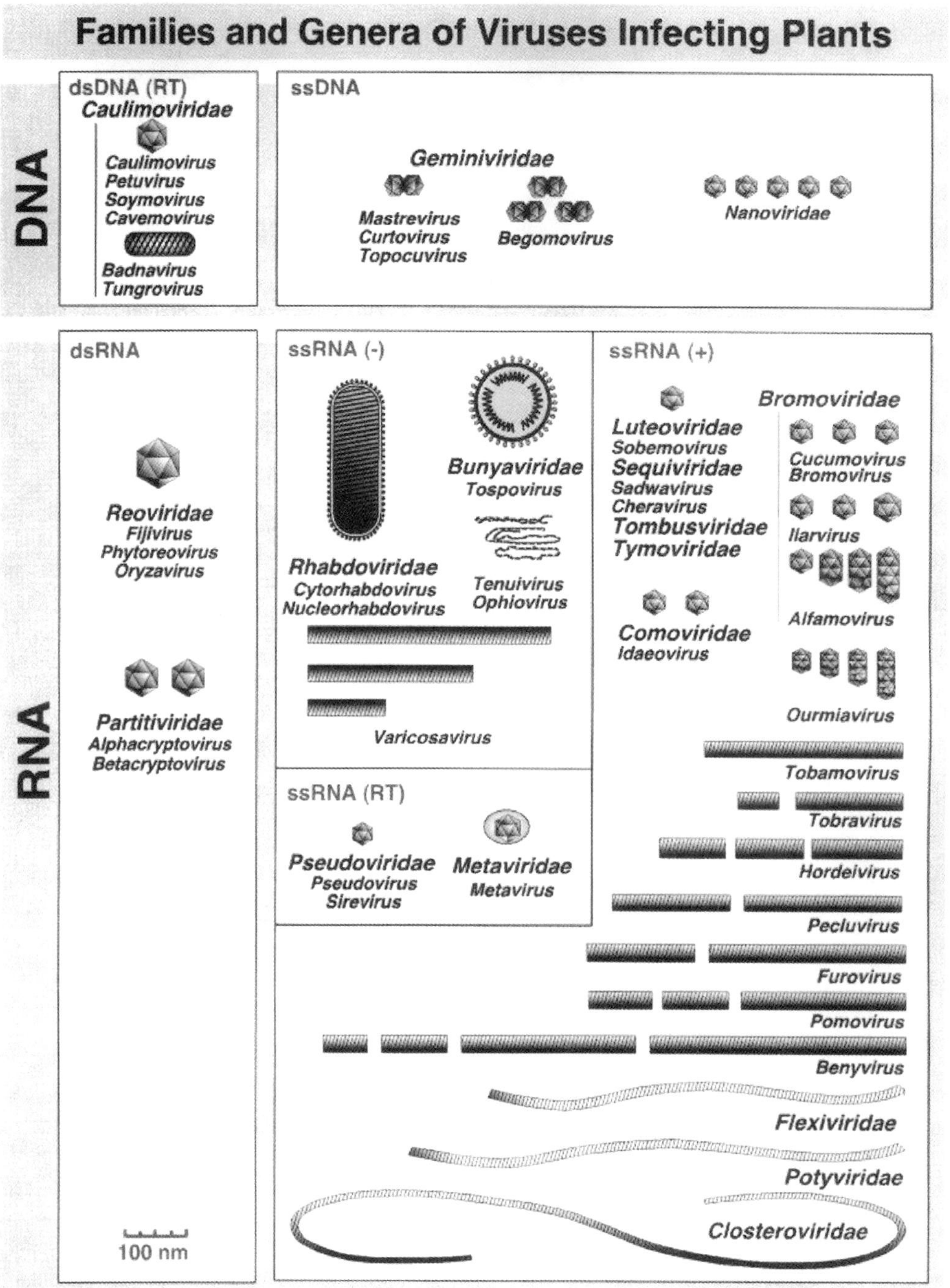

Families and Genera of Viruses Infecting Plants

FIGURE 1.5   Classification of plant viruses. From Virus Taxonomy, 8th Report of the National Committee on the Taxonomy of Viruses, Fauquet *et al.*, p. 18, Copyright Elsevier (2005).

BOX 1.4

QUASI-SPECIES

A *quasi-species* is a population structure in which collections of closely related genomes are subjected to a continuous process of genetic variation, competition, and selection. Usually, the distribution of mutants or variants is centred on one or several master sequences. The selection equilibrium is meta-stable and may collapse or change when an advantageous mutant appears in the distribution.

In this case, the previous quasi-species will be substituted by a new one characterised by a new master sequence and a new mutant spectrum. The stability of a quasi-species depends on the complexity of the genetic information in the viral genome, the copy fidelity on replication of the genome, and the superiority of the master sequence.

A quasi-species has a physical, chemical, and biological definition. In the physical definition, a quasi-species can be regarded as a cloud in sequence space, which is the theoretical representation of all the possible variants of a genomic sequence. For an ssRNA virus of 10 kb, the sequence space is 410,000. Thus, the quasi-species cloud represents only a very small proportion of the sequence space and is constrained by the requirements of gene and nucleic acid functions. Chemically, the quasi-species is a rated distribution of related nonidentical genomes. Biologically, a quasi-species is the phenotypic expression of the population, most likely dominated by that of the master sequence.

---

to farmers that the BWYV in their overwintering oilseed rape crop would not infect their beet crop the next year.

G. Use of Virus Names

The ICTV sets rules, which are regularly revised, on virus nomenclature and the orthography of taxonomic names (see the eighth ICTV report). The last word of a species is *virus*, and the suffix (ending) for a genus name is *-virus*. For a subfamily, it is *-virinae*; for a family, it is *-viridae*; and for an order, it is *-virales*. In formal taxonomic usage, the virus order, family, subfamily, genus, and species names are printed in italics (or underlined), with the first letter being capitalized; other words in species names are not capitalized unless they are proper nouns or parts of proper nouns. Also, in formal use, the name of the taxon should precede the name being used—for example, the family *Caulimoviridae*, the genus *Mastrevirus*, and the species *Potato virus Y*. An example of classification, nomenclature, and orthography is shown in Box 1.5.

In informal use, the family, subfamily, genus, and species names are written in lowercase Roman script, the taxon does not include the formal suffix, and the taxonomic unit follows the name being used—for example, the caulimovirus family, the mastrevirus genus, and the potato virus Y species. In even less formal circumstances, but still widely used, the taxonomic unit is omitted and the taxon for higher taxa can be in the plural—for example, caulimoviruses, mastreviruses, and potato virus Y.

EXAMPLE OF VIRUS CLASSIFICATION, NOMENCLATURE AND ORTHOGRAPHY

Taxa	Example	Suffix
Order	*Mononegavirales*	*-virales*
Family	*Rhabdoviridae*	*-viridae*
Subfamily		*-virinae*
Genus	*Nucleorhabdovirus*	*-virus*
Species	*Sonchus yellow net virus*	
Acronym	*SYNV*	

Informal usage arises from practicalities and can lead to the adoption of more formal use. For instance, the genus *Badnavirus* was not adopted in 1991 but was used widely in the literature and was adopted in the 1995 ICTV report. However, the year 2000 report limited its use to certain DNA viruses with bacilliform particles excluding *Rice tungro bacilliform virus*. As will be apparent in this book, it is necessary to distinguish the reverse transcribing DNA viruses that have isometric particles from those that have bacilliform particles; the informal usage will be *caulimoviruses* for the former and *badnaviruses* for the latter.

V. VIRUSES OF OTHER KINGDOMS

The eighth report from the ICTV (Fauquet *et al.*, 2005) noted over 2,700 accepted, tentative, and unassigned virus species classified into 3 orders, 73 families (4 of these divided into subfamilies), and 287 genera. Most of these taxonomic groupings at the genus level are specific to viruses of plants, vertebrates, invertebrates, or prokaryotes, but some genera of viruses infect more than one kingdom. The overall classification is based on genome type, some very obvious differences exist between the genome types of plant, vertebrate, invertebrate, and prokaryotic viruses (Table 1.5).

TABLE 1.5   Numbers of Virus Species in Various Kingdoms

	Plant		Vertebrate		Invertebrate		Prokaryote		Fungi and Algae	
	Number	% Total	Number	% Total	Number	% Total	Number	% Total	Number	% Total
dSDNA	0	0	263	28.4	150	60.7	203	62.5	127	67.9
sSDNA	198	19.1	66	7.1	27	10.9	92	28.3	0	0
RT	66	6.4	62	6.7	24	9.7	0	0	13	7.0
dSRNA	48	4.6	75	8.1	38	15.4	24	7.4	33	17.6
sS-RNA	48	4.6	227	24.5	0	0	0	0	0	0
sS+RNA	677	65.3	234	25.2	8	3.2	6	1.8	14	7.5
Total	1,037		927		247		325		187	

Data from Fauquet *et al.* (2005), using numbers of assigned, unassigned, and tentative virus species.

VI. SUMMARY

- Plant viruses are important pathogens.
- The study of plant viruses has made important contributions to the understanding of viruses in general—for example, the recognition of viruses as pathogens, the structure of virus particles, and the infectious nature of RNA.
- This chapter defines a virus, contrasts it with similar agents, and discusses how viruses are classified.

References

Ackermann, H.-W. (2008). [History of virology]—Bacteriophage. *Encyclopedia of virology*, Vol. 2, 442–449.

Caspar, D.L.D. and Klug, A. (1962). Physical principles in the construction of plant viruses. *Cold Spring Harbor Symp. Quant. Biol.* **27**, 1–24.

Fauquet, C.M. and Mayo, M.A. (1999). Abbreviations for plant virus names—1999. *Arch Virol.* **144**, 1249–1272.

Fauquet, C.M., Mayo, M.A., Maniloff, J., Desselberger, U., and Ball, L.A. (2005). *Virus taxonomy. Eighth Report of the International Committee on Taxonomy of Viruses.* Elsevier, San Diego.

Fenner, F.J. (2008) History of viruses/Vertebrate viruses. *Encyclopedia of Virology*, Vol. 2, 455–458.

Hull, R. (2001). *Matthew's plant virology*, 4th ed. Academic Press, San Diego.

Murphy, F.A., Fauquet, C.M., Bishop, D.H.L., Ghabrial, S.A., Jarvis, A.W., Martelli, G.P., Mayo, M.A., and Summer, M.D. (1995). *Virus taxonomy. Sixth Report of the International Committee on Taxonomy of Viruses.* Springer-Verlag, Wien, Austria.

Rivers, T.M. (1937). Viruses and Koch's postulates. *J. Bacteriol.* **33**, 1–12.

Further Reading

Fauquet, C.M. (2008). Taxonomy, classification, and nomenclature of viruses. *Encyclopedia of Virology*, Vol. 5, 9–23.

Grandbastien, M.-A. (2008). Retrotransposons of plants. *Encyclopedia of Virology*, Vol. 4, 428–435.

Hull, R. (2002) *Matthew's plant virology*, 4th ed. Academic Press, London.

Hull, R. (2008). History of virology: Plant viruses. *Encyclopedia of Virology*, Vol. 2, 450–454.

Levin, H.L. (2008). Metaviruses. *Encyclopedia of Virology*, Vol. 3, 301–310.

Pavord, A. (1999). *The tulip.* Bloomsbury, London.

Sanjuan, R. (2008). Quasispecies. *Encyclopedia of Virology*, Vol. 4, 359–366.

Van Regenmortel, M.H.V. Virus species. *Encyclopedia of Virology*, Vol. 5, 401–405.

Voytas, D.F. (2008). Pseudoviruses. *Encyclopedia of Virology*, Vol. 4, 352–358.

Overview of Plant Viruses

Plant viruses are important both economically and as model systems to explore the interactions of viruses and their host cells. They have some properties in common with viruses of other kingdoms and some properties specific to the properties of their plant hosts.

I. INTRODUCTION

Most of the plant virus names that are commonly used today include terms that describe an important symptom in a major host or the host from which the virus was first described. Some viruses, under appropriate conditions, may infect a plant without producing any obvious signs of disease. Others may lead to rapid death of the whole plant. Between these extremes, a wide variety of diseases can be produced.

Virus infection does not necessarily cause disease at all times in all parts of an infected plant. We can distinguish six situations in which obvious disease may be absent:

1. Infection with a very mild strain of the virus
2. A tolerant host
3. "Recovery" from disease symptoms in newly formed leaves (see Chapter 11)
4. Leaves that escape infection because of their age and position on the plant

5. Dark green areas in a mosaic pattern (see Chapter 11)
6. Plants that are infected with cryptic viruses

II. ECONOMIC LOSSES DUE TO PLANT VIRUSES

One of the main driving forces for the detailed studies of plant viruses is the impact that the resulting diseases have on crop productivity worldwide, although it is difficult to obtain firm data on the actual losses themselves. The losses due to fungal and bacterial pathogens are well documented, and lists of loss estimates attributable to specific named fungi or bacteria are easily obtainable. In these compendia, the losses due to viruses are often lumped together in categories such as "Virus Diseases," "All Other," or "Miscellaneous" diseases. However, viruses are responsible for far greater economic losses than are generally recognised. This lack of recognition are due to several factors, especially their insidious nature. Virus diseases are frequently less conspicuous than those caused by other plant pathogens, and they last much longer. This are especially true for perennial crops and those that are vegetatively propagated.

One further problem with attempting to assess losses due to virus diseases on a global basis is that most of the data are from small comparative trials that do not necessarily give information that can be used for more global estimates of losses. This is due to factors such as variation in losses by a particular virus in a particular crop from year to year, variation from region to region and climatic zone to climatic zone, differences in loss assessment methodologies, identification of the viral aetiology of the disease, variation in the definition of the term "losses," and complications with other loss factors.

In addition to the obvious detrimental effects such as reduced yields and visual product quality, virus infections often do not induce noticeable

TABLE 2.1 Some Types of Direct and Indirect Damage Associated with Plant Virus Infections

Reduction in growth
  Yield reduction (including symptomless infection)
  Crop failure
Reduction in vigour
  Increased sensitivity to frost and drought
  Increased predisposition to attack by other pathogens and pests
Reduction in quality or market value
  Defects of visual attraction: size, shape, colour
  Reduced keeping quality
  Reduced consumer appeal: grading, taste, texture, composition
  Reduced fitness for propagation
Cost of attempting to maintain crop health
  Cultural hygiene on farm, including vector control
  Production of virus-free propagation materials
  Checking propagules and commodities on export/import (quarantine programmes)
  Eradication programmes
  Breeding for resistance
  Research, extension, and education.

From Waterworth and Hadidi (1998).

disease but influence their effects on plants in a variety of more subtle ways. Table 2.1 identifies some of the ways that viruses can damage crop plants. We can see that the effects of virus infection extend into areas far beyond the actual reduction in yield and quality. Loss estimates do not take account of these indirect factors.

In spite of all these limitations, various collections of loss data (Table 2.2) have been compiled. Various newly emerging virus problems in crops are being exacerbated by changes in agricultural practices, global trade, and the climate (see Anderson *et al.*, 2004).

III. VIRUS PROFILES

Appendix A provides profiles of 32 plant viruses that feature strongly in this book. These profiles give you basic information on these important viruses and are referred to in this and subsequent chapters.

TABLE 2.2   Some Examples of Crop Losses Due to Viruses

Crop	Virus	Countries	Loss/Year
Rice	Tungro	SE Asia	1.5×10^9
	Ragged stunt	SE Asia	1.4×10^8
	Hoja blanca	S. and C. America	9.0×10^6
Barley	Barley yellow dwarf	UK	£6 $\times 10^6$
Wheat	Barley yellow dwarf	UK	£5 $\times 10^6$
Potato	Potato leafroll	UK	£3-5 $\times 10^7$
	Potato virusY		
	Potato virus X		
Sugar beet	Beet yellows	UK	£5-50 $\times 10^6$
	Beet mild yellows		
Citrus	Citrus tristeza	Worldwide	£9-24 $\times 10^6$
Cassava	African cassava mosaic	Africa	2×10^9
Many crops	Tomato spotted wilt[a]	Worldwide	1×10^9
Cocoa	Cocoa swollen shoot	Ghana	1.9×10^8 trees[b]

[a]Data from Prins and Goldbach (1996); references to other data given in Hull and Davies (1992).
[b]Number of trees eradicated over about 40 years.

IV. MACROSCOPIC SYMPTOMS

Symptoms on plants may be local on the inoculated leaves and/or systemic on spread to other parts of the plant from inoculated leaves.

A. Local Symptoms

Localised lesions that develop near the site of entry on leaves are not usually of any economic significance but are important for biological assay. Three types of local response to infection can result: necrotic local lesions in which the infected cells die (Profile 14; see Appendix) and that vary from small pinpoints to large irregular spreading necrotic patches; chlorotic local lesions in which the infected cells lose chlorophyll and other pigments; and ring spot lesions that typically consist of a central group of dead cells beyond which develop one or more superficial concentric rings of necrotic or chlorotic cells with normal green tissue between them (Profile 6; see Appendix). Some viruses in certain hosts show no visible local lesions in the intact leaf, but when the leaf is cleared in ethanol and stained with iodine, "starch lesions" may become apparent (see Chapter 9).

Viruses that produce local lesions when inoculated mechanically onto leaves may not do so when introduced by other means. For example, *Beet yellows virus* produces necrotic local lesions on *Chenopodium capitatum*, but it does not do so when the virus is introduced by the aphid *Myzus persicae* feeding on parenchyma cells.

B. Systemic Symptoms

Various symptoms often appear in combination in particular diseases, and the pattern of disease development for a particular host-virus combination often involves a sequential development of different kinds of symptoms.

1. *Effects on Plant Size*

Reduction in plant size is the most general symptom induced by virus infection (see Box 2.1). The degree of stunting is generally correlated with the severity of other symptoms, particularly where loss of chlorophyll from the leaves is concerned. Stunting is usually almost entirely due to reduction in leaf size and internode length. Leaf number may be little affected.

In perennial deciduous plants, such as grapes, there may be a delayed initiation of growth in the spring. Root initiation in cuttings from virus-infected plants may be reduced, as in chrysanthemums. In vegetatively propagated plants, stunting is often a progressive process. For example, virus-infected strawberry plants and tulip bulbs may become smaller in each successive year.

2. *Mosaic Patterns and Related Symptoms*

One of the most common obvious effects of virus infection is the development of a pattern of light and dark green areas, which creates a mosaic effect in infected leaves. The detailed nature of the pattern varies widely for different host-virus combinations. In dicotyledons, the areas that make up the mosaic are generally irregular in outline. For example, only two shades of colour—like dark green and a pale or yellow-green—may be involved, (see Profiles 3 and 14; see Appendix), or there may be many different shades of green, yellow, or even white, as with *Turnip yellow mosaic virus* (TYMV) in Chinese cabbage (Profile 18;

see Appendix). The junctions between areas of different colour may be sharp, and such diseases resemble quite closely the mosaics produced by inherited genetic defects in the chloroplasts. *Abutilon mosaic virus* is a good example of this type (Profile 7; see Appendix).

In mosaic diseases that infect herbaceous plants, usually a fairly well-defined sequence in the development of systemic symptoms occurs. The virus moves up from the inoculated leaf into the growing shoot and into partly expanded leaves. In these leaves, the first symptoms are a "clearing" or yellowing of the veins; it is suggested that this symptom is an optical illusion. However, chlorotic vein-banding is a true symptom and may be very faint or may give striking emphasis to the pattern of veins [see *Cauliflower mosaic virus* (CaMV), Profile 4, and *Beet yellows virus* (BYV), Profile 5; see Appendix]. Vein-banding may persist as a major feature of the disease.

No mosaic pattern develops in leaves that are past the cell division stage of leaf expansion when they become infected (about 4–6 cm long for leaves such as tobacco). These leaves become uniformly paler than normal. In the oldest leaves to show mosaic, a large number of small islands of dark green tissue usually appear against a background of paler colour (for molecular aspects of dark green islands, see Chapter 11). The mosaic areas may be confined to the youngest part of the leaf blade— that is, the basal and central region. Although considerable variation in different plants may appear, successively younger systemically infected leaves show, on the average, mosaics consisting of fewer and larger areas. The mosaic pattern is laid down at a very early stage of leaf development and may remain unchanged, except for general enlargement, for most of the life of the leaf.

In monocotyledons, a common result of virus infection is the production of stripes or streaks of tissue lighter in colour than the rest of the leaf.

BOX 2.1

GROUNDNUT ROSETTE DISEASE

Groundnut rosette is an important disease of groundnuts (peanut) (*Arachis hypogaea*) in West and East Africa. It is transmitted by the aphid *Aphis craccivora* in a circulative nonpropagative manner. It causes the leaves to become light green or chlorotic, stunts the plant, and reduces yield. The figure here compares a disease plant (upper) with a healthy plant (lower).

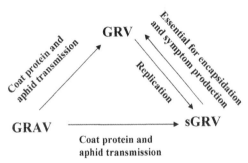

The disease is caused by a complex of two viruses—an umbravirus, *Groundnut rosette virus* (GRV), and a luteovirus, *Groundnut rosette assistor virus* (GRAV)—and a satellite RNA (sGRV). Umbraviruses do not encode a coat protein, and that for GRV is provided by GRAV. sGRV is essential for this encapsidation and for the symptom production by GRV. GRV's role in this complex is to potentiate the replication of sGRV. Thus, in the final complex, all three entities interact as shown in the diagram.

The shades of colour vary from pale green to yellow or white, and the more or less angular streaks or stripes run parallel to the length of the leaf (see *Banana streak virus* (BSV), Profile 4, and *Maize streak virus*, Profile 7). The development of the stripe diseases found in monocotyledons follows a similar general pattern to that found for mosaic diseases in dicotyledons.

A variegation or "breaking" in the colour of petals commonly accompanies mosaic or streak symptoms in leaves. The breaking usually consists of flecks, streaks, or sectors of tissue with a colour different from normal (see Box 1.1). The breaking of petal colour is frequently due to loss of anthocyanin pigments, which reveals any underlying colouration due to plastid pigments. Flower colour-breaking may sometimes be confused with genetic variegation, but it is usually a good diagnostic feature for infection by viruses that produce mosaic symptoms. In a few plants, virus-induced variegation has been valued commercially. As with the development of mosaic patterns in leaves, colour-breaking in the petals may develop only in flowers that are smaller than a certain size when infected. Virus infection may reduce pollen production and decrease seed set, seed size, and germination.

Fruits that form on plants showing mosaic disease in the leaves may show a mottling—for example, zucchinis infected with *Cucumber mosaic virus* (CMV; Profile 3). In other diseases, severe stunting and distortion of fruit may occur. Seed coats of infected seed may be mottled.

3. *Yellow Diseases*

Viruses that cause a general yellowing of the leaves are not as numerous as those that cause mosaic diseases, but some, such as the viruses that cause yellows in sugar beet, are of considerable economic importance. The first sign of infection is usually a clearing or yellowing of the veins in the younger leaves followed by a general yellowing or reddening of the leaves [see *Barley yellow dwarf virus*, Profile 8 (see Appendix); *Rice tungro bacilliform virus* (RTBV), Profile 4; and BYV, Profile 5]. This yellowing may be slight or severe. No mosaic is produced, but in some leaves sectors of yellowed and normal tissue may appear.

4. *Leaf Rolling*

Virus infection can result in leaf rolling, which is usually upward but occasionally downward (see *Potato leaf roll virus*, Profile 8). Pronounced epinasty (downward bending of leaves) may sometimes be a prominent feature.

5. *Ring Spot Diseases*

Ring spots are a pattern of concentric rings and irregular lines on the leaves and sometimes also on the fruit (see *Tobacco ring spot virus*, Profile 6, and *Papaya ring spot virus*, Profile 10; see Appendix). The lines may consist of yellowed tissue or may be due to death of superficial layers of cells, giving an etched appearance. In severe diseases, complete necrosis through the full thickness of the leaf lamina may occur. Ring spot patterns may also occur on other organs, such as bulbs and tubers (Profile 15; see Appendix).

6. *Necrotic Diseases*

The death of tissues, organs, or the whole plant is the main feature of some diseases. Necrotic patterns may follow the veins as the virus moves into the leaf (Profile 16; see Appendix). In some diseases, the entire leaf is killed. Necrosis may extend fairly rapidly throughout the plant. For example, with joint infections of *Potato virus X* (PVX) and *Potato virus Y* (PVY) in tomatoes, necrotic streaks appear in the stem (see Box 11.3 in Chapter 11). Necrosis spreads rapidly to the growing point, which is killed, and subsequently all leaves may collapse and die. Wilting of the parts that are about to become necrotic often precedes such systemic necrotic disease.

7. *Developmental Abnormalities*

Besides being generally smaller than normal, virus-infected plants may show a wide range of developmental abnormalities. Such changes may be the major feature of the disease or may accompany other symptoms. For example, uneven growth of the leaf lamina is often found in mosaic diseases. Dark green areas may be raised to give a blistering effect, and the margin of the leaf may be irregular and twisted.

In some diseases, the leaf blade may be more or less completely suppressed—for example, in tomatoes infected with CMV (Profile 3).

Some viruses cause swellings in the stem, which may be substantial in woody plants—for example, in *Cocoa swollen shoot virus* (CSSV) disease. Another group of growth abnormalities is known as enations, which are outgrowths from the upper or lower surface of the leaf, usually associated with veins.

Viruses may cause a variety of tumour-like growths. The most studied tumours are those produced by *Wound tumor virus* (WTV). In a systemically infected plant, external tumours appear on leaves or stems where wounds are made. In infected roots they appear spontaneously, beginning development close to cells in the pericycle that are wounded when developing side roots break through the cortex. Stem deformation such as stem splitting and scarring is caused by some viruses in some woody plants. Virus infection of either the rootstock or scion can cause necrosis and/or failure of the graft union. One of the unusual symptoms of BSV in some *Musa* cultivars is that the fruit bunch emerges from the side of the pseudostem instead of the top of it. This is due to necrosis of the cigar leaf.

8. Wilting

Some virus infections induce wilting of the aerial parts, leading to the death of the whole plant (see *Citrus tristeza virus*, Profile 5).

9. Recovery from Disease

Not uncommonly, a plant shows disease symptoms for a period, and then new growth appears in which symptoms are milder or absent, although virus is still present. This commonly occurs with *Nepovirus* infections. Many factors influence this recovery phenomenon. The stage of development at which a plant is infected may have a marked effect on the extent to which symptoms are produced. The environment can also affect recovery from disease, as can host species

or variety and virus strain. The molecular aspects of disease recovery are discussed in Chapter 11.

10. Genetic Effects

Infection with *Barley stripe mosaic virus* (BSMV) induces an increase in mutation rate in *Zea mays* and also a genetic abnormality known as an aberrant ratio (AR). This AR effect was observed only when the original pollen parent was infected and showing virus symptoms on the upper leaves. The AR effect is inherited in a stable manner in plants in which the virus can no longer be detected, with a low frequency of reversion to normal ratios.

C. The Cryptoviruses

Cryptoviruses escaped detection for many years because most of them cause no visible symptoms or, in a few situations, very mild symptoms. They are not transmissible mechanically or by vectors, but they are transmitted efficiently in pollen and seed. They occur in very low concentrations in infected plants. Nevertheless, they have molecular characteristics that might be expected of disease-producing viruses.

The genome of cryptoviruses consists of two dsRNA segments, and these viruses share some properties with the reoviruses. There is no indication, other than the low concentration at which they occur, as to why they cause symptomless infection.

D. Diseases Caused by Viral Complexes

The joint infection of two viruses can cause symptoms that are more severe than those of either of the two viruses; this is termed synergism. An example is tomato streak caused by the joint infection with PVX and PVY, which was just discussed (see Box 11.3).

Some virus diseases are caused by the complexes of two (or more) viruses, each contributing features to the disease. Rice tungro disease is caused by the joint infection of RTBV and

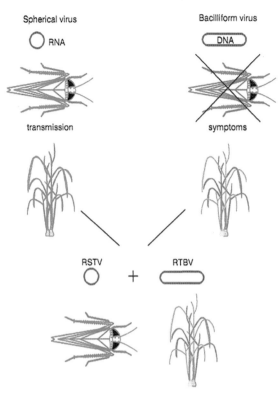

FIGURE 2.1 Diagram showing interaction of *Rice tungro bacilliform virus* and *Rice tungro spherical virus* in tungro disease complex.

Rice tungro spherical virus (RTSV; Figure 2.1). RTBV causes severe symptoms in rice but is not leafhopper transmitted on its own; RTSV is leafhopper transmitted but causes few, if any, symptoms. The complex is leafhopper transmitted with RTSV, giving the transmission properties and RTBV the severe disease symptoms. Similarly, umbraviruses require a luteovirus for aphid transmission. An even more complicated complex is show in groundnut rosette disease (see Box 2.1).

E. Agents Inducing Virus-Like Symptoms

Disease symptoms, similar to those produced by viruses, can be caused by a range of biological, physical, and chemical agents. These include phytoplasmas, spiroplasmas, rickettsia-like organisms (all three of these are discussed in Chapter 3), bacteria, toxins produced by arthropods, nutritional deficiencies, high temperatures, hormones (weed killers), and insecticide damage and air pollutants. Furthermore, genetic variants, some of which may be caused by transposons, can resemble mosaic and leaf variegations caused by viruses. Thus, care must be taken in diagnosing the viral aetiology of a new problem.

V. HISTOLOGICAL CHANGES

The basic structure of a plant and a plant cell is shown in Box 2.2. The macroscopic symptoms induced by viruses frequently reflect histological changes within the plant. These changes are of three main types—necrosis, hypoplasia, and hyperplasia—that may occur singly or together in any particular disease.

A. Necrosis

Necrosis, as a macroscopic symptom, was discussed previously. It may be general or may be limited to specific tissues, such as localised areas of the roots (e.g., *Tobacco necrosis virus*), internally in immature tomato fruits with late infection by *Tobacco mosaic virus* (TMV), or limited to the phloem, as in potato leaf roll disease.

B. Hypoplasia

Leaves with mosaic symptoms frequently show hypoplasia (localised retarded growth frequently leading to thinner areas on leaves) in the yellow areas. The lamina is thinner than in the dark green areas, and the mesophyll cells are less differentiated with fewer chloroplasts and fewer or no intercellular spaces (Figure 2.2).

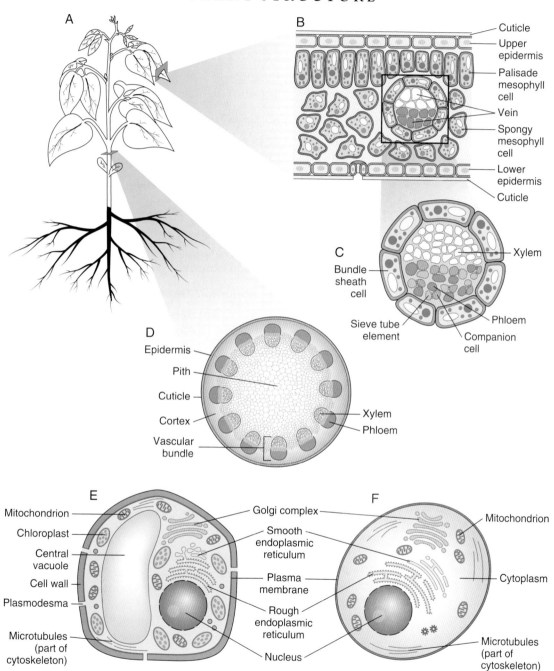

BOX 2.2

PLANT STRUCTURE

A

B
- Cuticle
- Upper epidermis
- Palisade mesophyll cell
- Vein
- Spongy mesophyll cell
- Lower epidermis
- Cuticle

C
- Bundle sheath cell
- Sieve tube element
- Xylem
- Phloem
- Companion cell

D
- Epidermis
- Pith
- Cuticle
- Cortex
- Vascular bundle
- Xylem
- Phloem

E
- Mitochondrion
- Chloroplast
- Central vacuole
- Cell wall
- Plasmodesma
- Microtubules (part of cytoskeleton)
- Golgi complex
- Smooth endoplasmic reticulum
- Plasma membrane
- Rough endoplasmic reticulum
- Nucleus

F
- Golgi complex
- Mitochondrion
- Cytoplasm
- Microtubules (part of cytoskeleton)

Fig. Basic structure of plant tissues. A. Diagrammatic dicotyledonous plant; B. Diagram of a section of a leaf showing the various tissues; C. Details of the cell types in a vein. D. Diagram of a section of a stem; E. Structure of a plant cell; F. Structure of an animal cell.

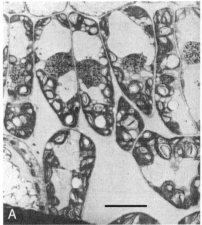
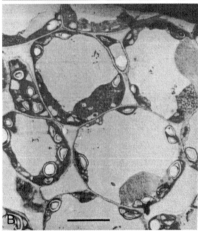

FIGURE 2.2 Hypoplasia in tobacco cells infected with TMV. A. Section through palisade cells of a dark green area of a leaf with mosaic showing that the cells are essentially normal; B. Section through a nearby yellow-green area showing that the cells are large and undifferentiated in shape. (Courtesy of P.H. Atkinson)

The major anatomical effect of *Apple stem-grooving virus* in apple stems is the disappearance of the cambium in the region of the groove. Normal phloem and xylem elements are replaced by a largely undifferentiated parenchyma.

C. Hyperplasia

Hyperplasia is growth of abnormally large cells or excessive cell division.

1. Cell Size

Vein-clearing symptoms are due, with some viruses at least, to enlargement of cells near the veins. The intercellular spaces are obliterated, and since there is little chlorophyll present, the tissue may become abnormally translucent.

2. Cell Division in Differentiated Cells

Some viruses such as PVX may produce islands of necrotic cells in potato tubers. The tuber may respond with a typical wound reaction in a zone of cells around the necrotic area. Starch grains disappear, and an active cambial layer develops. Similarly, in a white halo zone surrounding necrotic local lesions induced by TMV in *N. glutinosa* leaves (Profile 14), cell division occurred in mature palisade cells.

3. Abnormal Division of Cambial Cells

The vascular tissues appear to be particularly prone to virus-induced hyperplasia. For example, in the CSSV diseased shoots, abnormal amounts of xylem tissue are produced, but the cells appear structurally normal. In crimson clover infected by WTV, abnormal development of phloem cambium cells is present, and phloem parenchyma forms meristematic tumour cells.

VI. CYTOLOGICAL EFFECTS

A. Effects on Cell Structures

1. Nuclei

Many viruses have no detectable cytological effects on nuclei, but others give rise to intranuclear inclusions of various sorts and may affect the nucleolus or the size and shape of the nucleus, even though they appear not to replicate in this organelle. Here are some examples:

• In pea leaves and pods infected with *Pea enation mosaic virus* (PEMV), particles accumulate first in the nucleus, and then the nucleolus disintegrates. PEMV also causes vesiculation in the perinuclear space.

- Crystalline platelike inclusions can be seen by light microscopy in the nuclei of cells infected with several potyviruses.
- Geminivirus infection can cause marked hypertrophy of the nucleolus, which may come to occupy three-quarters of the nuclear volume. Fibrillar rings of deoxyribonucleoprotein appear, and masses of virus particles accumulate in the nucleus.

2. Mitochondria

The long rods of *Tobacco rattle virus* may be associated with the mitochondria in infected cells (see Figure 2.3A), as are the isometric particles of *Broad bean wilt virus-1* (see Figure 2.3B).

3. Chloroplasts

Infection with TYMV induces small peripheral vesicles (Figure 2.4) and other changes in and near the chloroplasts, as well as many other cytological changes in the chloroplasts, most of which appear to constitute a structural and biochemical degeneration of the organelles.

In many infections, the size and number of starch grains seen in leaf cells are abnormal.

In mosaic diseases, there is generally speaking less starch than normal, but in some diseases (e.g., sugar beet curly top and potato leaf roll) excessive amounts of starch may accumulate. Similarly in local lesions induced by TMV in cucumber cotyledons, chloroplasts become greatly enlarged and filled with starch grains.

4. Cell Walls

The plant cell wall tends to be regarded mainly as a physical supporting and barrier structure. In fact, it is a distinct biochemical and physiological compartment that contains a substantial proportion of the total activity of certain enzymes in the leaf. Three kinds of abnormality have been observed in or near the walls of virus-diseased cells:

- Abnormal thickening, due to the deposition of callose, may occur in cells near the edge of virus-induced lesions.
- Cell wall protrusions involving the plasmodesmata have been reported for several unrelated viruses. These are often involved in cell-to-cell movement of viruses (see Chapter 9).
- Depositions of electron-dense material between the cell wall and the plasma membrane (sometimes called paramural bodies) may extend over substantial areas of the cell wall.

FIGURE 2.3 Association of virus particles with mitochondria. A. Section of *Nicotiana clevelandii* showing mitochondrion with fringe of *Tobacco rattle virus* particles. [From Reichmann *et al.* (*Journal of General Virology* **73**, 1–16, 1992)] B. Section of *N. clevelandii* showing particles of *Broad bean wilt virus-1* arranged between mitochondria. [From Hull and Plaskitt (1974; *Intervirology* **2**, 352–359, S. Karger AG, Basel).]

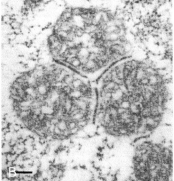

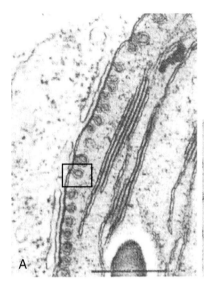

FIGURE 2.4 TYMV-induced peripheral vesicles in the chloroplasts of Chinese cabbage. A. General view; B. Details of the boxed vesicles in A showing continuity of the chloroplast membranes in the vesicle and fibrillar material inside the vesicle. Reprinted from *Virology*, **59**, T. Hatta and R.E.F. Matthews, The sequence of early cytological changes in Chinese cabbage leaf cells following systemic infection with *turnip yellow mosaic virus*, pp. 37–50, Copyright (1974), with permission from Elsevier.

5. Cell Death

Drastic cytological changes occur in cells as they approach death. These changes have been studied by both light and electron microscopy, but they do not tell us how virus infection actually kills the cell.

B. Virus-Induced Structures in the Cytoplasm

Various types of virus-induced structures are found in the cytoplasm of infected plants. They are either accumulations of virus particles, aggregates of virus-encoded proteins, or modifications of the cell organelles associated with virus replication (often termed viroplasms; see Box 2.3).

1. Accumulations of Virus Particles

Virus particles may accumulate in an infected cell in sufficient numbers and exist under suitable conditions to form three-dimensional crystalline arrays. These may grow into crystals large enough to be seen with the light microscope, or they may remain as small arrays that

---

BOX 2.3

VIROPLASMS

As we will see in Chapter 8, virus replication often involves membranes that causes cellular disturbance. This disturbance can be seen as clustering of membranes and/or associated organelles that are visible under the electron microscope (and sometimes even the light microscope)—for instance, in TMV infections. These different structures are sometimes called viroplasms, although this term is also used to describe CaMV and rhabdovirus inclusion bodies.

---

can be detected only by electron microscopy. The ability to form crystals within the host cell depends on properties of the virus itself and is not related to the overall concentration reached in the tissue or to the ability of the purified virus to form crystals. For example, TYMV can readily crystallize *in vitro*. It reaches high concentration in infected tissue but does not normally form crystals there. By contrast, *Satellite tobacco necrosis virus* occurring in much lower concentrations frequently forms intracellular crystals.

Rod-shaped viruses belonging to various groups may aggregate in the cell into more or less ordered arrays. In tobacco leaves showing typical mosaic symptoms caused by TMV, leaf-hair and epidermal cells over yellow-green areas may almost all contain crystalline inclusions of virus particles, whereas those in fully dark green areas contain none (see Figure 2.5A). Sometimes long, curved, fibrous inclusions, or spike-like or spindle-shaped inclusions made up largely of virus particles, can be seen by light microscopy. Different strains of the virus may form different kinds of paracrystalline arrays. Most crystalline inclusions have been found only in the cytoplasm, but some have been detected in nuclei.

Many small icosahedral viruses form crystalline arrays in infected cells (see Figure 2.5B), sometimes in ordered structures such as tubules. Plant cells infected with viruses that belong to the *Reoviridae* or *Rhabdoviridae* frequently contain masses of virus particles in regular arrays in the cytoplasm. With some rhabdoviruses the bullet-shaped particles accumulate in more or less regular arrays in the perinuclear space (Figure 2.5C).

2. *Aggregates of Virus-Encoded Proteins*

Potyviruses induce the formation of characteristic cylindrical inclusions in the cytoplasm of infected cells (Box 2.4).

3. *Caulimovirus Inclusions*

Two forms of inclusion bodies (also termed viroplasms) have been recognised in the cytoplasm of plants infected with CaMV and other "caulimoviruses" (see Figure 2.6). Both forms contain virus particles. Electron-dense inclusions are made up of open reading frame (ORF) VI product and are considered to be the sites of virus synthesis and assembly (see Chapter 8). Electron-lucent inclusion bodies are made up of ORF II product, one of the proteins involved in aphid transmission (see Chapter 12).

FIGURE 2.5   Crystals of virus particles in infected leaves. A. Freeze-etched preparation of TMV crystal within a tobacco mesophyll cell showing herringbone arrangement of particles; bar = 1 µm. [This article was published in *J. Ultrastr. Res.*, **54**, J.H. Willison, The hexagonal lattice spacing of intracellular crystalline tobacco mosaic virus, 176–182, Copyright Elsevier (1976).] B. Transverse section of tubes composed of particles of *Broad bean wilt virus-1* in *Nicotiana clevelandii* leaf; bar = 250 nm. [From Hull and Plaskitt (1974; *Intervirology* **2**, 352–359, S. Karger AG, Basel).] C. Thin section of maize leaf showing particles of *Maize mosaic virus* apparently budding through the inner nuclear membrane (INM). Cy, cytoplasm; N, nucleus; ONM, outer nuclear membrane. Bar = 0.3 µm. [From McDaniel *et al.* (1985; *Phytopathology* **75**, 1167–1172).]

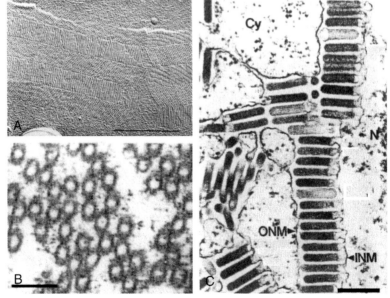

BOX 2.4

POTYVIRAL INCLUSION BODIES

Potyviruses induce the formation of characteristic cylindrical inclusions in the cytoplasm of infected cells. The most striking feature of these inclusions viewed in thin cross-section is the presence of a central tubule from which radiate curved "arms" to give a pinwheel effect. Reconstruction from serial sections shows that the inclusions consist of a series of plates and curved scrolls with a finely striated substructure with a periodicity of about 5 nm. The bundles, cylinders, tubes, and pinwheels seen in section are aspects of geometrically complex structures (Figure).

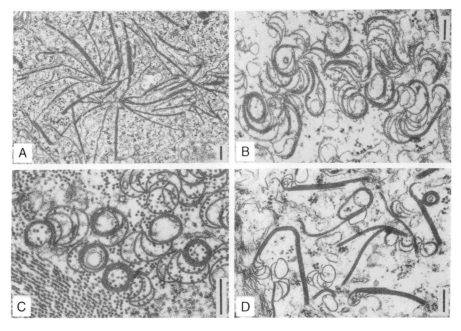

Fig. A. Scrolls induced in *Nicotiana clevelandii* by an unidentified potyvirus; B. Laminated aggregates induced by *Statice virus Y* in *Chenopodium quinoa*; C. Scrolls and laminated aggregates induced by *Turnip mosaic virus* in *N. clevelandii*; D. Scrolls and short curved laminated aggregates in *N. tabacum* infected with PVY. Bars = 200 nm. [From Lesemann (1988; in *The plant viruses*, Vol. 4, *The filamentous plant viruses*, R.G. Milne, Ed., pp. 179–235, Plenum Press, New York).]

Pinwheel inclusions originate and develop in association with the plasma membrane at sites lying over plasmodesmata. The central tubule of the pinwheel is located directly over the plasmodesmata, and it is possible that the membranes may be continuous from one cell to the next. The core and the sheets extend out into the cytoplasm as the inclusion grows. Later in infection, they may become dissociated from the plasmodesmata and come to lie free in the cytoplasm. Virus particles may be intimately associated with the pinwheel arms at all times and particularly at early stages of infection. The 71 kDa virus-coded protein that is found in these inclusions is part of the viral replicase.

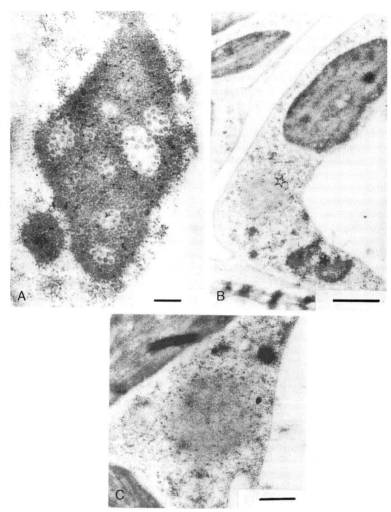

FIGURE 2.6 Electron micrographs of CaMV inclusion bodies in infected turnip leaves immunogold-labelled and anti-P62 (ORF VI product) antiserum. A. Electron-dense inclusion bodies with gold particles preferentially labelling the matrix (bar = 200 nm); B. Cell showing heavily labelled electron-dense inclusion body (filled-in star) and unlabelled electron-lucent inclusion (open star; bar = 1 μm); C. Electron-lucent inclusion (bar = 500 nm). [This article was published in *Virology*, **185**, A.M. Espinoza, V. Medina, R. Hull, P.G. Markham, *Cauliflower mosaic virus* gene II product forms distinct inclusion bodies in infected plant cells, 337–344, Copyright Elsevier (1991).]

C. Why Inclusion Bodies?

Viruses can go through numerous rounds of replication, which if allowed in an uncontrolled manner could result in cell death. Thus, it is likely that aggregation of a virus is a mechanism of removing it from the replication pool. Also, the aggregates of virus gene products are considered to be a means of sequestering potentially deleterious proteins from the cell metabolism. For instance, the potyviral genome is expressed from a polyprotein, which gives rise to as many copies of protease and replicase molecules as of coat protein molecules (see Chapter 7). It is likely that the proteases and replicases could be deleterious to the cell, so they are sequestered into inclusion bodies.

D. Cytological Structures

Some normal structures in cells could be mistaken for virus-induced effects—for example, crystalline or membrane-bound inclusions in plastids. Phosphorus-deficient bean leaves show degenerative changes in the chloroplasts like those seen in some virus infections. Similarly, in sulphur-deficient *Zea mays*, chloroplasts may contain many osmiophilic granules and small vesicles.

Nuclei of healthy cells sometimes contain crystalline structures that might be mistaken for viral inclusions. Such virus-induced effects as disorganisation of membrane systems, presence of numerous osmiophilic granules, and disintegration of organelles are similar to normal degenerative processes associated with aging or degeneration induced by other agents.

VII. THE HOST RANGE OF VIRUSES

Since the early years of the last century, plant virologists have used host range as a criterion for attempting to identify and classify viruses. In a typical experiment, the virus under study would be inoculated by mechanical means to a range of plant species that would then be observed for the development of virus-like disease symptoms. Back-inoculation to a host known to develop disease might be used to check for symptomless infections. In retrospect, it can be seen that reliance on such a procedure gives an oversimplified view of the problem of virus host ranges. Over the past few years, our ideas of what we might mean by "infection" have been considerably refined, and some possible molecular mechanisms that might make a plant a host or a nonhost for a particular virus have emerged.

The term *host* is sometimes used rather loosely. Technically, it is defined as "an organism or cell culture in which a given virus can replicate." This would mean that a plant species in which the virus can replicate in the initially infected cell (subliminal infection) is a host. However, this is impractical, and in this book the term *local host* is used for a species in which

the virus is restrained to the inoculated leaf, and *systemic host* is used for a species in which the virus spreads from the inoculated leaf to other, but not necessarily all, parts of the plant.

A. Limitations in Host Range Studies

Almost all the plant viruses so far described have been found infecting species among the angiosperms. Only a minute proportion of the possible host-virus combinations has been tested experimentally. The following arithmetic indicates the scale of our ignorance. In a major study on host range, 24 viruses were tested on 456 angiosperm species revealing 1,312 new host-virus combinations, or 12 percent of those tested. There may be about 250,000 species of angiosperms, and over 1,000 plant viruses have been recorded. If the 12 percent rate applied on average to all these plants and viruses, then there may be more than 27×10^6 new compatible host-virus combinations awaiting discovery. In relation to this figure, the number of combinations already tested must be almost negligible.

Our present knowledge of the occurrence and distribution of viruses among the various groups of plants is both fragmentary and biased. There are four probable reasons for this.

- Plant virologists who work on diseases as they occur in the field have been primarily concerned with viruses that cause economic losses in cultivated plants. They have usually been interested in other plant species only to the extent that they might be acting as reservoirs of a virus or its vector affecting a cultivated species. Thus, until fairly recently all the known plant viruses were confined to the angiosperms. Within this group, most of the known virus hosts are plants used in agriculture or horticulture or are weed species that grow in cultivated areas.

- It seems likely that widespread and severe disease in plants due to virus infection is largely a consequence of human agricultural manipulations. Under natural conditions, viruses are probably closely adapted to their hosts and cause very little in the way of obvious disease (see Chapter 4). Thus, casual inspection of plants growing in their natural habitat may give little indication of the viruses that might be present.
- The selection of "standard" test plants for viruses is to a great extent governed by those species that are easy to grow in glasshouses and to handle for mechanical and insect vector inoculation.
- The genera and species chosen for a host range study may not form a taxonomically balanced selection. Most virologists who work in the north temperate zone use mainly festucoid grasses in host range studies, whereas in other parts of the world, nonfestucoid groups predominate in the flora and in agricultural importance.

B. Patterns of Host Range

In spite of the preceding limitations, some general points can be made. Different viruses may vary widely in their host range. At one extreme BSMV is virtually confined to barley as a host in nature. At the other, CMV can infect more than 1,300 species in more than 100 botanical families, and *Tomato spotted wilt virus* (TSWV) has a host range of over 800 species belonging to 80 botanical families.

C. The Determinants of Host Range

On the basis of present knowledge there are four possible stages where a virus might be blocked from infecting a plant and causing systemic disease: during initial events—the uncoating stage; during attempted replication in the initially infected cell; during movement from the first cell in which the virus replicated;

and by stimulation of the host's cellular defenses in the region of the initial infection.

1. Initial Events

The initial event for any infection is the recognition of a suitable host cell or organelle. Bacterial viruses and most of those infecting vertebrates and invertebrates have specific proteins on their surface that act to recognise a protein receptor on the surface of a susceptible host cell. The surface proteins on plant rhabdo-, reo-, and tospo-viruses may have such a cell recognition function that is unlikely to be of use in plants but is likely to be important in recognising receptors of insect vectors. No evidence exists for plant cell recognition receptors on the surface of any of the ssRNA plant viruses; however, there are receptors on the surface of RNA viruses that have a circulative interaction with their biological vector (see Chapter 12). The evidence available for these small viruses suggests that host range is usually a property of the RNA rather than the protein coat. When it has been tested, the host range of a plant virus is the same whether intact virus or the RNA is used as inoculum. Surface recognition proteins may be of little use to a virus in the process of infecting a plant because they must enter cells through wounds on the plant surface. Various lines of evidence suggest that there is little or no host specificity in the uncoating process. Thus, both TMV and TYMV are uncoated as readily in nonhosts as in host species.

2. Expression and Replication

It has already been noted that the majority of plant viruses have single-stranded (+)-strand genomes and, as will be discussed in Chapter 7, the first events on entering a cell are that the virus particles are uncoated and the genomic RNA is translated. Following inoculation of TMV to plant species considered as nonhosts, viral RNA has been found in polyribosomes; furthermore, TMV particles uncoat and express their RNA in *Xenopus* oocytes.

However, evidence exists that specific viral genes involved in replication may also be involved as host range determinants. Various host proteins have been found in replication complexes of several viruses (see Chapter 8). The replication of many viruses takes place in association with particular cell organelles, and therefore recognition of a particular organelle or site within the cell by a virus (or by some subviral component or product) must be a frequent occurrence. Plant viruses may have evolved a recognition system basically different from that of viruses of vertebrates and invertebrates that normally encounter and recognise their host cells in a liquid medium or at a plasma membrane surface. This recognition can be effected in two ways: by the virus genome-encoding products that recognise specific cell components or by complementation between two coinfecting viruses. The latter case is exemplified by squash leaf curl disease in the United States, which is caused by two distinct but highly homologous bipartite geminiviruses. The host range of one virus is a subset of the other. Virus replication is involved in the host restriction of one of the viruses, and the replication of the restricted virus is rescued in trans by coinfection with the nonrestricted virus. Sequence analysis revealed that the restricted virus had a 13-nucleotide deletion in the common region. In other respects, the sequences of the two common regions were almost identical.

3. Cell-to-Cell Movement

As we will see in Chapter 9, plant viruses move from cell to cell through cytoplasmic connections (plasmodesmata), and this process requires a functional cell-to-cell movement protein. Thus, this is one of the factors that determines whether a particular virus can give rise to readily detectable virus replication in a given host species or cultivar. The situation when a virus can replicate in the initially infected cell but not move to adjacent cells is termed subliminal infection. A functional cell-to-cell movement protein may not be the only requirement for systemic movement of a virus, and other gene products may be involved at this stage.

4. Stimulation of Host-Cell Defenses

Plants contain various defense systems against viruses, such as RNA silencing and hypersensitive response (see Chapters 10 and 11) which the virus must overcome to establish a successful infection.

VIII. VIRUSES IN OTHER KINGDOMS

Many obvious differences that are related to the natures of the organisms are apparent in virus infections in members of different kingdoms. Factors such as host range are easier to understand than others.

IX. SUMMARY

- Plant viruses cause considerable economic losses.
- Local symptoms occur in inoculated leaves and are usually either necrotic or chlorotic lesions or ringspots
- Systemic symptoms include mosaics and mottles, general chorosis, distortions of leaves and stems, wilting and stunting, and necrosis.
- Virus infection of plants causes various histological changes of both plant cells and intracellular structure.
- Viruses can form recognisable changes to the cell, including aggregates of virus particle and viral proteins.
- Plant viral host ranges are controlled by various factors, including no virus-plant host cell receptor recognition system and little host specificity in the initial uncoating

process; mechanisms that block virus replication in the cells where the virus first gained entry; the ability of the virus to move from the first infected cell; and the host defence system.

References

Anderson, P.K., Cunningham, A.A., Patel, A.G., Morales, F.J., Epstein, P.R., and Daszak, P. (2004). Emerging infectious diseases in plants: pathogen pollution, climate change, and agrotechnology drivers. *Trends Ecol. Evolut.* **19**, 535–544.

Hull, R. and Plaskitt, A. (1974). The *in vivo* behaviour of *Broad bean wilt virus* and three of its strains. *Intervirol.* **2**, 352–359, S. Karger AG, Basel.

Hull, R. and Davies, J.W. (1992). Approaches to nonconventional control of plant virus diseases. *Crit. Rev. Plant Sci.* **11**, 17–33.

Prins, M. and Goldbach, R. (1996). RNA-mediated resistance in transgenic plants. *Arch Virol.* **141**, 2259–2276.

Waterworth, H.E. and Hadidi, A. (1988). Economic losses due to plant viruses. In *Plant virus disease control* (A. Hadidi, R.K. Khetarpal, and H. Koganezawa, Eds.), pp. 1–13. APS Press, St. Paul.

Further Reading

Hull, R. (2002). *Matthew's plant virology.* Academic Press, San Diego.

Agents That Resemble or Alter Plant Virus Diseases

Various agents that resemble virus infections can induce diseases in plants, and other agents can alter virus diseases. The former include viroids and phytoplasma, and the latter comprise small nucleic acids associated with plant pathogenic viruses. Essentially, most fall into two groups—viroids and phytoplasma—that can replicate autonomously and satellite nucleic acids and defective interfering (DI) nucleic acids that require a functional virus for their replication.

OUTLINE

I. VIROIDS

Various important virus-like diseases in plants have been shown to be caused by pathogenic RNAs known as viroids, which have the following basic properties:

- Viroids are small circular molecules, a few hundred nucleotides long, with a high degree of secondary structure.

- Viroids do not code for any polypeptides and replicate independently of any associated plant virus.
- Viroids are the smallest known self-replicating genetic unit.

Viroids are of practical importance as they cause several economically significant diseases and are of general biological interest as being among the smallest known agents of infectious disease. The most studied viroid is *Potato spindle tuber viroid* (PSTVd). Viroid names are abbreviated to initials with a "d" added to distinguish them from abbreviations for virus names.

A. Classification of Viroids

Based on the sequence and predicted structures of their RNAs, the present 29 known viroids are classified into two families: the *Pospiviroidae* and the *Asunviroidae;* each family has several genera (Table 3.1).

B. Pathology of Viroids

1. Macroscopic Disease Symptoms

Viroids infect both dicotyledonous and monocotyledonous plants. As a group, there is nothing that distinguishes the disease symptoms produced by them from those caused by viruses. Their symptoms include stunting, mottling, leaf distortion, and necrosis and range from the slowly developing lethal disease in coconut palms caused by *Coconut cadang-cadang viroid* (CCCVd) to the worldwide symptomless infection of *Hop latent viroid* (HLVd).

2. Cytopathic Effects

Various effects of viroid infection on cellular structures have been reported. For example, in some infections changes have been observed in membranous structures called plasmalemmasomes. Several workers have described pronounced corrugations and irregular thickness in cell walls of viroid-infected tissue. Various

TABLE 3.1   Classification of Viroids[a]

Family	Genus	Viroid	Abbreviation	Size (nt)
Pospiviroidae	*Pospiviroid*	Potato spindle tuber	PSTVd	356–359
		Chrysanthemum stunt	CSVd	355–356
		Citrus exocortis	CEVd	369–373
		Tomato apical stunt	TASVd	360
		Tomato planta macho	TPMV	360
	Hostuviroid	Hop stunt	HSVd	296–303
	Cocadviroid	Coconut cadang-cadang	CCCVd	246
		Citrus viroid IV	CVd-IV	284
		Hop latent	HLVd	256
	Apscaviroid	Apple scar skin	ASSVd	329–331
		Apple dimple fruit	ADFVd	306
		Citrus viroid III	CVd-III	294
		Grapevine yellow speckle viroid 1	GYSVd-1	367
		Grapevine yellow speckle viroid 2	GYSVd-2	363
	Coleviroid	Coleus blumei viroid 1	CbVd-1	248–251
		Coleus blumei viroid 2	CbVd-2	301–302
		Coleus blumei viroid 3	CbVd-3	361–364
Avsunviroidae	*Avsunviroid*	Avocado sunblotch	ASBVd	247
	Pelamoviroid	Peach latent mosaic	PLMVd	337–338
		Chrysanthemum chlorotic mottle	CChMVd	399

[a]Only some representative viroid species listed. For more recent full lists, see Fauquet *et al.* (2005) and Flores *et al.* (2005).

degenerative abnormalities have been found in the chloroplasts of viroid-infected cells.

3. Location of Viroids in Plants

Using confocal laser scanning microscopy and transmission electron microscopy in conjunction with in situ hybridisation, both *Citrus exocortis viroid* (CEVd) and CCCVd were found in vascular tissues as well as mesophyll cells.

From experiments involving fractionating components of viroid-infected cells, it has become generally accepted that most viroids are located in the nucleus. The main exception is *Avocado sun blotch viroid* (ASBVd), which is found in chloroplasts. Within nuclei, PSTVd and CCCVd are located in nucleoli, whereas CEVd accumulates to higher concentrations in the nucleoplasm.

4. Movement in the Plant

Viruses with defective coat proteins and naked RNAs move slowly through the plant by cell-to-cell movement (see Chapter 9). By contrast, viroids move rapidly from cell to cell of a host plant in the manner of competent viruses. The cell-to-cell movement is via plasmodesmata and is mediated by specific sequences or structural motifs. Long-distance movement of viroids is almost certainly through the phloem. The relative resistance of viroid RNA to nuclease attack probably facilitates their long-distance movement. It is also possible that viroid particles undergoes RNA translocation while bound to some host protein.

5. Transmission

Viroids are readily transmitted by mechanical means in most of their hosts. Transmission in the field is probably mainly by contaminated tools or similar means. This ease of transmission of an RNA molecule in the presence of nucleases is probably due to viroid secondary structure and to the complexing of viroids to host components during the transmission process. Several viroids have been shown to be pollen and seed transmitted in tomato, potato, and grapes.

6. Epidemiology

The main methods by which viroids are spread through crops are by vegetative propagation, mechanical contamination, and pollen and seed. The relative importance of these methods varies with different viroids and hosts. For example, vegetative propagation is dominant for PSTVd in potatoes and *Chrysanthemum stunt viroid* in chrysanthemums. Mechanical transmission is a significant factor for others, such as CEVd in citrus and HSVd in hops. Seed and pollen transmission are factors in the spread of ASBVd in avocados.

For most viroid diseases, the reservoir of inoculum appears to be within the crop itself, which raises the question as to where the viroid diseases came from. The evidence suggests that many viroid diseases are of relatively recent origin. None of the recognised viroid diseases was known to exist before 1900, and many were first described since 1940. The sudden appearance and rapid spread of a new viroid disease can probably be accounted for by viroids being readily transmitted by mechanical means and many modern crops being grown as large-scale monocultures. Thus, from time to time a viroid present in a natural host and probably causing no disease might escape into a nearby susceptible commercial crop and spread rapidly within it. If the viroid and crop plant had not evolved together, disease would be a likely outcome. There is direct evidence for such a sequence of events with the tomato planta macho disease in Mexico.

C. Properties of Viroid RNAs

The properties of the RNAs of the two families of viroids are summarised in Table 3.2.

1. Sequence and Structure

The nucleotide sequences of most members of the viroid group and those of numerous viroid variants are now known. They range in size from 246 to 375 nucleotides (see Table 3.1).

TABLE 3.2   Properties of Viroid RNAs

	Pospiviroidae	*Avsunviroidae*
Shape of molecule	Circular	Circular
Secondary structure	Rodlike, about 37 nm long	Branched
Central conserved region	Present	Absent
Site of replication	Nucleus	Chloroplast
Mechanism of replication	Rolling circle, asymmetrical pathway	Rolling circle, symmetrical pathway
Replication enzyme	Nuclear DNA-dependent RNA polymerase II	Chloroplastic DNA-dependent RNA polymerase
Cleavage of product of replication	Host nuclear RNase	Hammer-head ribozyme

Members of both families have circular RNA molecules, but those of the *Pospiviroidae* are rodlike, whereas those of the *Avsunviroidae* are branched (Figure 3.1).

It should be remembered that these structures have been derived either from computer predictions or from *in vitro* experiments and that, *in vivo*, viroids may be associated with

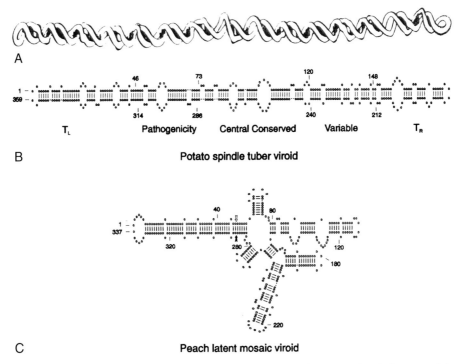

FIGURE 3.1   Structure of viroids. A. Three-dimensional representation of viroid molecule [Reprinted from *Matthews' Plant Virology*, 4th ed., R. Hull, Viroids, Satellite Viruses and Satellite RNAs, pp. 593–626, Copyright (2002), with permission from Elsevier]; B. Secondary structure of PSTVd showing the functional domains; C. Proposed secondary structure of *Peach latent mosaic viroid*. The predicted self-cleavage sites are indicated, (+) strand filled arrow, (−) strand open arrow. [This article was published in *Encyclopedia of Virology*, R. Owens (A. Granoff and R.G. Webster, Eds.), Viroids, pp. 1928–1937, Copyright Elsevier (1999).]

host proteins and have other structures. However, other evidence points to at least a partial rod-shaped structure *in vivo* in that viable duplications or deletions preserve this type of structure. Viroids have tertiary structure that is thought to be important in interactions with host proteins.

The predicted rodlike structures of the *Pospiviroidae* have five structural-functional domains that are common to all members (Figure 3.1B; see Box 3.1). These were thought to have specific functions, but the situation is now considered to be more complex. For instance, symptom expression is thought to be controlled by determinants located within the T_L, P, V, and T_R domains.

2. Replication

Even if it was assumed that the three out-of-phase potential reading frames were fully utilized, viroids do not contain enough information to code for an RNA replicase. It is now generally accepted that viroids are not translated to give any polypeptides. Table 3.2 shows that members of the *Pospiviroidae* replicate in the nucleus and members of the *Avsunviroidae* in chloroplasts. It is likely that viroids of both families have sequences and/or structure motif(s) for import into their replication organelle.

Viroids replicate via an RNA template, most probably by a rolling circle mechanism. Figure 3.2 illustrates two rolling circle models. In the asymmetric pathway, the infecting circular (+)-strand monomer is transcribed into linear multimeric (−) strands, which then are the template for the synthesis of linear multimeric (+) strands. In the symmetric pathway, the linear multimeric (−) strands are processed and ligated to give (−) monomer circles that are the template for linear multimeric (+)-strand synthesis. In both cases, the multimeric (+) strand is processed to give monomeric circles. As the symmetric pathway involves both (+)- and (−)-strand circular forms and the asymmetric pathway only (+)-strand circular forms, the two mechanisms can be distinguished by the presence or absence of the (−)-strand circular form. This RNA species has not been found in plants infected with PSTVd, and thus the replication of this viroid is considered to follow the asymmetric pathway. In contrast, (−)-strand circular monomer RNA forms have been found in ASBVd infections, which suggests that replication of this viroid follows the symmetric pathway.

The rolling circle replication of viroids involves the following features:

- Nuclear replication of members of the *Pospiviroidae* is by DNA-dependent RNA polymerase II. Chloroplastic replication of members of the *Avsunviroidae* is by a chloroplastic DNA-dependent RNA polymerase.
- Initiation of replication of PSTVd maps to the left terminal loop (see Figure 3.1B). The replication initiation site for members of the *Avsunviroidae* has still to be determined.

BOX 3.1

DOMAINS IN POSPIVIROIDAE RNA

The domains on PSTVd are illustrated in Figure 3.1.

C (central domain): This contains a central conserved region (CCR) of about 95 nucleotides.

P (pathogenicity domain): Implicated in pathogenesis.

T_L and T_R (left and right terminal domains): Implicated in replication.

V (variable domain): Varies between species often showing <50 percent homology between otherwise closely related viroids.

ASYMMETRIC

SYMMETRIC

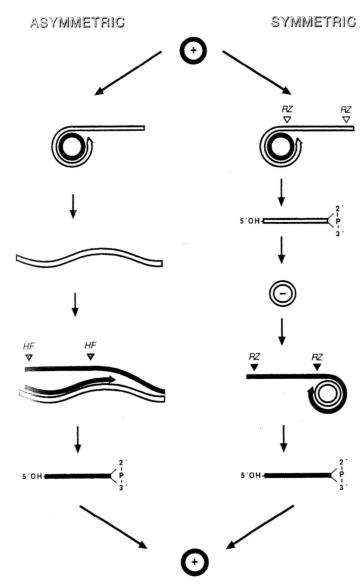

FIGURE 3.2   Models for viroid replication; see text for details. [This article was published in *Sem. Virol.*, **8,** R. Flores, F. Di Serio, and C. Hernandez, Viroids: the non-coding genomes, pp. 65–73, Copyright Elsevier (1997).]

- Processing of the long products of rolling circle replication of members of the *Pospiviroidae* to give monomeric (+) RNAs is effected by a host RNase activity in the nucleolus.
- Cleavage of the long replication products of members of the *Asunviroidae* is by a hammerhead ribozyme (see Box 3.2)

- Circularisation of the monomeric molecules of nuclear viroids is thought to be calalysed by host RNA ligase. It is unclear if a chloroplastic RNA ligase exists for members of the *Avsunviroidae* or whether the reaction is autocatalytic.

BOX 3.2

HAMMERHEAD RIBOZYMES

A hammerhead ribozyme is a small catalytic RNA motif that catalyzes self-cleavage reaction. Its name comes from its secondary structure, which resembles a carpenter' hammer (see figure). The hammerhead ribozyme is involved in the replication of the *Avsunviroidae* viroid and some satellite RNAs.

MINUS POLARITY HAMMERHEADS

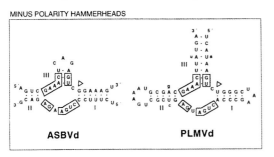

PLUS POLARITY HAMMERHEADS

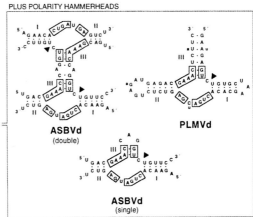

Fig. Hammerhead structures formed by *Avsunviroidae* that cleave replication (−)-sense and (+)-sense replication intermediates. Conserved nucleotides are boxed and cleavage site indicated by arrowhead. This article was published in *Sem. Virol.*, 8, R. Flores, F. Di Serio, and C. Hernandez, Viroids: the non-coding genomes, pp. 65–73, Copyright Elsevier (1997).

All hammerhead ribozymes contain three base-paired stems and a highly conserved core of residues required for cleavage. The cleavage reaction proceeds by an attack of a 2′ hydroxyl oxygen of a catalytic site cytosine on the phosphorus atom attached to the 3′ carbon of the same residue, breaking the sugar phosphate backbone and producing a 2′, 3′ cyclic phosphate. As for protein ribonucleases, a metal ion bound in the active site (Mg^{++}) stabilises the ionized form of the 2′ hydroxyl oxygen, promoting the catalytic attack.

Artificial ribozymes can be made to enable RNA to be cleaved at a predetermined site. The hammerhead motif is prepared *in vitro* by synthesising two separate strands and combining them together to give rise to the self-cleavage reaction.

3. Recombination Between Viroids

The nucleotide sequence data make it highly probable that recombination events have taken place in the past between different viroids, presumably during replication in mixed infections. For example, *Tomato apical stunt viroid* appears to be a recombinant viroid made up of most of the sequence of a CEVd-like viroid but with the T2 domain replaced by a T2 domain from a PSTVd-like viroid.

4. Interference Between Viroids

Inoculation of tomato plants with a very mild strain of PSTVd gives substantial protection against a second inoculation with a severe strain applied two weeks after the first. However, the mild strain must be established first. This protection, termed cross-protection (see Chapter 10), occurs not only between strains of a particular viroid but also between different viroids. For example, when PSTVd RNA transcripts from a cloned PSTVd DNA are inoculated together with HSVd RNA, the PSTVd reduces the level of HSVd RNA in infected plants. Plants inoculated with dual transcripts—two copies of a severe PSTVd strain linked to two of HSVd—develop PSTVd symptoms and only PSTVd progeny RNA can be detected.

D. Molecular Basis for Biological Activity

Because of their very small size, their autonomous replication, the known structure of many variants, and the lack of any viroid-specific polypeptides, it has been hoped by many workers that viroids might provide a simple model system that would give insights as to how variations in the structure of a pathogen modulate disease expression. Very small changes in nucleotide sequence may give rise to dramatic changes in the kind of disease induced by a viroid, and, therefore, disease induction must involve specific recognition of the viroid sequence by some host macromolecule(s). Until the nature of the host macromolecule(s) is known, the interpretation of correlations between nucleotide sequence and biological properties of viroids will remain speculative. At present, the only practicable biological properties that can be observed are infectivity and severity of disease, although viroid-binding proteins are starting to be recognised.

Sequence comparison between viroid variants and site-directed mutagenesis has identified the P domain but has shown the importance of viroid structure, with many mutations rendering it noninfectious. Comparisons with host RNA sequences reveal similarities between viroid sequences and ribosomal RNA processing sequences, suggesting the potential

for interference. Some host proteins have been found to bind to viroids, though their role in pathogenesis has not yet been elucidated. A protein of 602 amino acids from tomato has been suggested to be involved either in the transport of PSTVd RNA into or from the nucleus or in the formation of the replication complex. The discovery of viroid-specific siRNAs has led to the suggestion that viroids could be inducers of RNA silencing (see Chapter 11 for RNA silencing). It is thought that RNA silencing could mediate the pathogenicity of viroids.

E. Diagnostic Procedures for Viroids

Since viroids produce no specific proteins, the immunological methods applied so successfully to viruses cannot be used for the diagnosis of diseases that they cause. Similarly, because no characteristic particles can be detected, electron microscopic techniques are inappropriate. For these reasons diagnostic procedures have been confined to biological tests, gel electrophoresis, PCR, and nucleic acid hybridisation tests.

II. PHYTOPLASMA

Some important virus-like diseases are caused by bacteria that lack a cell wall and are obligate parasites of the phloem tissue. There are two genera of such organisms: *Phytoplasma* and *Spiroplasma*.

Phytoplasma were formerly known as Mycoplasma-like organisms, or MLOs. They have a pleomorphic or filamentous shape with a diameter of less than 1 μm and very small genomes. Taxonomically they belong to the order *Acholeplasmatales*; the genus name of *Phytoplasma* is not yet formally recognised and is currently at Candidatus stage.

Two common symptoms of phytoplasma infection are phyllody—the production of leaf-like structures in place of the floral parts and yellowing of the leaves due to the presence of the pathogen in the phloem disrupting the transport of carbohydrates. Many phytoplasma-infected plants have a bushy appearance due to changes in growth patterns that induce proliferation of axillary side shoots and increase in size of internodes, giving witches' broom symptoms. This symptom can be of commercial use—for instance, it can enable the production of poinsettia plants that have more than one flower. Phytoplasma are restricted to phloem tissue, and it is considered that the symptoms of infected plants are due to disturbance of phloem functions, including upsetting the hormone balance as well as inducing host defence proteins.

Phytoplasma are mainly transmitted by leafhoppers (family *Cicadellidae*) and planthoppers (family *Fulgoridae*) in a propagative circulative manner (see Chapter 12 for virus-vector relationships). They infect the vector and thus are pathogens of both the plant and insect. Phytoplasma are also transmitted by grafting infected material into healthy plants and by taking cuttings from infected plants.

The genome sizes of sequenced phytoplasma range from 530 kb to 1,350 kb, encoding between 671 and 754 genes; they also contain plasmids. They lack many of the genes essential for cell metabolism, such as those for de novo synthesis of amino acids, fatty acids, or nucleotides, and thus they rely on the uptake of nutrients from both their plant and insect hosts.

Diagnosis of phytoplasma infection is usually by molecular techniques such as PCR. Control is most frequently by breeding and deployment of resistant varieties and by control of the insect vector. For long-term valuable perennials such as trees, the application of tetracycline has proved effective but very expensive.

Spiroplasma resemble phytoplasma in many respects but have a distinct helical morphology. They can also be cultured *in vitro*, requiring a rich culture medium, and they grow well at 30°C but not at 37°C. Taxonomically they belong to the order *Entomoplasmatales*, genus *Spiroplasma*. The Spiroplasma genome size is 780–2,200 kb and, as with Phytoplasma, lacks many of the genes requires for full cell metabolism. They contain genes that confer helicity and motility, but these properties are not required for pathogenicity or transmission.

Two important diseases caused by spiroplasma are citrus stubborn disease due to infection by *Spiroplasma citri* and corn stunt caused by *Spiroplasma kunkelii*. The symptoms of citrus stubborn are shorter and broader cupped leaves and the development of witches' broom symptoms. Fruiting is suppressed and fruit quality affected. In corn stunt infections the plants are stunted, have broad chlorotic stripes on the leaves, and have poor filling of the ears. Both these diseases are transmitted by leafhoppers, and diagnosis and control are similar to that for phytoplasma.

III. SATELLITE VIRUSES AND SATELLITE RNAs

Satellites are subviral agents that lack the genes encoding functions that are necessary for their replication. They depend for their multiplication on coinfection of a host cell with a helper virus. The genomes of satellites differ substantially or totally from those of their helper virus.

Two major classes of satellite agents can be distinguished according to the source of the coat protein used to encapsulate the nucleic acid. In **satellite viruses**, the satellite nucleic acid codes for its own coat protein. In **satellite RNAs** or **DNAs**, the nucleic acid becomes packaged in protein shells made from coat protein of the helper virus. Satellite viruses and satellite nucleic acids have the following properties in common:

- Their genetic material is a nucleic acid molecule of small size. The nucleic acid is not part of the helper virus genome and usually has little sequence similarity to it apart from the terminal regions.
- Replication of the nucleic acid is dependent on a specific helper virus.

- The agent affects disease symptoms, at least in some hosts.
- Replication of the satellite interferes to some degree with replication of the helper.
- Satellites are replicated on their own nucleic acid template.

Both RNA and DNA viruses have satellites associated with them—those of RNA viruses being RNA and those of DNA viruses being DNA. Satellite viruses and nucleic acids can be categorised into six groups based on their properties (Table 3.3). Not included in the table are defective RNAs and DNAs, which have sequences derived from those of the helper virus; these are described in detail later in this chapter. The coat-dependent RNA replicons are included because their sequences differ from those of the helper virus.

Several satellite RNAs associated with a particular group of viruses have been shown to have viroid-like structural properties. These agents have been termed virusoids, but this term is no longer favoured. There appears to be no taxonomic correlation between the

viruses that are associated with satellites, and satellitism seems to have arisen a number of times during virus evolution. Some viruses are associated with more than one satellite, and satellites can even require a second satellite as well as the helper virus for replication.

A. Satellite Plant Viruses (A-type)

Satellite tobacco necrosis virus (STNV) was the first satellite virus to be recognised; the term satellite virus was coined to denote the 17 nm isometric particles associated with the 30 nm *Tobacco necrosis virus* (TNV) isometric particles. The helper *Necrovirus* TNV replicates independently of other viruses and normally infects plant roots in the field. There is significant specificity in the relationship between satellite and helper. Strains of both viruses have been isolated, and only certain strains of the helper virus will activate particular strains of the satellite.

Both STNV and TNV are transmitted by the zoospores of the fungus *Olpidium brassicae*

TABLE 3.3   Satellite Viruses, Nucleic Acids, and Coat-Dependent RNA Replicons

Satellite Type	Satellite	Helper Virus[a]	Independent Replication	Directs Protein Synthesis	Coat Protein
Satellite virus genomic RNA (A type)	STNV genomic RNA	TNV	No	Yes	Own
Messenger satellite RNA (B type)	BsatTomBRV RNA	TomBRV	No	No	Helper virus
Nonmessenger satellite RNA (C type)	SCMV RNA	CMV	No	No	Helper virus
Nonmessenger satellite RNA (D type)	STRSV RNA	TRSV	No	No	Helper virus
Satellite DNA	Begomovirus DNAβ		No	No	Helper virus
Coat-dependent RNA replicon	ST9aRNA		Yes	Yes	Helper virus

[a]Abbreviations of viruses: CMV = *Cucumber mosaic virus*; TNV = *Tobacco necrosis virus*; TomBRV = *Tomato black ring virus*; TRSV = *Tobacco ringspot virus*.

(Chapter 12). Transmission depends on an appropriate combination of four factors: satellite and helper virus strains, race of fungus, and species of host plant.

The complete nucleotide sequence of STNV RNA was one of the first viral sequences to be determined and has no significant similarity with that of the TNV genome. The STNV genome contains only one open reading frame (ORF): that for its coat protein. The 5′ terminus of STNV is unlike that of most ss (+)-sense RNA plant viruses (see Chapter 6 for RNA terminal structures). There is neither a 5′ cap nor a VPg, and the 5′ termination is 5′-ppApGpUp-. The 3′-terminal region can be folded to give a tRNA-like structure with an AUG anticodon, but there is no evidence that this structure can accept methionine.

Relatively little is known about the replication of STNV *in vivo*, but it is widely assumed that the RNA replication must be carried out by an RNA-dependent RNA polymerase coded for, at least in part, by the helper virus. It is assumed that replication is through a (−) strand in a manner similar to that of (+)-strand RNA viruses (see Chapter 8). Replication of STNV substantially suppresses TNV replication, and it is thought likely that this involves competition for the replicase. The presence of STNV in the inoculum reduces the size of the local lesions produced by the helper virus, possibly by reduction in TNV replication.

B. Satellite RNAs (satRNAs)

As Table 3.3 shows, there are three classes of satellite RNAs. These do not encode their own coat protein and are encapsidated in the helper virus coat protein. They are grouped on size, the large ones with mRNA activity being termed B type and the smaller ones without mRNA activity being divided on form: those with linear molecules (C type) and those with circular molecules (D type).

1. Large Satellite RNAs (B-type)

As with the other satellite RNAs, B-type satellites are dependent on the helper virus for replication and encapsidation, but in contrast to the other types they direct protein synthesis both *in vitro* and *in vivo* from a single ORF. The RNAs of the B-type satellites thus far studied are 0.7 kb or larger. Most of the B-type satellites (given the prefix "Bsat") are found associated with nepoviruses, and one has been found associated with a potexvirus.

The B satellite of *Tomato black ring virus* (BsatTBRV) is an RNA of 1,374 nucleotides that is packaged in varying numbers in particles made of the helper virus coat protein, giving rise to a series of components of differing buoyant density in solutions of CsCl. Like the helper virus, the satellite RNA is transmitted through the seed and also by the nematode vector of the virus. The satellite does not appear to affect replication of the helper virus or to modify symptoms, except that the number of local lesions on *Chenopodium amaranticolor* is reduced.

BsatTBRV RNA contains one ORF with a coding capacity for a protein of 419–424 amino acids, has a VPg at the 5′ terminus, and is polyadenylated at the 3′ end (see Chapter 6 for RNA terminal structures). Mutagenesis studies suggest that this protein region and the 5′ and 3′ untranslated regions are all necessary for the replication of the Bsat. It is most likely that B satellites replicate in the same manner as the helper virus.

2. Small Linear Satellite RNAs (C-type)

C-type satellites have linear RNA molecules that are generally smaller than 0.7 kb and do not have mRNA activity. They are found associated with cucumoviruses, tombusviruses, and carmoviruses and can cause severe disease outbreaks (Box 3.3).

There are many natural variants of the satellite associated with *Cucumber mosaic virus* (satCMV), all differing in sequence from CMV

BOX 3.3
_____

SEVERE DISEASE IN TOMATO CAUSED BY A SATELLITE OF CUCUMBER MOSAIC VIRUS (CMV)

In 1972, a devastating outbreak of a lethal necrotic disease of field-grown tomatoes occurred in the Alsace region of France. It was discovered that CMV was involved in this outbreak, but it was not clear why necrosis occurred instead of the usual fern-leaf symptoms. An additional small RNA, called satCMV or CARNA5, that was not part of the viral genome was identified and when added to CMV isolates caused lethal necrotic disease in tomatoes. Similar outbreaks in tomatoes in southern Italy have been shown to be due to a necrogenic isolate of satCMV. However, most CMV satellites attenuate the symptoms, and the overall effect depends on the combination of satellite RNA and helper virus strain.

---

itself. They vary in length from about 330 to 390 nucleotides. It is probable that most satellite RNA preparations consist of populations of molecules with closely related sequences all dependent on the CMV genome. Packaging in CMV coat protein enables the satellite RNA to be transmitted by aphids that transmit CMV.

Like the helper genomic CMV RNAs, they are all capped with M^7Gppp at their 5′ termini and have a hydroxyl group at the 3′ end (see Chapter 6 for RNA terminal structures). A 3′-terminal tRNA-like structure is possible, but attempts to aminoacylate the RNA have been unsuccessful. The RNA has a high degree of secondary structure, which may account for the stability of the RNA both *in vitro* and *in vivo* and for its relatively high specific infectivity.

The carmovirus *Turnip crinkle virus* (TCV) supports a family of satellite RNAs. Two of the satellites, D and F, do not affect symptoms, while C is virulent, intensifying TCV symptoms in turnip. The virulent satellite C is of particular interest, since, unlike other known satellites, it appears to be a recombinant between D satellite and a defective interfering RNA (see following) of the helper virus TCV (Box 3.4). As with B satellites, it is thought that C satellites replicate in the same manner as the helper virus.

3. Small Circular Satellite RNAs (D-type)

D-type satellite RNAs are small, about 350–450 nucleotides, and occur as circular as well as linear molecules. They are found associated with nepoviruses, luteoviruses, and sobemoviruses. Several nepoviruses support the replication of satellite RNAs, which become encapsulated in particles made of the helper virus coat protein. As well as the large B-type satellites described previously, small ones with circular RNA molecules also exist. The satellite of *Tobacco ringspot virus* (sTRSV) consists of a small RNA species of about 359 nucleotides in a protein shell identical to that of the helper virus. Twelve to 25 satellite RNA molecules become packaged in a single particle. The satellite cannot replicate on its own, and it interferes with the replication of TRSV. There are no ORFs, and the RNA is not translated *in vitro*.

The satellite RNA associated with the luteovirus *Cereal yellow dwarf virus* (sRPV) reduces the accumulation and symptom production of the helper virus in oats. This satellite also has multimeric linear and circular forms together with a hammerhead ribozyme sequence (see Box 3.2).

BOX 3.4

TURNIP CRINKLE VIRUS SATELLITE C

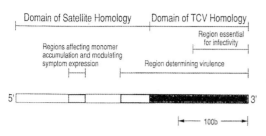

Fig. Structure of TCV satellite RNA C showing the various domains. From Hull (2002).

TCV satellite C is 355 bases (see figure). The 189 bases at the 5′ end are homologous to the entire sequence of a smaller nonvirulent TCV satellite D RNA. The rest of the RNA C molecule (166 bases in the 3′ region) is nearly identical in sequence to two regions at the 5′ end of the TCV helper genome. Thus, in a structural sense at least, TCV satellite RNA C is a hybrid between a satellite RNA and a DI RNA. When plants are inoculated with a mixture of TCV, satellite RNA D, and satellite RNA C transcripts containing nonviable mutations in the 5′ domain, recombinant satellite RNAs are recovered. Sequence analysis around 20 recombinant junctions supports a copy-choice model for this recombination. In this model, while in the process of replicating (−) strands, the replicase can leave the template together with the nascent (+) strand and can reinitiate synthesis at one of two recognition sequences on the same or a different template.

Four viruses described from Australia, which belong in the *Sobemovirus* genus, have been shown to contain small RNAs that occur in both circular and linear forms. The small RNAs have been shown to have the biological properties of satellite RNAs. The helper virus can replicate independently, while the small RNA cannot. The satellite RNAs, which range in size from about 325 to 390 nucleotides, are encapsidated in both linear and closed circular forms. They show no sequence similarity with the helper RNAs.

There is quite strong evidence that D satellites are replicated by a rolling circle mechanism as just described for viroids. The cleavage of the multimeric forms appears to be autolytic by a phospho-transfer reaction between an adenylate residue and the 2′ hydroxyl of a neighboring cytidylate, to give rise to a 5′ hydroxyl terminal adenosine and a 3′ cytosine 2′,3′-cyclic phosphodiester group. The viroid-like satellite RNAs are predicted to form hammerhead-shaped self-cleavage domains that resemble those of ASBVd.

4. Satellite-Like RNAs

Most satellites are not essential for the biological success of the helper virus. However, it is a relatively small evolutionary step from this situation to one where the satellite is involved in the disease spread and expression. Since such a molecule cannot be described as nonessential for the helper virus, it cannot technically be considered a satellite; hence, they are grouped together as "satellite-like RNAs."

a. A Satellite RNA of Groundnut Rosette Virus (GRV).
The satellite associated with GRV (satGRV) is a linear ssRNA of 895–903 nucleotides that relies on GRV for its replication. Although different variants of sGRV contain up to five potential ORFs in either the

(+)- or (−)-sense strands, none of these ORFs is essential for replication of the satellite RNA.

As discussed in Box 2.1, satGRV is involved with the two viruses causing groundnut rosette disease and is essential for encapsidation and symptom production within the complex.

b. Ancillary RNAs of Beet Necrotic Yellow Vein Virus (BNYVV).

Field isolates of BNYVV contain four or five RNA components, but laboratory isolates maintained by mechanical inoculation of *Chenopodium quinoa* or *Tetragonia expansa* may lack any or all of the three smaller RNAs. Thus, the BNYVV genome comprises RNAs 1 and 2 (Profile 2; see Appendix) together with RNAs 3, 4, and 5 that resemble satellite RNAs. Each of the satellite-like RNAs encodes one ORF and contributes to the field pathology of the parent virus. RNA3 contributes to a large degree to the pathology of the virus, whereas RNA4 greatly increases the efficiency of virus transmission by its vector. RNA5 seems to play a role in both the symptom production and transmission.

5. The Molecular Basis for Symptom Modulation

As noted in Box 3.3, satellite RNAs can be responsible for outbreaks of severe disease in the field. In addition, because of their small size and the availability of many sequence variants, either isolated from nature or produced in the laboratory, they are amenable to detailed molecular study. For these reasons it has been hoped that studies correlating molecular structure of the satellite RNAs with their effects on disease in the plant would give us some insight into the molecular basis of disease induction. Results so far have been difficult to interpret. This is not surprising, since multiple factors are involved, including the strain of the helper virus, the coat protein of the helper virus (TCV and sat RNA C), the host (satCMV), the strain of satellite, specific sequences within the satellite RNA, and perhaps the environment. This is exemplified by comparison of the nucleotide sequences of CMV satellite RNAs, suggesting that only a few nucleotide changes may be necessary to change the host response and that different kinds of disease response (e.g., yellowing or necrosis) may be associated with different domains of the satellite sequence. The domain for chlorosis in tomato is in the 5′ 185 nucleotides of these RNAs, while the domain for necrosis is in the 3′ 150 nucleotides. The sequence of the WL1-sat, which attenuates CMV symptoms on tomatoes, differs from all necrogenic satellite RNAs at three nucleotides positions in the conserved 3′ region of the RNA. This has implications in the use of satellite RNAs as a control measure to mitigate symptom expression (see Chapter 14) in that relatively few mutations could convert a symptom attenuation satellite strain to a necrogenic strain.

A satellite RNA of particular interest in relation to disease modulation is the virulent satellite RNA of TCV illustrated in Box 3.4. Other avirulent strains of the satellite (e.g., strain F) lack the 3′ sequence derived from the helper TCV genome. The 3′ sequence derived from TCV contains a region essential for infectivity and a larger overlapping region determining virulence. The domain of satellite homology contains regions affecting monomer accumulation and modulating symptom expression.

The only general conclusion to be drawn is that changes in a disease induced by the presence of a satellite RNA depend on changes in its nucleotide sequence. So far, there is no convincing evidence that such changes are mediated by a polypeptide translated from the satellite RNA. Indeed, differences in disease modulation occur between RNAs that have no significant ORFs. Disease modulation is most probably brought about by specific macromolecular interactions between the satellite RNA and (1) helper virus RNAs, (2) host RNAs, (3) virus-coded proteins, (4) host proteins, or (5) any combination of 1–4. Until the kinds of interaction that are important in disease induction have been established on a molecular basis, differences in nucleotide sequence between related satellites will remain largely uninterpretable.

C. Satellite DNAs

DNA extracted from preparations of monopartite begomoviruses often contains molecules that are not the viral genomic DNA (Figure 3.3). Some of these are considered to be satellite DNA and others defective interfering DNAs.

The most studied satellite DNA is termed DNAß, which forms a complex with most, if not all, monopartite begomoviruses, causing significant disease symptoms. DNAßs are approximately half the size of the genomic DNA (Figure 3.3), have a 5' stem-loop structure, one ORF (ßC1), an A-rich region, and a 3' sequence conserved between all DNAßs. They show no significant homology to their helper begomovirus, on which they are dependent for replication, encapsidation, and movement both within and between hosts. They enhance the titre of the helper virus and are often involved with symptom production that involves ORF ßC1.

The replication of DNAß is considered to be by the rolling circle mechanism used by its helper virus (see Chapter 8 for geminivirus replication). The replication of one DNAß species is supported by several begomoviruses, but the symptom enhancement is only found in the original helper virus/DNAß combination.

The first satellite to be recognised to be associated with a DNA virus was a 262 nucleotide circular ssDNA associated with the begomovirus *Tomato leaf curl virus* (ToLCV) from northern Australia. Replication of ToLCV-sat is dependent on the helper virus replication-associated protein (see Chapter 8 for geminivirus replication-associated protein), and the satellite DNA is encapsidated by the helper virus coat protein. It has no significant ORFs, and the only significant sequence similarity to that of the

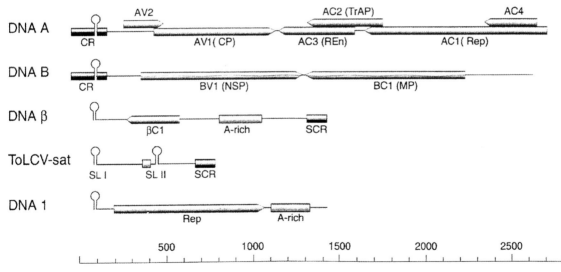

FIGURE 3.3 Genome organisation of begomoviruses and their associated subviral components. All components are circular but are shown as linear maps for clarity. The common region (CR) is highly conserved between DNA-A and DNA-B components of a particular begomovirus. The satellite-conserved region (SCR) is highly conserved between all known DNA-ß components. Genes encoding the coat protein (CP), replication-associated protein (Rep), transcriptional transactivator protein (TrAP), replication enhancer (REn), movement protein (MP), and nuclear shuttle protein (NSP) are indicated. This article was published in *Virology*, **344**, R.W. Briddon and J. Stanley, Subviral agents associated with plant single-stranded DNA viruses, pp. 198–210, Copyright Elsevier (2006).

helper virus is in two short motifs in separate putative stem-loop structures: TAATATTAC, which is conserved in all geminiviruses, and AATCGGTGTC, which is identical to a putative replication-associated protein binding motif in ToLCV (Figure 3.3).

D. Discussion

Satellite viruses and RNAs form specific associations with their helper viruses. This raises two interlinked questions: How did they evolve? and What is their natural function? It is thought that satellite viruses arose from an independent virus by degenerative loss of functions in a mixed infection with the helper virus. It seems likely that satellite RNAs arose from at least three different lineages, as shown in Figure 3.4.

As to function, there are two possibilities: Satellites are either "molecular parasites" or they have a beneficial role in the biology of the helper viruses. As molecular parasites, satellites compete for the replication machinery with the helper virus. However, this may not be to the disadvantage of the helper virus.

It is possible that the attenuation of symptoms by many satellites could enhance the survival of the host to the benefit of the helper virus. The variants that induce more severe symptoms, such as some CARNA 5 isolates, would be selected against under natural conditions because they kill their host and thus limit the spread of the helper virus.

IV. DEFECTIVE AND DEFECTIVE INTERFERING NUCLEIC ACIDS

Truncated, and often rearranged, versions of genomic viral nucleic acids are associated with many plant and animal RNA viruses and some plant DNA viruses. These subviral nucleic acids are termed defective interfering (DI) if they modulate the helper virus replication or defective (D) if they have no effect. They are dispensable for helper virus propagation, completely dependent on the helper virus for their replication, and are derived mainly by one or more premature termination and reinitiation events during the replication of the helper virus genomic nucleic acid. These subviral nucleic acids contain all the *cis*-acting elements necessary for replication by the replication system of the parent virus but usually lack genes required for encapsidation and movement.

DI nucleic acids can cause major alterations to the normal disease progression of the helper virus. Most moderate the symptoms of the helper virus but some (e.g., TCV) can enhance symptom severity. The D and DI nucleic acids and viruses of plants are listed in Table 3.4.

The deletion patterns in the DI and D RNAs and viruses fall into two groups: Group 1, in which the defective RNA is derived from a single internal deletion, and Group 2, which consists of a mosaic of the parental viral genome (Table 3.4 and Figure. 3.5).

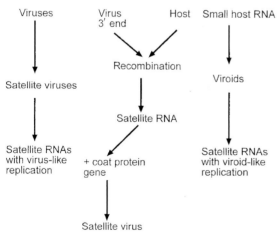

FIGURE 3.4 Hypothetical schemes for the derivation of satellite RNAs. [From Hull (2002).]

TABLE 3.4 Examples of Defective and Defective Interfering Nucleic Acids (Modified from Hull, 2002)

Family/ Genus	Virus[a]	Deleted Segment	Type of Defective Element	Comments
RNA Viruses				
Reoviridae	WTV	RNAs 2 and 5	D virus	Loss of vector transmission
Tospovirus	TSWV	L RNA	DI RNA	Encodes polymerase protein
		M RNA	D virus	Loss of viral envelope and probable loss of vector transmission
Bromoviridae	BBMV	RNA2	DI RNA	Exacerbates symptoms in some hosts, encodes viral polymerase.
	CMV	RNA3	D RNA	Deletion in 3a protein; encodes coat protein.
Tobravirus	TRV	RNA2	D and DI Virus	Vector transmission eliminates DI
Furovirus	SBWMV	RNA2	D Virus	Loss of vector transmission
		RNAs 3 and 4	D Virus	Loss of vector transmission and ability to infect roots
Tombusviridae	TCV	Various	DI RNA	See text
DNA Viruses				
Begomovirus	CLCuMV	See DNA-1 in Fig. 3.3	DI DNA	See text

[a]Abbreviations for viruses: BBMV = *Broad bean mottle virus*; CLCuMV = *Cotton leaf curl Multan virus*; CMV = *Cucumber mosaic virus*; SBWMV = *Soil-borne wheat mosaic virus*; TCV = *Turnip crinkle virus*; TRV = *Tobacco rattle virus*; TSWV = *Tomato spotted wilt virus*; WTV = *Wound tumor virus*.

FIGURE 3.5 Defective and defective interfering RNAs. A. Group 1 single-deletion D RNA. CMV genomic RNA3 (top) and the D RNA 3ß derived from it (bottom; the region deleted from the 3a ORF is indicated. [This article was published in *Sem. Virol.*, **7**, M.V. Graves, J. Pogany, J., and J. Romero, Defective interfering RNAs and defective viruses associated with multipartite RNA viruses of plants, pp. 399–408, Copyright Elsevier (1996).] B. Group 2 DI RNA. The upper part shows the genome organisation of TBSV. Below are two size classes of DI RNAs with the portions derived from the TBSV genomic RNA indicated by shaded boxes with the deleted intervening regions shown as lines. [This article was published in *Sem. Virol.*, **7**, K.A. White, Formation and evolution of Tombovirus devective interfering RNAs, pp. 409–416. Copyright Elsevier (1996).]

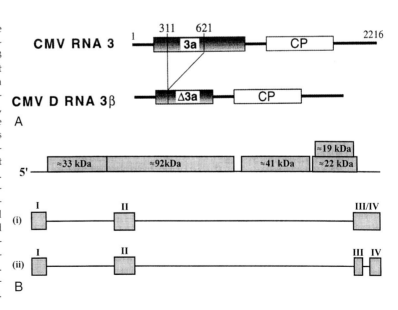

A. Group 1: Single Deletion D RNAs

The defective RNAs associated with CMV have no apparent effect on symptom production or virus accumulation. They are derived from RNA3 (Figure 3.5A) by a single deletion that removes a segment of the 3a ORF but maintains the reading frame downstream of the deletion. This leaves a defective cell-to-cell movement protein and a functional coat protein. CMV D RNAs accumulate in various *Nicotiana* species, but in tomato, zucchini squash, and muskmelon, the D RNAs only accumulate in inoculated tissue but do not move systemically. Furthermore, the D RNA accumulates and is encapsidated in both inoculated cotyledons and leaves of tomato and zucchini squash and accumulates but is not encapsidated only in the inoculated cotyledons of muskmelon. This indicates that host and tissue specificity is involved in replication, cell-to-cell movement, systemic movement, and encapsidation of CMV D RNAs.

B. Group 2: Multiple Deletion D and DI RNAs

Multiple deletion DI and D RNAs are characteristic of several members of the *Tombusviridae* family (Table 3.4 and Figure 3.5). These DI RNAs are about 10 to 20 percent the size of the parent genome and have common structural features that retain terminal regions and an internal segment but lack all coding capacity (Figure 3.5B).

DI RNAs can be generated de novo by high-multiplicity passages of transcripts from cloned helper virus infectious cDNA. Although the DI RNA can represent 60 percent of virus-specific RNA in leaf extracts, less than 5 percent of the encapsidated RNA is DI RNA. The presence of DI RNAs in infected plants greatly reduces helper virus accumulation and persistently attenuates symptoms. Three mechanisms have been suggested for DI RNAs to give these effects: competition with the helper virus for transacting factors necessary for replication, specific interaction with virus-encoded products, and activation of the RNAi defence system. It is possible that more than one of these mechanisms operates in a helper virus/DI RNA interaction.

C. Defective DNAs Associated with DNA Viruses

Besides having satellite DNAs, some monopartite begomoviruses also have DI DNAs. These ameliorate the symptoms and reduce virus accumulation of the helper virus. These circular ssDNA molecules have been termed DNA-1 and are about half the size of the helper begomovirus. They encode the Rep gene (see Chapter 8 for geminivirus replication) and can replicate autonomously; however they depend on the helper begomovirus for encapsidation and movement both within and between plants. Thus, technically they are neither satellites nor true DI DNA, as they do not fit the strict definitions of either. It has been suggested that they are similar to the nanovirus component encoding the Rep and have been modified to overcome the size constraints of geminivirus encapsidation.

V. VIRUSES OF OTHER KINGDOMS

Many other agents cause virus-like diseases in other kingdoms. For instance, prions were originally considered to be similar to viroids but are now recognised to be completely different protein entities. Spiroplasms have been associated with transmissible spongiform encepatholopies.

Satellite viruses have been found in association with viruses of both vertebrates and invertebrates. *Hepatitis delta virus* (HDV) is a satellite of *Hepatitis B virus* (HBV). Like its helper virus, HDV has an RNA genome, but it encoded only one protein: the delta antigen that encapsidates it. It requires HBV for it replication. In invertebrates *Chronic bee-paralysis associated satellite virus* has an RNA genome encoding just its coat protein and depends on *Chronic bee-paralysis virus* for its replication. Satellite RNAs and DNAs have only been found associated with plant viruses, but D and DI RNAs and DNAs are found with both plant and animal viruses.

VI. SUMMARY

- The effects of various agents resemble or alter plant virus diseases.
- Viroids are small pathogenic molecules that do not encode any proteins and are the smallest known self-replicating nucleic acid.
- Phytoplasma are obligate parasites of plants that lack a cell wall.
- Satellite viruses encode their own coat protein but depend on the helper virus for replication.
- Various forms of satellite RNAs and DNAs exist, and each requires the helper virus for replication, although their nucleic acid differs from that of the helper virus.
- Defective and defective interfering nucleic acids are derived from that of the helper virus by deletion and thus have sequence similarities to that of the helper virus.
- DI RNAs and DNAs modulate the symptoms caused by the parent virus.

References

Fauquet, C.M., Mayo, M.A., Maniloff, J., Desselberger, U., and Ball, L.A. (2005). *Virus taxonomy. Eighth Report of the International Committee on Taxonomy of Viruses.* Elsevier, San Diego, California.

Flores, R., Hernández, C., Martinéz de Alba, A.E., Daròs J.-A., and Di Serio, F. (2005). Viroids and viroid-host interactions. *Annu. Rev. Phytopathol.* **43**, 117–139.

Hull, R. (2001). *Matthews' plant virology.* Academic press, San Diego.

Further Reading

Bové, J.M., Renaudin, J., Saillard, C., Foissac, X., and Garnier, M. (2003). *Spiroplasma citri*, a plant pathogenic mollicure: Relationships with its two hosts, the plant and the leafhopper vector. *Annu. Rev. Phytopathol.* **41**, 483–500.

Briddon, R.W. and Mansoor, S. (2008). Beta ss DNA satellites. *Encyclopedia of Virology*, Vol. 1, 314–320.

Christensen, N.M., Axelsen, K.B., Nicolaisen, M., and Schulz, A. (2005). Phytoplasmas and their interactions with hosts. *Trends in Plant Science.* **10**, 526–533.

Ding, B. and Itaya, A. (2007). Viroid: A useful model for studying the basic principles of infection and RNA biology. *Molec. Plant Microbe Interact.* **20**, 7–20.

Flores, R. and Owens, R.A. (2008). Viroids. *Encyclopedia of Virology*, Vol. 5, 332–341.

Hogenhout, S.A., Oshima, K., Ammar, E.-D., Kabizawa, S., Kingdom, H.N., and Namba, S. (2008). Photoplasmas: bacteria that manipulate plants and insects. *Molec. plant. Pathol.* **9**, 403–423.

Lee, I.-M., Vavis, R.E., and Gundersen-Rindal, D.E. (2000). Phytoplasma: Phytopathogenic mollicutes. *Annu. Rev. Microbiol.* **54**, 221–255.

Palukaitis, P., Rezaian, A., and Garcia-Arenal, F. (2008). Satellite RNAs, satellite DNAs, and satellite viruses. *Encyclopedia of Virology*, Vol. 4, 526–534.

Rocheleau, L. and Pelchat, M. (2006). The subviral RNA database: A toolbox for viroids, the hepatitis delta virus, and satellite RNAs research. *BMC Microbiology* 6:24. http://www.biomedcentral.com/1471-2180/6/24.

Roux, L. (2008). Defective interfering viruses. *Encyclopedia of Virology*, Vol. 2, 1–4.

Simon, A.E., Roosinck, M.J., and Havelda, Z. (2004). Plant virus satellite and defective interfering RNAs: New paradigms for a new century. *Annu. Rev. Phytopathol.* **42**, 415–437.

Taylor, J.M. (2008). Hepatitis delta virus (Deltavirus). *Encyclopedia of Virology*, Vol. 2, 375–376.

Westhof, E. and Lescoute, A. (2008). Ribozymes. *Encyclopedia of Virology*, Vol. 4, 475–480.

Plant Virus Origins and Evolution

Like other living entities, viruses substantially resemble the parent during their replication, but they can change to create new types, or *strains*. This inherent variation enables viruses to adapt to new and changing situations. Over longer periods of time, new viruses evolve, but there must have been a time at which the archetypical virus arose.

I. INTRODUCTION

For many groups of organisms, an understanding of the evolutionary pathways can be obtained from the fossil record over geological time, but viruses do not leave the conventional form of fossils; insects preserved in amber and seeds from archaeological sites may have some evidence of viruses. However, the increasing molecular information that is accruing on plant (and other) viruses is revealing what can be termed molecular fossil information. Even so, our knowledge of the pathways of virus evolution is quite fragmentary, although there is no doubt that viruses have undergone, and continue to undergo, evolutionary change that is sometimes quite rapid.

II. VIRUS EVOLUTION

In discussing evolution of viruses, we must recognise that it is distinct from the evolution of virus diseases. A new disease may be the consequence of the "evolution" of the causal virus, but it can also result from no change in the virus. For instance, a new disease can result from the movement of an "old" virus into a new situation. It is likely that the epidemics of swollen shoot in cacao in West Africa, maize streak in Africa, and tungro in rice in Southeast Asia were due to the spread of the viruses from asymptomatic natural hosts into either a new species in that area or a changed agronomic situation.

As just noted, viruses do not form conventional fossils, but their genome sequences contain molecular fossils. Analysis of nucleic acids and proteins of bacteria, fungi, and higher organisms is providing much information on relatedness and even on evolutionary pathways. The high turnover rate of viral genomes coupled with their potential for great variation would seem to preclude this source of information on viral evolution. However, most basic functions are controlled by highly conserved sequences that are essential, say, for enzyme catalytic sites; significant variation in these motifs would be lethal for the organism. Such motifs are conserved in viruses and form an important source of information for gaining an insight into viral evolution and even origins.

Because of this variation and high replication rates, viruses are the most rapidly evolving genetic agents of all biological entities. However, this has to be viewed in the light of the current environment in which they evolve. One can consider that there are two "ages" of virus evolution: the long "prehuman age" when the environment for viruses changed slowly and the recent "human age" when the environment (e.g., human populations and movement, growing monocultures of crops and movement of crop species) is changing rapidly and when

viruses are probably evolving more rapidly. Most, if not all, of the available data on virus evolution relates to the latter "age."

A. Origins of Viruses

For many years, three hypotheses for the origins of viruses (Box 4.1) were accepted. However, the drawbacks to each of these hypotheses and modern thinking are that the basic development of the "virus world" may have predated the common ancestors of the three major domains of Bacteria, Archaea, and Eukarya. It is suggested that the virus world

---

BOX 4.1

HYPOTHESES ON THE ORIGIN OF VIRUSES

The virus-first hypothesis: Viruses are descended from primitive precellular life forms. This hypothesis was based on the ideas that the earliest prebiotic polymers were RNA, which had enzymatic properties, such as that shown by ribozymes (see Box 3.2). These prebiotic RNAs later parasitised the earliest cells.

The reduction hypothesis. Viruses developed from the normal constituents of cells. The suggestion is that viruses arose from some cell constituent that escaped the normal control mechanism and became self-replicating entities. Examples of the normal cell constituent include transposable elements and mRNAs.

The escape hypothesis. Viruses are derived from degenerate cells that eventually parasitised normal cells. This has been suggested as a possible origin for large vertebrate viruses, such as poxviruses.

might even have played a significant role in the origin and evolution of these domains.

In considering the evolution from ancient forms of viruses to viruses of modern organisms, it must be remembered that in geological times five major great extinctions occurred, when up to 90 percent of living organisms died out. If viruses *did* arise early in evolution, they must have passed through these extinctions and subsequently diversified after each one.

Further support for an ancient origin of present-day viruses is the geographic distribution of some groups. For instance, tymoviruses have no known means of intercontinental spread other than by humans. Different tymoviruses occur in Europe, North and South America, Asia, Africa, and Australia, which were still one landmass at the end of the Jurassic period. Thus, the present-day tymoviruses are thought to have diverged from a common stock; the continents drifted apart between 138 and 80 million years ago.

B. Virus Variation

The two main forms of variation of viruses are mutation (nucleotide changes) and recombination (nucleic acid sequence rearrangement). (See Figure 4.1.) Mutations (see Chapter 8) usually occur during nucleic acid replication as a result of lack of proofreading of the newly synthesised strand; they can also be induced by external agents such as radiation. Single base changes that occur in a coding region may lead to replacement of one amino acid by another in the protein product, the introduction of a new stop codon that results in early termination of translation, and a shorter polypeptide or replacement of a codon that has either greater or lesser usage in the particular host. Deletion or addition of a single nucleotide in a coding region will give rise to a frameshift, with consequent downstream amino acid changes. Such deletions or additions will usually be lethal unless compensated for by a second change (addition or deletion) that restores the original reading frame. Nucleotide changes in noncoding regions will vary in their effects, depending on the regulatory or recognition functions of the sequence involved. There is no doubt that a single base change giving rise to a single amino acid substitution in the protein concerned is a frequent source of virus variability under natural conditions *in vivo*. Mutations that involve more than one nucleotide change can give rise to more major changes in the encoded protein or in the regulatory sequence.

Recombination occurs in viruses with genomes consisting of either DNA or RNA. (Recombination in both DNA and RNA viruses is discussed in detail in Chapter 8.) Recombinational events can lead to changes in the genomes of both DNA and RNA viruses and can lead to deletions, additions, or exchanges of sequence between two genomes. Deletions often give rise to defective or defective interfering nucleic acids, which are discussed in Chapter 3.

A further form of recombination, termed pseudo-recombination or reassortment, occurs in viruses that have their genomes divided between several nucleic acid strands. In joint infections between two strains of such viruses, new variants can arise by the encapsidation of genome segments from the two strains. This reassortment or shuffling of genomic segments is found in most divided genome viruses including bromoviruses, geminiviruses, and orthomyxoviruses (e.g., influenza virus).

C. Types of Evolution

The long-term evolution of viruses was just discussed. Now we will examine some of the factors involved in virus evolution in modern eukaryotic cells.

1. Microevolution and Macroevolution

The variation of viruses gives the material on which selection pressures can act, which results in their evolution. The different forms of variation have different importance in the

```
Wild type ORF
 M  D  D  Q  S  R  M  L  Q  T  L  A  G  V  N  L
atggacgatcaatccaggatgctgcagactctggccggggtgaacctg


Silent (third base) mutation
 M  D  D  Q  S  R  M  L  Q  T  L  A  G  V  N  L
atggacgatcaatccaggatgctgcaaactctggccggggtgaacctg


Point mutation (missense)
 M  D  D  Q  S  R  M  L  K  T  L  A  G  V  N  L
atggacgatcaatccaggatgctgaagactctggccggggtgaacctg


Point mutation (nonsense)
 M  D  D  Q  S  R  M  L stop
atggacgatcaatccaggatgctgtagactctggccggggtgaacctg


Single base deletion leading to frame-shift
 M  D  D  Q  S  R  M  L R  L  W  P  G stop
atggacgatcaatccaggatgctg-agactctggccggggtgaacctg
```

A

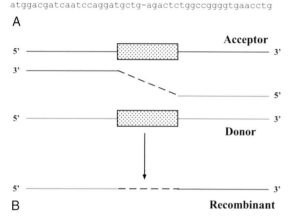

FIGURE 4.1   Variation in a sequence. A. Mutation. The top lines show the wild type ORF with the nucleotide sequence in lowercase and the amino acid sequence in uppercase. Mutations are shown by red nucleotides and deletion by –. The mutations focus on the codon cag encoding glutamine (Q); the third base mutation does not change the amino acid; a change in the first base of the codon (c > a) changes the polar glutamine to lysine (K), which has a positive charge; changing the c to a t introduces a stop codon; deleting the c leads to a frameshift and different downstream amino acids. B. A simple model of recombination between the donor virus (green) and the acceptor (red). The box indicates a region of homology between the donor and acceptor. During transcription of the donor the polymerase switch strands in the homology region, giving the recombinant.

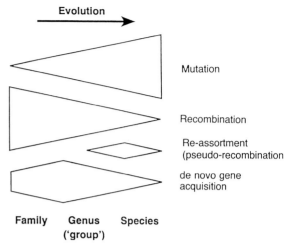

FIGURE 4.2   The apparent relative importance of some sources of genetic novelty that influence virus evolution. [From Gibbs and Keese (1995; in *Molecular basis for virus evolution*, A.J. Gibbs, C.H. Calisher, and F. Garcia-Arenal, Eds., pp. 76–90, reprinted with permission of Cambridge University Press).]

quasi-species cloud (for quasi-species, see Box 1.3). This can be termed **microevolution**. Larger and more radical changes caused by recombination and/or acquisition of new genes, termed **macroevolution**, lead to the generation of new genera or families. This essentially starts a new quasi-species cloud that is selected upon in further microevolutionary diversification.

Because of the population structure of a virus isolate, microevolution can be a continuous process. However, extremely high mutation rates do not necessitate rapid evolution. Evidence exists that some virus genera (e.g., begomoviruses) appear to be diversifying rapidly at present, whereas others (e.g., tobamoviruses) appear to be more stable. It is likely that, as with the evolution of higher organisms, viruses go through a stage of relatively rapid diversification when presented with a changing environment and then enter a relatively stable phase where they have adapted to the new environment(s). The selection pressures are discussed later in this chapter.

level or type of evolution (Figure 4.2). Thus, strains are differentiated mainly by mutations and small insertions or deletions that are selected, changing the master sequence in the

Macroevolution is a much rarer and a step-wise process. The great majority of recombination events between viral sequences or those leading to the acquisition of new genes will be lethal or deleterious to the virus. However, it is the very rare event that leads to the formation of a viral genome with new properties that enable it to be more successful than its progenitors. Examples of this will be discussed later in this chapter.

Microevolution and macroevolution are not necessarily independent systems. It is quite possible that microevolutionary changes of one virus could lead to the formation of "hot-spots" for recombination with sequences in other viruses. Similarly, not all recombination events will lead to major changes. Recombination between near homologous sequences will create new sequence combinations that have only minor differences to the parental sequences. Thus, microevolution and macroevolution are all a matter of degree.

2. Sequence Divergence or Convergence

Sequence similarity between two genes does not necessarily indicate evolutionary relationship (homology). Without other evidence, it may be impossible to establish whether sequence similarity between two genes is due to a common evolutionary origin or to convergence. There is much debate as to whether the evolution of viral RNA-dependent RNA polymerases is convergent or divergent.

3. Modular Evolution

A process of modular evolution was initially proposed for DNA bacteriophage but is now considered to also apply to all viruses. It is suggested that viruses have evolved by recombinational rearrangements or reassortments of interchangeable elements or modules (Box 4.2).

The basic module is for replicating the genome, and it consists of the replicase and, in at least the case of reverse transcribing viruses, the coat protein. To this are added modules that

enable the basic replicon to fill its "niche"—that is, to provide functions for the virus to infect its host and to move between hosts.

4. Sources of Viral Genes

The current understanding is that a significant part of viral evolution involves the acquisition and exchange of functional modules that make up the virus genome. As to the overall origins of genes or modules, two theories predominate: a common origin by a molecular "big bang" and continuous creation. Mechanisms by which new coding sequences can be generated in a continuous creation scenario include "overprinting," in which an existing nucleotide sequence is translated de novo in a different reading frame or from noncoding ORFs and gene duplication (for example, in Closteroviruses).

a. Replicases. *RNA replicases.* The structure of core RNA replicases is discussed in Chapter 8, where it is pointed out that they comprise several functional units: the RNA-dependent RNA polymerase (RdRp), a helicase, and a methyl-transferase; the RdRps and helicases fall into several supergroups or superfamilies (see Boxes 8.2 and 8.3). There is disagreement as to whether a common origin exists for each of these basic activities giving a monophyletic evolutionary pathway or whether their evolution is polyphyletic from several origins. However, it has been suggested that the earliest form of nucleic acid was RNA, which would indicate RNA replication is very ancient.

Notwithstanding the divergence of opinion about the monophyletic or polyphyletic origins of the component parts of core RNA replicases, reasonable evidence exists that their arrangement in modern viral genomes has arisen by modular shuffling of these components (Figure 4.3).

Reverse transcriptase. The replication of genomes of members of the several virus families, including the *Retroviridae* and *Caulimoviridae*, is

---

BOX 4.2
==========

VIRUS MODULES

Modules are defined as interchangeable genetic elements, each of which carries out a particular biological function; a basic modular structure for a plant virus is shown in the figure.

| Replicon | Coat protein | Cell-to-cell movement | Plant-to-plant movement | Suppressor of RNAi |

These modules, or parts of them, can be interchanged, which enables their independent evolution under a wide variety of selective conditions. Such modular mobility can overcome the evolutionary constraints that would occur if all the modules had to coevolve within a single genomic unit.

The following are the essentials of modular evolution:

- The product of evolution is a favourable combination of modules selected to work optimally individually and together to fill a particular niche.
- Joint infection of the host by two or more viruses is essential for the assembly of new combinations of modules. The viruses do not necessarily have to give full systemic infection of the host; they just have to replicate in the same cell. This can lead to changes in virus host range of the recombinant virus when compared with the donor viruses.
- Viruses in the same "interbreeding" population can differ widely in any characteristic because these are aspects of the function of individual modules.
- Evolution acts primarily at the level of the individual module and not at the level of the intact virus. Selection of modules is for a good execution of function, retention of the appropriate regulatory sequences, and functional compatibility with most, if not all, other modules in that genome.

---

by reverse transcriptase (RT), which converts an RNA template to DNA (see Chapter 8). Several arguments point to the basic elements of RT being very ancient. The suggestion that RNA preceded DNA evolutionarily, coupled with sequence similarities between RdRps and RTs, has been taken to indicate that RT is among the earliest of enzymes. The amino acid homologies between the RTs of all retro-elements suggest that the enzyme has evolved only once.

DNA replicases. The DNA > DNA replication of various plant virus groups is described for geminiviruses in Chapter 8. These viruses use host DNA replicase functions, and because they replicate in differentiated cells, the cell cycle must be modified. Thus, one of the viral gene functions is to bind to retinoblastoma (Rb) proteins, which directs the cell into the S phase. It is likely that the Rb-binding protein(s) has been acquired by the viruses from host sources.

b. Proteinases. Three classes of virus-coded proteinases are found among plant viruses (Box 4.3). The functions of these were identified by comparison with amino acid sequences from plants and animals, hence the terms *chymotrypsin-like* and *papain-like*. This would suggest a host origin. In most cases, the protease is a unique gene product, but it is sometimes associated with a protein with another

FIGURE 4.3 Conservation and variability in the arrangement of "core" genes in proteins involved in replication of (+)-strand RNA viruses. POL 1, 2, 3, RNA-dependent RNA polymerase of subgroups 1, 2, and 3, respectively; HEL 1, 2, 3, (putative) helicase of superfamily 1, 2, and 3, respectively; S-PRO, serine chymotrypsin-like proteinase; C-PRO, cysteine chymotrypsin-like proteinase; P-PRO, papain-like proteinase; MTR 1, 2, methyl transferases of types 1 and 2, respectively; Boxes are not to scale. [From Koonin, E.V., Dolja, V.V., and Morris, T.J. (1993). Evolution and taxonomy of positive-strand RNA viruses: implications of comparative analysis of amino acid sequences. *Biochem. Mol. Biol.* **28**, 357–430, reprinted by permission of publisher (Taylor & Francis Ltd., http://www.tandf.co.uk/journals).]

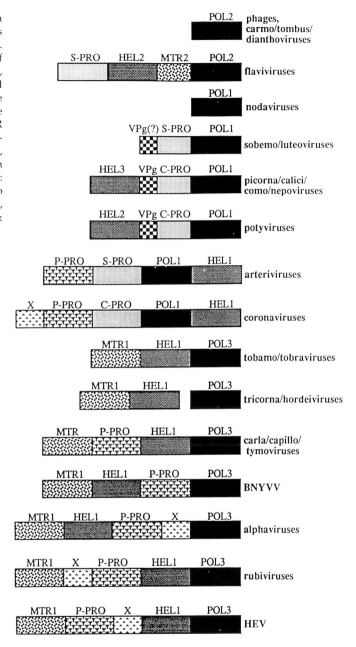

function—for example, the HC-Pro of potyviruses, in which it is associated with the insect transmission helper factor (see Chapter 12) and also the suppressor of host gene silencing (see Chapter 11). Whether this represents a multiple function for the original proteinase or the acquisition of various functional modules into the proteinase molecule is unknown.

BOX 4.3

VIRUS CODED PROTEINASES

Four classes of virus-coded proteinases are currently recognised: serine, cysteine, aspartic, and metallo-proteinases, named usually after their catalytic site. Three of these four types of proteinase are found among plant viruses.

Proteinases encoded by some plant viruses

Virus group	Viralprotease
Caulimoviridae	Aspartate
Potyviridae	Serine, cysteine, serine-like[a]
Comoviridae	Serine-like
Tymovirus	Cysteine
Closterovirus	Cysteine, aspartate

[a] Has a cysteine at its active site but a structure similar to a serine proteinase.

Serine proteinases are also termed 3C proteases (from their expression in picornaviruses) or chymotrypsin-like proteases. Most have a catalytic triad of amino acids (His, Asp, Ser), but in some, the Ser is replaced by Cys. The serine residue is unusually active and acts as a nucleophile during catalysis by donating an electron to the carbonyl carbon of the peptide bond to be hydrolyzed. An acyl serine is formed, and a proton is donated to the departing amyl group by the active-site histidine residue. The acyl enzyme is then hydrolyzed, the carboxylic acid product is released, and the active site is regenerated. Serine proteases cleave primarily at Gln-Gly, Gln-Ser, Gln-Ala, and Gln-Asn.

Cysteine proteinases, also known as papain-like or thiol proteinases, have a catalytic dyad comprising Cys and His residues in close proximity that interact with each other. During proteolysis, the Cys sulfhydryl group acts as a nucleophile to initiate attack on the carbonyl carbon of the peptide bond to be hydrolyzed. An acyl enzyme is formed through the carbonyl carbon of the substrate and the sulfhydryl group of the active-site His. The carbonyl carbon is then hydrolyzed from the thiol group of the proteinase, and the active-site residues are regenerated.

Aspartic or acid proteinases are composed of a catalytic dyad of two Asp residues. They most likely do not form covalent enzyme-substrate intermediates and are thought to operate by an acid-base catalysis.

Viral proteinases are highly specific for their cognate substrates, a specificity that depends on the three-dimensional structures of both the proteinase and substrate. For instance, *Cowpea mosaic virus* proteinase does not recognise primary translation products of mRNAs from other comoviruses.

c. Coat Proteins The structure of the virus particle is determined by the structure of the coat protein subunit (see Chapter 5). Because of packing considerations, proteins making up rod-shaped particles have to be wedge shaped, whereas those making up spherical particles have to be conical or have a three-dimensional wedge shape. The shape of these proteins is determined by the secondary and tertiary structures of their polypeptide chains. The coat proteins of most, if not all, viruses with isometric particles have a basic structure comprising eight antiparallel ß sheets, termed the ß barrel or jellyroll motif (see Chapter 5). This suggests a single origin for the isometric virus coat protein. However, the coat protein of *Sindbis virus* has close relationships to chymotrypsin-like proteinases, which would indicate a separate origin.

d. Cell-to-Cell Movement Proteins. The cell-to-cell movement of many plant viruses is potentiated by one or more movement proteins (MP; see Chapter 9). At least one type of these proteins—the "30K superfamily," which includes the *Tobacco mosaic virus* (TMV) MP—appears to function by gating the plasmodesma open to enable the viral infection unit to pass to an uninfected cell. A plant paralogue of TMV MP (a transcription factor, the KNOTTED1 homeodomain) has been recognised. Thus, it is likely that this functional module has been acquired by viruses from the plant genome. As there are several ways in which cell-to-cell movement is effected, it is likely that different genes have been acquired by various viruses. For instance, the triple gene block of potexviruses and hordeiviruses would appear to be a combination of three different genes rather than the separation of functional domains of one gene product.

e. Suppressors of Gene Silencing. The plant (and other organisms) gene silencing defence system against "foreign" nucleic acids is described in Chapter 11. Most, if not all, plant viruses have genes that suppress this defence system, and some viruses of vertebrates do as well. The origin of these suppression genes is not known, but there is a possibility that they have been acquired from the host genome. One time that plants and animals have to tolerate "foreign" nucleic acids is during the fertilisation process. Proteins involved in the fertilisation process might be the source of the viral silencing suppression gene(s).

D. Selection Pressures for Evolution

Virus evolution, like that of other organisms, is a trade-off between costs and benefits and is driven by the impact of selection pressures on the inherent variation. Much of the molecular information on viruses comes from laboratory experiments, which, although they can provide an insight into the mechanisms involved, cannot give a picture of what happens in the "natural situation." In this section, we examine some of the factors that drive and limit the evolution of plant viruses.

1. Adaptation to Niches

A virus is closely adapted to specific hosts and within those hosts to specific cellular sites. These can be thought of as ecological niches. There are two sorts of niches: macro-niches and micro-niches. Macro-niches are the hosts and geographic situations in which the virus can establish a successful infection. Thus, viruses that infect animals require different genes (modules) for spreading from cell to cell than those that infect plants (see Chapter 9). An example of geographic situations is rice tungro disease, the occurrence of which is determined by the distribution of its insect vector, the rice green leafhopper, *Nephotettix virescens* (Figure 4.4). Micro-niches are the sites within the host cells in which the virus replicates and moves from cell to cell and are usually defined by organelles, membranes, and the cytoskeleton (see Chapters 8 and 9).

2. Maximising the Variation

Earlier in this chapter, we discussed the sources of variation, mutation, pseudo-recombination, and recombination that provide the material upon which selection pressures can act. The rapid replication of viruses and the lack of proofreading or repair mechanisms give the potential for much mutagenic variation in both RNA and DNA viruses. Thus, selection pressures operate on quasi-species, which comprise a dominant sequence and a multitude of minor variant sequences (see Box 1.4). Three factors contribute to the success of recombination, giving rise to new viruses: mixed infections, high levels of viral replication, and increased host range of the vector.

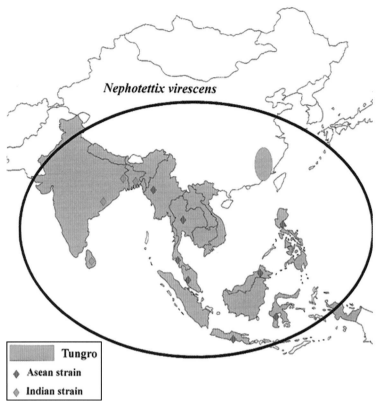

FIGURE 4.4 Distribution of rice tungro disease; the circle shows the distribution of the main leafhopper vector (*Nephotettix virescens*), and the diamonds show the sites from which the two *Rice tungro bacilliform virus* strains were obtained. [This article was published in *Encyclopedia of Virology*, vol. 2, R. Hull, Rice tungro disease, p. 483, Copyright Elsevier (2008).]

3. Controlling the Variation

Viruses are absolutely dependent on their hosts for their propagation and survival. Any viral variant that damages the host significantly would be selected against, and thus the evolutionary pressure is toward an equilibrium state between the virus and its host. There are very few examples of the impact of factors that control variation in the natural situation, as most viruses that are studied cause overt diseases in organisms of interest to man. In most cases these are new interactions between the virus and the host, and there has not been sufficient time for evolution to select an equilibrium. One example of the natural situation (Table 4.1) is found in a study of viruses in 92 plants of the wild species *Plantago lanceolata*. Eight viruses were found, five of which caused no symptoms in *P. lanceolata*.

In light of this high level of variation, some constraints must be present that control the preservation of the identity of a virus species or strain.

TABLE 4.1 Viruses of *Plantago lanceolata*

Plantain virus X	Potexvirus	Symptoms
Plantain virus A	Rhabdovirus	Symptoms
Ribgrass mosaic virus	Tobamovirus	Slight symptoms
Plantago virus 4	Caulimovirus	No symptoms
Plantago virus 5	? Necrovirus	No symptoms
Plantago virus 6	? Tombusvirus	No symptoms
Plantago virus 7	? Potyvirus	No symptoms
Plantago virus 8	? Carlavirus	No symptoms

Data from J. Hammond (1980). PhD Thesis, University of East Anglia, UK.

a. Muller's Ratchet. It is likely that most mutations in a viral genome will be either neutral or deleterious. Mutations that lead to loss of a critical function would not be propagated in a population unless they were complemented by other members of that population. However, mutations that caused only a decrease in "fitness" would be more likely to be retained. Theoretical considerations have led to the concept of "Muller's ratchet" (Box 4.4).

BOX 4.4

MULLER'S RATCHET AND VIRUSES

Theoretical considerations have led to the concept of Muller's ratchet, which indicates that high mutation rates can have significant impact on populations, especially if they are small in size. The concept of Muller's ratchet is that if the average mutation is deleterious, there will be a drift to decrease of population fitness leading to "mutational meltdown" (see Lynch *et al.*, 1993). Muller's ratchet is particularly applicable to small populations, and for many viruses, transmission and infection form a bottleneck in which the population is small.

Back mutations at the specific site of a deleterious mutation or compensatory mutations are likely to occur at a lower rate than forward mutations. In populations of higher organisms, this drift is limited by sex which recreates, through genetic exchange, genomes with fewer or no mutations. Obviously, this process does not occur in the conventional sense in viruses, but it is likely that recombination or genetic reassortment within the quasi-species could play a part in controlling the effects of Muller's ratchet.

---

b. Muller's Ratchet and Plant Viruses. Evidence exists that Muller's ratchet has the following effects.

- *Effects on plant virus populations.* The tobamoviruses in herbarium samples and living samples of *Nicotiana glauca* in New South Wales, Australia, covering a period from 1899 to 1993 were analyzed. Before 1950, many plants were infected with both TMV and *Tobacco mild green mosaic virus* (TMGMV), but after that date only TMGMV was found. In experimental joint infections of *N. glauca*, TMV accumulated to about 10 percent of the level of that in single inoculations; the level of TMGMV was not affected. It was concluded that TMV colonised *N. glauca* in New South Wales earlier or faster than TMGMV, but in joint infections the latter virus caused a decrease of the TMV population below the threshold at which deleterious mutations were eliminated.

- *Methods to prevent it.* Nematode transmission of *Tobacco rattle virus* (TRV) serves as a bottleneck to clear the virus population of defective interfering (DI) RNAs (for DI RNA, see Chapter 3). These DI RNAs, derived from RNA2, have a modified coat protein gene and interfere with viral replication. It was suggested that TRV RNA2 and the DI RNA are encapsidated in *cis* by their encoded coat proteins, which are, respectively, functional and nonfunctional in transmission, eliminating the DI RNA at the transmission bottleneck.

Variants that reduce the fitness of the virus population have to be removed. From their success, it is obvious that plant viruses can overcome the constraint of Muller's ratchet, although the mechanism is not known for most viruses.

4. Role of Selection Pressure

Functional viral genes will be retained only if they are needed for survival of the virus. For example, RNAs 3 and 4 of *Beet necrotic yellow vein virus* (Profile 2) appear to suffer deletions when not under the requirement to be transmitted by the vector.

5. Selection Pressure by Host Plants

Conditions within a given host species or variety will exert pressure on an infecting virus against rapid and drastic change. Viral genomes and gene products must interact in highly specific ways with host macromolecules during virus replication and movement. These host molecules, changing at a rate that is slow compared with the potential for change in a virus, will act as a brake on virus evolution.

The evolution of variants in populations of TMV originating from an *in vitro* transcript of a cDNA clone was studied by assessing the ability to cause necrotic lesions (nl) in *Nicotiana sylvestris*. The proportion of nl variants in tobacco, tomato, *Solanum nigrum,* and *Petunia hybrida* was similar. However, in *Physalis floridiana* there was strong selection pressure against nl variants, which were reduced to almost undetectable levels. Nevertheless, in all hosts tested large and apparently random changes in the proportion of nl variants in individual plants were observed, showing that viral populations can evolve rapidly on a timescale of days. Perhaps these variations in a strain population occur during the early stages of infection in new host species. Over a period of many transfers in the same species, the populations may stabilise. This idea is supported by the fact that TMV obtained from W.M. Stanley in the 1930s and independently subcultured many times over decades in tobacco varieties in the United States and Germany was shown in both countries to have the same coat protein amino acid sequence.

E. Timeline for Evolution

The direction and timing of the evolution of most organisms is derived from the stratigraphical position of the conventional fossil ancestors.

However, it is not possible to use this approach for the molecular fossils of viruses. The only potential sources of information are the molecular clocks of change of sequence, but these usually do not give the direction of change.

1. Nonconstant Rates of Evolution

The molecular clock hypothesis, which assumes a constant rate of change in sequence over evolutionary time, has often been used in the interpretation of phylogenetic "trees." However, there is considerable uncertainty as to whether molecular clocks can be applied to the evolution of viral genomes. Furthermore, different proteins, or parts of a protein coded by a viral genome, may evolve at different rates. In addition, some noncoding sequences in the genome may be highly conserved, particularly those recognition sequences essential for genome replication.

2. Estimated Rates of Evolution

The rate of point mutation for RNA viruses has been estimated to be approximately 10^{-3}–10^{-4} per nucleotide per round of replication with some variation between different viruses contrasting with estimates of about 10^{-11} for DNA polymerases (see Figure 4.5). However, it is very difficult to relate mutation rates to the actual rates of change in viruses that might be occurring in the field at present or over past evolutionary time because of (1) variation in rates of change in different parts of a viral genome, (2) uneven rates of change over a time period, and (3) lack of precise historical information.

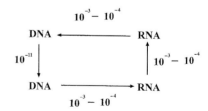

FIGURE 4.5   Error rates of transcription within and between RNA and DNA. [From Hull (2002).]

III. EVIDENCE FOR VIRUS EVOLUTION

As just noted, evidence for virus evolution must come from the study of present-day viruses in present-day hosts. It is the comparisons of genome organisations and the details of sequences that are giving evidence for the evolutionary pathways of many viruses. The major factors involved can be shown by three examples of plant viruses but can also be applied to many other virus groups.

A. Geminiviruses

The DNA > DNA replication of geminiviruses should result in a less mutagenic variation than RNA > RNA replication (see Figure 4.5). However, there is evidence for significant genomic variation in at least two geminivirus genera: the begomoviruses and the mastreviruses. Sequence analysis of three *Maize streak virus* (MSV) isolates obtained by serial passage of a wild-type isolate through a semi-resistant cultivar of maize indicated that the original infection had a quasi-species structure with mutations distributed throughout the genome. Mutation frequencies were estimated to be between 3.8×10^{-4} and 10.5×10^{-4}, levels similar to those found with RNA > RNA replication. The mutagenic variation of geminiviruses is unexpected in view of the proofreading normally associated with DNA > DNA replication and may reflect a lack of postreplicative repair. Geminivirus DNA does not appear to be methylated, as its replication is inhibited by methylation. Thus, although these viruses use the host system for their replication, the normal host mechanisms for mismatch repair probably do not function. There is increasing evidence for natural recombination within geminiviruses, on occasions leading to new diseases (Box 4.5).

B. Closteroviruses

The genome organisations of members of the family *Closteroviridae* are described in Profile 5 in the Appendix. These viruses have the largest (+)-strand ssRNA genomes among plant viruses. It has been suggested that the closterovirus genome arose from a common ancestor, with rearrangement of that genome and acquisition of other modules by recombination (Box 4.6).

Recombination is an essential part of this proposed evolutionary pathway. Analyses of multiple species of *Citrus tristeza virus* (CTV) defective RNAs show that they appear to have arisen by recombination of a subgenomic RNA (sgRNA) with distant parts from the 5′ end of the CTV genome. It is suggested that closteroviruses are able to use the sgRNA and/or their promoter signals for modular exchange and rearrangement of their genomes.

In spite of the potential for variation, a study showed that mild CTV isolates maintained in different citrus hosts, from several geographical locations (Spain, Taiwan, Colombia, Florida, and California) and isolated at different times were remarkably similar indicating a high degree of evolutionary stasis in some CTV populations.

C. Luteoviruses

The organisation of the (+)-sense ssRNA genomes of members of the family *Luteoviridae* is shown in Profile 8 in the Appendix. The genome organisation of the viruses in the families *Luteoviridae* and *Tombusviridae* (Profile 16) and the genus *Sobemovirus* comprises two basic modules: the 5′ replicase proteins and the 3′ proteins, which include the virion coat protein. Phylogenetic analyzes of the RdRp and the coat protein suggest that there have been gene transfer events between the modules of members of this supergroup and suggested two main clusters termed the *enamo* and *carmo* clusters. Likely gene transfers were recognised both between and within these main clusters.

BOX 4.5

EXAMPLE OF GEMINIVIRUS EVOLUTION

Tomato yellow leaf curl disease spread and diversified rapidly through the Mediterranean basin during the twentieth century, as shown in the figure, which gives the first dates that the disease was recognised in various countries.

Three viruses are currently known to cause the disease: *Tomato yellow leaf curl virus* (TYLCV; see Profile 7), *Tomato yellow leaf curl Sardinia virus* (TYLCSV), and *Tomato yellow leaf curl Sudan virus* (TYLCSDV); these viruses are all begomoviruses but differ in sequence.

Various recombinants have now been detected between these viruses. TYLCV has recombined with TYLCSV to give a new virus, *Tomato yellow leaf curl Malaga virus*, which has subsequently recombined with TYLCSDV and also formed further recombinants with its parents. Both TYLCV and TYLCSV have also recombined with TYLCSDV. Thus, there is rapid variation in this group of viruses.

The current classification of the *Luteoviridae* recognises three genera: the *Luteovirus*, which falls in the carmo cluster, and the *Polerovirus* and *Enamovirus*, which are placed in the enamo cluster. The proposed evolutionary pathway is shown in Box 4.7.

There is further considerable evidence for recombination within the *Luteoviridae*. One can distinguish three types of events: recombination within a gene, recombination of large parts of the genome within a genus, and recombination between large parts of the genome between genera.

Thus, recombination appears to be rampant both within the *Luteoviridae* and between members of this family and those of some other groups of small ssRNA viruses. There is also evidence that *Potato leaf roll virus* (PLRV) can recombine with host sequences; the 5′-terminal 119 nucleotides of some RNAs of a Scottish isolate of PLRV are very similar to an exon of tobacco chloroplast DNA.

CLOSTEROVIRUS EVOLUTION

The progenitor of the alphavirus supergroup has been proposed to comprise a complex of genes encoding methyl transferase (MTR), helicase (HEL), papain-like protease (P-PRO), polymerase (POL) (for description of these modules in the replicase see Chapter 8, Section IV, B), and capsid protein. The proposed evolutionary pathyway is shown in the Figure:

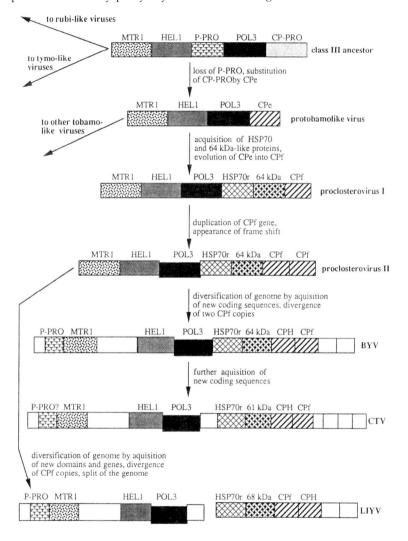

Fig. A tentative scenario for the evolution of closteroviruses. CPe designates an elongated particle capsid protein; CPH, capsid protein homolog; HEL1, RNA helicase of superfamily 1; HSP70r, HSP70-related protein; MTR 1, type 1 methyl transferase; POL 3, polymerase of supergroup (see Box 8.2) 3; P-PRO, papain-like protease. [From Dolja *et al.* (1994; *Annu. Rev. Phytopathol.* **32**, pp. 261–285); Reprinted, with permission, from the *Annual Review of Phytopathology* Volume 32 ©1994 by Annual Reviews www.annualreviews.org.]

(continued)

BOX 4.6   *(continued)*

This evolutionary pathway comprises the following steps:

1. Deletion of P-PRO from the core of the replication genes.
2. Substitution of the postulated alphavirus type capsid protein by a capsid protein capable of forming elongated virions.
3. Invention of the frameshift mechanism of POL expression (see Chapter 7, Section V, B, 12).
4. Acquisition of the heat-shock protein HSP70 from the cellular genome.
5. Duplication of the capsid protein and the functional switch for one of the tandem copies to facilitate aphid transmission.
6. Insertion of long coding sequences between the MET and HEL cistrons.
7. Secondary acquisition of the leader P-PRO, perhaps from a potyvirus or a related virus.
8. Additional diversification and acquisition of the 3'-terminal genes.
9. Split of the genome into two components giving the crinivirus (a genus in the *Closteroviridae* with a divided genome) genome organisation.

It is suggested that steps 1 and 2 occured early in evolution and gave a common ancestor of the whole tobamovirus cluster. The order of the other events is rather arbitrary.

BOX 4.7

PROPOSED EVOLUTION OF THE *LUTEOVIRIDAE*

The most likely model for the origin of the genomes of the luteovirus and polerovirus genomes (Profile 8) (Figure) suggests that recombination arose by strand switching at subgenomic RNA start sites during RNA replication in cells jointly infected by the two parental viruses. For the derivation of the luteovirus genome, the sgRNA start site on diantho-like viruses has homology to that of poleroviruses. Recombination at this site would create a hybrid virus with dianthovirus polymerase and polerovirus coat protein and neighbouring genes. A recombination event at the sgRNA start site downstream of ORF5 would give the complete luteovirus genome organisation. A single recombination between the 5' region of a sobemovirus and the 3' part of a luteovirus followed by premature termination would give the polerovirus genome organisation.

(continued)

BOX 4.7   *(continued)*

Origin of a subgroup I luteovirus by recombination

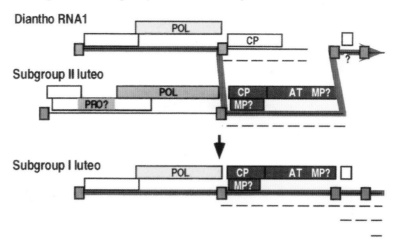

Origin of a subgroup II luteovirus by recombination

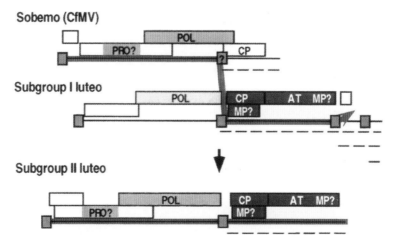

Fig. Model for the origin of genomes of luteoviruses and poleroviruses. Solid black lines represent viral genomic RNA; dashed lines indicate sub-genomic RNAs; Boxes indicate the genes; blue shading, genes with sequence similarity to umbra-diantho- and carmo-viruses; green. Sequence similarity fo sobemoviruses. Grey boxes represent putative origins of replication and sub-genomic RNA promoters. POL, RNA-dependent RNA polymerase; PRO?, putative protease; CP, coat protein; MP?, putative movement protein; AT, read-through domain of coat protein gene. Pink line shows the proposed path of replicase as it switched strands during copying viral RNA in a mixed infection. From Miller *et al.* (1997; *Plant Dis.* **81**, 700-710).

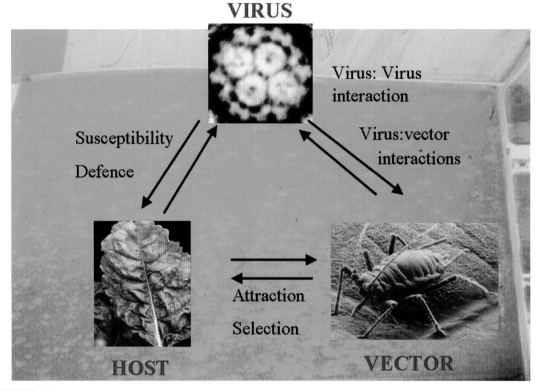

FIGURE 4.6   Interactions between a virus, its host, and its vector.

IV. COEVOLUTION OF VIRUSES WITH THEIR HOSTS AND VECTORS

Fahrenholtz's rule postulates that parasites and their hosts speciate in synchrony (Eichler, 1948). Thus, there is a prediction that phylogenetic trees of parasites and their hosts should be topologically identical. In view of the known wide host ranges of many present-day plant viruses, it is not to be expected that Fahrenholtz's rule will be followed closely for viruses and their hosts. Nevertheless, it is now widely accepted that viruses have had a long evolutionary history and have coevolved with their host organisms. The concept of coevolution does not imply that viruses have evolved to a state of higher complexity following their plant hosts in this respect. On the contrary, the evidence available at present shows that the largest and most complex virus infecting photosynthetic organisms is found in the simplest host, a Chlorella-like green alga.

When considering coevolution, one must consider the interactions between the virus and the host, between the virus and its vector, and between the host and the vector (Figure 4.6). These interactions can be complex and are also affected by external factors such as climate and competition from other organisms. However, in the natural situation it is essential for the virus to be able to replicate efficiently but not affect its host significantly and also to be efficiently transmitted to other hosts and establish infections therein.

V. VIRUSES OF OTHER KINGDOMS

There is extensive molecular evidence that has been used to suggest evolutionary pathways for many groups of viruses. The examples just given are from plant viruses, but the same principles apply to viruses from other kingdoms.

VI. SUMMARY

- Viruses do not produce conventional fossils, but viral sequences contain "molecular fossils."
- Viruses probably originated early in the evolution of living organisms.
- Virus evolution is through variation and selection.
- Virus variation is due to mutations that can create new virus strains and some species and recombination that gives rise to new virus genera and families.
- Viral genomes are composed of functional modules that can be interchanged by recombination.
- There are various selection pressures on virus evolution and, as viruses require living hosts for their replication, the ultimate state is an equilibrium state between the virus and its host.
- Viruses have coevolved with their hosts and their vectors.

References

Eichler, W. (1948). Some rules on ectoparasitism. *Ann. Mag. Nat. Hist. Ser.* **12**, 588–598.

Lynch, M., Bürger, R., Butcher, D., and Gabriel, W. (1993). The mutational meltdown in asexual populations. *J. Hered.* **84**, 339–344.

Further Reading

Forterre, P. (2008). Origin of viruses. *Encyclopedia of Virology*, Vol. 3, 472–478.

Hull, R (2002). *Matthew's plant virology*. Academic Press, San Diego.

Karasev, A.V. (2000). Genetic diversity and evolution of closteroviruses. *Annu. Rev. Phytopathol.* **38**, 293–324.

Lovisolo, O., Hull, R., and Rosler, O. (2003). Coevolution of viruses with hosts and vectors and possible paleontology. *Adv. Virus Res.* **62**, 326–379.

Roossinck, M.J. (Ed.) (2008). *Plant virus evolution*. Springer, Berlin.

Villarreal, L.P. (2008). Evolution of viruses. *Encyclopedia of Virology*, Vol. 2, 174–184.

WHAT IS A VIRUS MADE OF?

Architecture and Assembly
of Virus Particles

The genomic nucleic acids of plant viruses are protected by virus-coded proteins which form defined virus particles. The particles of most viruses have a simple structure, but others are more complex.

OUTLINE

I. INTRODUCTION

Knowledge of the detailed structure of virus particles is an essential prerequisite to our understanding of many aspects of virology— for example, how viruses survive outside the cell, how they infect and replicate within the cell, and how they are related to one another.

Knowledge of virus architecture has increased greatly in recent years, due both to more detailed chemical information and to the application of more refined electron microscopic, optical diffraction, and X-ray crystallographic procedures.

Many different terms are used in virus structure (see Box 5.1). Figures 1.3 and 1.5 indicate

---

BOX 5.1

TERMS USED IN VIRUS STRUCTURE

Capsid: The closed shell or tube of a virus

Capsomere: The clusters of subunits on the capsid as seen in electron micrographs; also termed **morphological subunit**

Encapsidation (or encapsulation): The process of enclosing the viral genomic nucleic acid in virus-encoded protein usually to form a virus particle

Multicomponent virus: Infectious virus genome divided between several nucleic acid segments that are separately encapsidated

Negative stain: The virus particle is embedded in an electron dense material such as phosphotungic acid or uranyl acetate, which forms a dense background against which the virus particle appears translucent in the electron microscope. This method is capable of providing information about structural details often finer than those visible in thin sections, replicas, or shadowed specimens and has the advantage of speed and simplicity

Nucleocapsid: The inner nucleoprotein core of membrane-bound viruses

Nucleoprotein: A complex between the viral genomic nucleic acid and virus-encoded protein, which may or may not have a defined structure

Particle (virus particle): The virus genome enclosed in a capsid and for some viruses also a lipid membrane

Protein subunit: Individual virus-encoded protein molecule that makes up the capsomere or nucleoprotein; also termed **structural subunit**

Virion: The mature virus that may be membrane bound; term interchangeable with **virus particle**

---

the range of sizes and shapes found among plant viruses. The three basic types of virus particle are rod-shaped, isometric (spherical), and complex.

II. METHODS

A. Chemical and Biochemical Studies

Knowledge of the size and nature of the viral nucleic acid and of the proteins and other components that occur in a virus particle is essential to an understanding of its architecture. Chemical and enzymatic studies may give various kinds of information about virus structure. For example, the fact that carboxypeptidase A removed the terminal threonine from intact *Tobacco mosaic virus* (TMV) indicated that the C-terminus of the polypeptide was exposed at the surface of the virus. For viruses with more complex structures, partial degradation by chemical or physical means—for example, removal of the outer envelope—can be used to establish where particular proteins are located within the particle. Studies on the stability of a virus under various pH and ionic or other conditions may give clues as to the structure and the nature of the bonds holding the structure together.

B. Methods for Studying Size and Fine Structure of Viruses

1. Hydrodynamic Measurements

The classic procedure for hydrodynamic measurements is to use the Svedberg equation (see Schachman, 1959). Laser light scattering has been used to determine the radii of several approximately spherical viruses with a high degree of precision. Virus particles bind water, and this method gives an estimate of the hydrated diameter, whereas the Svedberg equation and electron microscopy give a diameter for the dehydrated virus. Comparison of the two approaches can give a measure of the hydration of the virus particle.

2. Electron Microscopy

Measurements made on electron micrographs of isolated virus particles, or thin sections of infected cells, offer very convenient estimates of the size of viruses. For some of the large viruses and for the rod-shaped viruses, such measurements may be the best available, but they are subject to significant errors, such as magnification errors or flattening of the particles.

With helical rod-shaped particles, the central hollow canal is frequently revealed by negative staining. With the small spherical viruses, which often have associated with them empty protein shells, it was assumed by some workers that stained particles showing a dense inner region represented empty shells in the preparation. However, staining conditions may lead to loss of RNA from a proportion of the intact virus particles, allowing stain to penetrate, while stain may not enter some empty shells. Stains differ in the extent to which they destroy or alter a virus structure, and the extent of such changes depends closely on the conditions used. The particles of most viruses that are stabilised by protein:RNA bonds are disrupted by some negative stains at pHs above 7.0.

Some aspects of the structure of the enveloped viruses, particularly bilayer membranes, can be studied using thin sections of infected cells or of a pellet containing the virus. Cryo-electron microscopy, which involves the extremely rapid freezing of samples in an aqueous medium, allows the imaging of symmetrical particles in the absence of stain and under conditions that preserve their symmetry. This technique has been widely used for isometric viruses. For these particles, the images have to be reconstructed to give a three-dimensional object from a two-dimensional image using a range of approaches, leading to information that complements that from X-ray crystallography and also allows the detailed analysis of viruses that are not amenable to crystal formation.

3. X-Ray Crystallography

For viruses that can be obtained as stable crystalline preparations, X-ray crystallography can give accurate and unambiguous estimates of the radius and structure of the particles in the crystalline state. The technique is limited to viruses that are stable or can be made stable in the salt solutions necessary to produce crystals.

4. Neutron Small-Angle Scattering

Neutron scattering by virus solutions is a method by which low-resolution information can be obtained about the structure of small isometric viruses and in particular about the radial dimensions of the RNA or DNA and the protein shell. The effects of different conditions in solution on these virus dimensions can be determined readily. The method takes advantage of the fact that H_2O-D_2O mixtures can be used that match either the RNA or the protein in scattering power. Analysis of the neutron diffraction at small angles gives a set of data from which models can be built.

5. Atomic Force Microscopy

TMV particles, when dried on glass substrates, assemble into characteristic patterns that can be studied using atomic force microscopy. In highly oriented regions, the particle length was measured as 301 nm and the width as 14.7 nm; the latter measurement shows intercalation of packed particles. The particles are not flattened and their depth was 16.8–18.6 nm.

6. Mass Spectrometry

As well as measuring the mass of viral proteins and particles, mass spectrometry can be used to identify posttranslational modifications of viral protein such as myristoylation, phosphorylation, and disulfide bridging. When this technique is used in conjunction with X-ray crystallography, the mobility of the capsid can be studied. Similarly, nuclear-magnetic-resonance spectroscopy can detect mobile elements on the surface of virus particles.

7. Serological Methods

The reaction of specific antibodies with intact viruses or dissociated viral coat proteins has been used to obtain information that is relevant to virus structure. For instance, the terminal location of the minor coat protein on the flexuous rod-shaped particles of closteroviruses and criniviruses was recognised using polyclonal antibodies (Figure 5.1).

8. Stabilising Bonds

The primary structures of viral coat proteins and nucleic acids depend on covalent bonds. In the final structure of the simple geometric viruses, these two major components are held together in a precise manner by a variety of noncovalent bonds. Two kinds of interaction are involved: protein:protein and protein:nucleic acid. In addition, small molecules such as divalent metal ions (Ca^{2+} in particular) may have a marked effect on the stability of some viruses. Knowledge of these interactions is important for understanding the stability of the virus in various environments, how it might be assembled during virus synthesis, and how the nucleic acid might be released following infection of a cell. The stabilising interactions are hydrophobic bonds, hydrogen bonds, salt linkages, and various other long- and short-range interactions.

III. ARCHITECTURE OF ROD-SHAPED VIRUSES

A. Introduction

Crick and Watson (1956) put forward a hypothesis concerning the structure of small viruses, which has since been generally confirmed. Using the then recent knowledge that the viral RNA was enclosed in a coat of protein and that the naked RNA was infectious, they assumed that the basic structural requirement

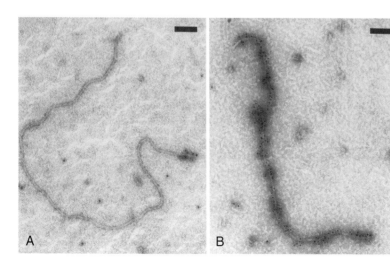

FIGURE 5.1   "Rattlesnake" tails on Citrus tristeza virus particles. A. Particle gold-labelled with p27 (minor coat protein) antibodies; B. Particles gold-labelled with coat protein (p25) antibodies. Bars = 100 nm. [From Febres et al. (1996; *Phytopathology* **86**, 1331–1335).]

for a small virus was the provision of a shell of protein to protect its ribonucleic acid. They considered that the protein coat might be made most efficiently by the virus that controlled the production in the cell of a large number of identical small protein molecules, rather than in one or a few very large ones.

They pointed out that if the same bonding arrangement is to be used repeatedly in the particle, the small protein molecules would aggregate around the RNA in a regular manner. There are only a limited number of ways in which the subunits can be arranged. The structures of all the geometric viruses are based on the principles that govern either rod-shaped or spherical particles.

In rod-shaped viruses, the protein subunits are arranged in a helical manner. There is no theoretical restriction on the number of protein subunits that can pack into a helical array in rod-shaped viruses.

B. Structure of TMV

1. General Features

The particle of TMV is a rigid helical rod, 300 nm long and 18 nm in diameter (Figure 5.2A). The composition of the particle is approximately 95 percent protein and 5 percent RNA. It is an extremely stable structure, having been reported to retain infectivity in

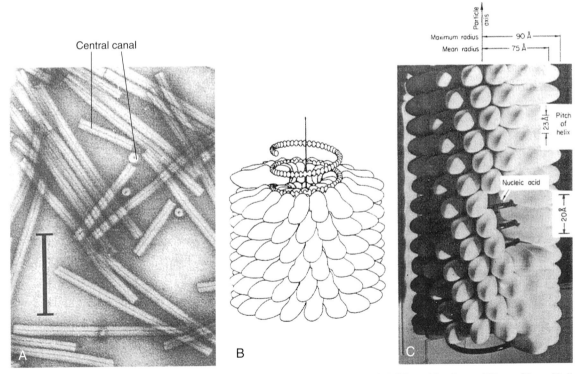

FIGURE 5.2   Structure of TMV. A Electron micrograph of negatively stained TMV particles; bar = 100 nm. [From Hull (2002).] B. Drawing showing relationship of the RNA and protein subunits; note that the RNA shown free of protein could not maintain that configuration in the absence of protein; C. Photograph of a model of TMV with the major dimensions indicated. B. and C. Reprinted from *Adv. Virus Res.*, 7, A. Klug and D.L.D Caspar, The structures of small viruses, pp. 225–325, Copyright (1961), with permission from Elsevier.

nonsterile extracts at room temperature for at least 50 years; however, the stability of naked TMV RNA is no greater than that of any other ssRNA. Thus, stability of the virus with respect to infectivity is a consequence of the interactions between neighbouring protein subunits and between the protein and the RNA.

X-ray diffraction analyses have given us a detailed picture of the arrangement of the protein subunits and the RNA in the virus rod. The particle comprises approximately 2,130 subunits that are closely packed in a right-handed helical array. The pitch of the helix is 2.3 nm, and the RNA chain is compactly coiled in a helix following that of the protein subunits (Figure 5.2B, C). There are three nucleotides of RNA associated with each protein subunit, and there are 49 nucleotides and $16\frac{1}{3}$ protein subunits per turn. The phosphates of the RNA are at about 4 nm from the rod axis. In a proportion of negatively stained particles, one end of the rod (the 5' end) can be seen as concave, and the other end (the 3' end) is convex. A central canal with a radius of about 2 nm becomes filled with stain in negatively stained preparations of the virus (Figure 5.2A).

2. Virus Structure

Because intact TMV does not form three-dimensional regular crystalline arrays in solution, fibre diffraction methods were used to solve the structure to a resolution of 2.9 Å and to produce a model for the virus (Figures 5.2C and 5.3).

The following are the important features of this model:

- *The outer surface.* The N- and C-termini of the coat protein are at the virus surface (Figure 5.3B). However, the very C- terminal residues 155–158 were not located on the density map and are therefore assumed to be somewhat disordered.

- *The inner surface.* The presence of the RNA stabilises the inner part of the protein subunit in the virus so that its position can be established. The highest peak in the radial density distribution is at about 2.3 nm.

- *The RNA binding site* (Figure 5.3C). The binding site is in two parts, being formed by the top of one subunit and the bottom of the next. The three bases associated with each protein subunit form a "claw" that grips the left radial helix of the top subunit.

- *Electrostatic interactions involved in assembly and disassembly.* Pairs of carboxyl groups with anomalous pK values (near pH 7.0) are present in TMV. These groups may play a critical role in assembly and disassembly of the virus. There are three sites where negative charges from different molecules are juxtaposed in subunit interfaces. These create an electrostatic potential that could be used to drive disassembly.

　A low-radius carboxyl-carboxylate pair appears to bind calcium
　A phosphate-carboxylate pair that also appears to bind calcium
　A high-radius carboxyl-carboxylate pair in the axial interface. This cannot bind calcium but can bind a proton and thus titrate with an anomalous pK.

- *Water structure.* Water molecules are distributed throughout the surface of the protein subunit, both on the inner and outer surfaces of the virus and in the subunit interfaces.

- *Specificity of TMV protein for RNA.* TMV protein does not assemble with DNA even if the origin of assembly sequence (see following) is included; thus, the specificity for RNA must involve interactions made by the ribose hydroxyl groups because all three base-binding sites could easily accommodate thymine.

FIGURE 5.3 Interactions between TMV subunits and RNA. A. From Figure 5.2B with the square indicating the region shown in panel B; the subunits marked * are the two subunits in panel B. B. Secondary structure of coat protein subunits. The four a-helices are indicated (LS, left-slewed; LR, left radial; RS, right-slewed; RR, right radial); N is the substantially obscured N-terminus; C is the C-terminus. C. Enlargement of the interaction of RNA with protein. The backbone structure of the protein subunits and the three RNA nucleotides (labelled 1, 2 and 3), represented as GAA, are illustrated. [This article was published in *J. Molec. Biol.*, **208**, K. Namba, R. Pattanayek, and G. Stubbs, Visualisation of protein-nucleic acid interactions in a virus. Refined structure of intact tobacco mosaic virus at 2.9 Å resolution by X-ray fiber diffraction, pp. 307–325, Copyright Elsevier (1989).]

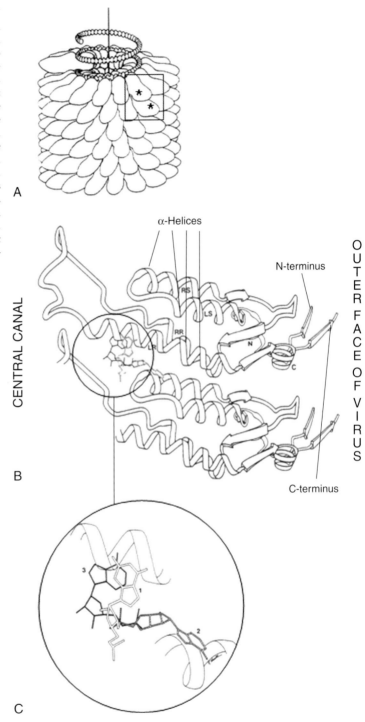

A

B

C

α-Helices

CENTRAL CANAL

N-terminus

C-terminus

OUTER FACE OF VIRUS

C. Assembly of TMV

TMV particles can be disassembled into coat protein subunits and RNA by dialysing virus preparations against alkaline buffers and separating the protein from the nucleic acid by ammonium sulphate precipitation. This has led to much study of the factors involved in *in vitro* virus disassembly and reassembly.

1. Properties of the Coat Protein

The coat protein comprises 158 amino acids, giving it a molecular weight of 17–18 kDa. Fibre diffraction studies have determined the structure to 2.9 Å resolution (Figure 5.3B). The protein has a high proportion of secondary structure with 50 percent of the residues forming four a-helices and 10 percent of the residues in ß-structures, in addition to numerous reverse turns. The four closely parallel or anti-parallel a-helices make up the core of the sub-unit, and the distal ends of the four helices are connected transversely by a narrow and twisted strip of ß-sheet. The central part of the subunit distal to the ß-sheet is a cluster of aromatic residues forming a hydrophobic patch.

2. Assembly of TMV Coat Protein

The protein monomer can aggregate in solution in various ways depending on pH, ionic strength, and temperature. The major forms are summarised in Figure 5.4A. Experiments with monoclonal antibodies show that both ends of stacked disks expose the same protein subunit surface. Thus, each two-layer unit in the stack must be bipolar (i.e., facing in opposite directions). The existence of these various aggregates has been important both for our understanding of how the virus is assembled and also for the X-ray analysis that has led to a detailed understanding of the virus structure. The helical protein rods that are produced at low pH are of two kinds: one with 16⅓ subunits per turn of the helix, as in the virus, and one with 17⅓. In both of these forms the

protein subunit structure is very similar to that in the virus.

3. Assembly of the TMV Rod

a. Assembly in Vitro. The classic experiments of Fraenkel-Conrat and Williams (1955) showed that it was possible to reassemble intact virus particles from TMV coat protein and TMV RNA. Since then, a detailed understanding has been gained on the assembly of the virus. The three-dimensional structure of the coat protein is known in atomic detail, and the complete nucleotide sequence of several strains of the virus and related viruses is known. The system therefore provides a useful model for studying interactions during the formation of a macromolecular assembly from protein and RNA.

Four aspects of rod assembly must be considered: the site on the RNA where rod formation begins; the initial nucleating event that begins rod formation; rod extension in the 5′ direction; and rod extension in the 3′ direction. The current understanding shows the following features.

The Assembly Origin in the RNA. Coat protein does not begin association with the viral RNA in a random manner, and the origin of assembly is between 900 and 1,300 nucleotides from the 3′ end. Nucleotide sequences near the initiating site can form quite extensive regions of internal base-pairing, as is illustrated in Figure 5.4B. Loop 1 in Figure 5.4B, with the sequence AAGAA-GUCG, combines first with a double disk of coat protein. The base sequence in the stem of loop 1 is not critical, but its overall stability is important as is the small loop at the base of the stem,

The Initial Nucleating Event. TMV rod assembly is initiated by the interaction between a double disk of coat protein (see Figure 5.4C, steps 1 and 2) with the long 5′ tail and the short 3′ tail of the RNA looping back down the axial hole. The transition between the double disk and helix is mainly controlled by a switching mechanism involving the abnormally titrating carboxyl groups. At low pH, the protein can form a helix on its own because the carboxyl

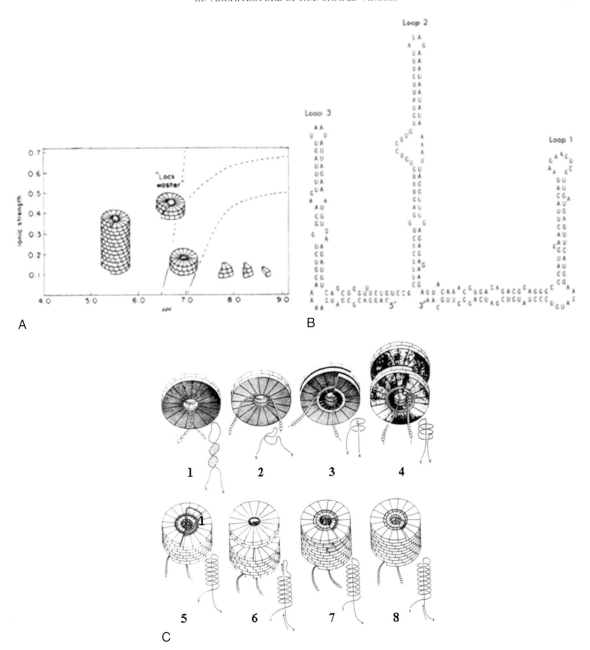

FIGURE 5.4 Assembly of TMV. A. Effects of pH and ionic strength on the formation of some aggregation states of TMV coat protein. The forms mentioned in the text that are involved in the assembly of the virus are identified. [Reprinted by permission from Macmillan Publishers Ltd: *Nature (London) New Biol.* **229,** 37–42 from Durham *et al.,* copyright 1971.] B. Proposed secondary structure of TMV origin of assembly extending from bases 5290 to 5527 of the viral sequence. [Reprinted by permission from Macmillan Publishers Ltd., from Zimmern (1983; *EMBO J.* **2,** 1901–1907).] C. Model for the assembly of TMV; the stages are described in the text. [Courtesy P.J.G. Butler.]

groups become protonated. It is thought that the inner part of the two-layered disk acts as a pair of jaws to bind specifically to the origin-of-assembly loop in the RNA and, in the process, converts each disk of the double disk into a "lock washer" and forming a protohelix (Figure 5.4C, step 3).

Rod Extension in the 5′ Direction. Following initiation of rod assembly, there is rapid growth of the rod in the 5′ direction "pulling" the RNA through the central hole until about 300 nucleotides are coated (Figure 5.4C, steps 4–8). It is generally agreed that rod extension is faster in the 5′ than in the 3′ direction by the addition of double disks. It is uncertain how the 5′ cap structure of the RNA is encapsidated. Disassembly by ribosomes (see Chapter 7) and *in vivo* would suggest that the structure at the extreme 5′ end might differ from that over most of the virus particle.

Rod Extension in the 3′ Direction. While 5′ extension of the rod is rapid and by addition of double disks, 3′ extension is much slower. It is thought that 3′ extension of the assembling rods is by the addition of small A protein aggregates.

b. Assembly in Vivo. Evidence exists that the process involved in the initiation of assembly outlined earlier is almost certainly used *in vivo*. However, there is no evidence that establishes the method by which the TMV rod elongates *in vivo*, but there is no reason to suppose that it differs from the mechanism that has been proposed for *in vitro* assembly.

IV. ARCHITECTURE OF ISOMETRIC VIRUSES

A. Introduction

From crystallographic considerations, Crick and Watson concluded that the protein shell of a small "spherical" or isometric virus could be constructed from identical protein subunits arranged with cubic symmetry, which has three types: tetrahedral (2:3), octahedral (4:3:2), and icosahedral (5:3:2); most isometric virus particles have icosahedral symmetry. The basic icosahedron has 60 identical subunits arranged identically on the surface of a sphere and display fivefold, threefold, and twofold rotational symmetry (Figure 5.5). A shell made up of many small identical protein molecules makes most efficient use of a virus's genetic material.

B. Possible Icosahedra

In developing potential structures for spherical viruses a problem arises with the limitation of 60 protein subunits. A very basic infectious genome would code for a capsid protein (about 1,200 nucleotides) and a polymerase (about 2,500 nucleotides), which would require a minimum internal "hole" of radius about 9 nm. A 60 subunit icosahedron made of coat protein of about 20–30 kDa (the usual size of a viral coat protein) would have an internal "hole" of about 6 nm and thus would not be sufficiently large to encapsidate most viral genomes. Caspar and Klug (1962) enumerated all the possible icosahedral surface lattices and the number of structural subunits involved. The basic icosahedron (Figure 5.5), with 12 groups of 5 structural subunits (termed pentamers) (or 20 × 3), giving a total of 60 structural subunits, can be subtriangulated (Box 5.2).

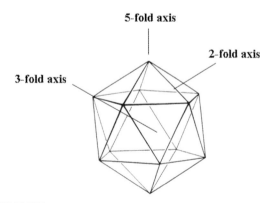

FIGURE 5.5   The regular icosahedron. The 5-fold, 3-fold, and 2-fold symmetry axes are identified.

BOX 5.2

BASIC ICOSAHEDRAL SYMMETRY STRUCTURE

The basic icosahedron of 60 structural subunits can be subtriangulated according to the formula:

$$T = PAf^2,$$

where T is called the triangulation number.

Parameter f is based on the fact that the basic triangular face can be subdivided by lines joining equally spaced divisions on each side.

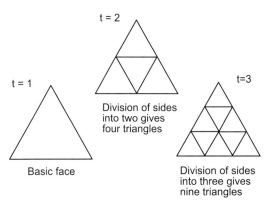

t = 2

t = 1

t = 3

Division of sides into two gives four triangles

Basic face

Division of sides into three gives nine triangles

Thus, f is the number of subdivisions of each side and is the number of smaller triangles formed. There is another way in which subtriangulation can be made, and this is represented by P. It is easier to consider a plane network of equilateral triangles.

Such a sheet can be folded down to give the basic icosahedron by cutting out one triangle from a hexagon (e.g., cross-hatching) and then joining the cut edges to give a vertex with five-fold symmetry.

However, if each vertex is joined to another by a line not passing through the nearest vertex, other triangulations of the surface are obtained. In the simplest case, the "next but one" vertices are joined.

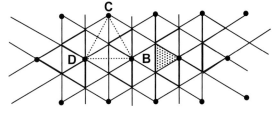

This gives a new array of equilateral triangles, and the plane net can be folded to give the solid shown in Figure 5.6A by removing the shaded triangle from each of the original vertices and then folding to give a vertex with 5-fold symmetry.

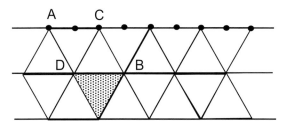

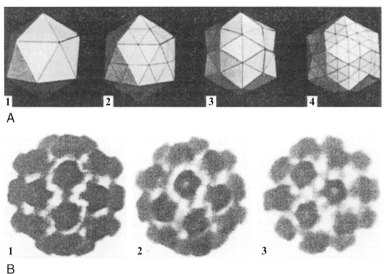

FIGURE 5.6 Icosahedral symmetry. A. Ways of subtriangulation of the triangular faces of the basic icosahedron shown in Figure 5.5; 1, The basic icosahedron with T = 1; 2, with T = 4; 3, with T = 3; 4, with T = 12. [From Caspar and Klug (1962), © Cold Spring Harbor Laboratory Press.] B. Structure of T = 3 particles of TYMV with reconstructed density distribution on particles showing 1, 2-fold axis; 2, 5-fold axis; 3, 3-fold axis. [This article was published in *J. Molec. Biol.*, **72**, J.E. Mellema and L.A. Amos, Three-dimensional image reconstruction of turnip yellow mosaic virus, pp. 819–822, Copyright Elsevier (1972).]

The process outlined in Box 5.2 leads to expansion of the particle by adding groups of six subunits (hexamers) in predetermined positions between the pentamer clusters (Figure 5.6A, panels 1–4). This gives a series of potential spheres of increasing size depending on the T number. The preceding formula gives limitations for the values of T, and the most common ones found in viruses are listed in Table 5.1. From this table it can be seen that with increasing particle size the number of

TABLE 5.1　Triangulation Numbers and Sizes of Isometric Virus Particles

Triangulation Number T	Number of Subunits	Number of Pentamers	Number of Hexamers	Approximate Diameter (nm)	Example
1	60	12	0	17	*Satellite tobacco necrosis virus*
				30	*Cowpea mosaic virus*[a]
				28	*Poliovirus*[a]
3	180	12	20	25–30	*Tomato bushy stunt virus*
				30	*Flock House virus*
				25	*Norwalk virus*
				27	*Enterobacteria phage MS2*
4	240	12	30	40	*Nudaurelia capensis ω virus*
				40	*Sindbis virus* core
7	420	12	60	50	*Cauliflower mosaic virus*
				45	*Simian virus 40*
13	780	12	120	75	*Wound tumor virus*
				75	*Bluetongue virus* core

[a] Pseudo T = 3 (see text).

pentamers remains constant (12) and that the number of hexamers increases.

C. Clustering of Subunits

The actual detailed structure of the virus surface will depend on how the physical subunits are packed together. For example, three clustering possibilities for the basic icosahedron are as follows:

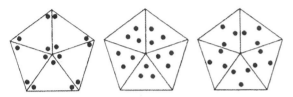

In fact many smaller viruses are based on the P = 3, f = 1, T = 3 icosahedron. In this structure, the structural subunits are commonly clustered about the vertices to give pentamers and hexamers of the subunits. These are the morphological subunits seen in electron micrographs of negatively stained particles (Figure 5.6B, 1–3).

D. Quasiequivalence

In the basic T = 1 icosahedron each of the subunits is in the same equivalent position to the adjacent subunits. However, when further subunits are introduced to form hexamers, it is not possible for them to occupy equivalent positions. Caspar and Klug (1962) assumed that not all the chemical subunits in the shell need to be arrayed in a strictly mathematically equivalent way. They also assumed that the shell is held together by the same type of bonds throughout but that the bonds may be deformed in slightly different ways in different nonsymmetry-related environments. This is termed **quasiequivalence**. Quasiequivalence would occur in all icosahedra except the basic structure (Fiure 5.5). Basically, pentamers give the 3-dimensional curvature, 12 giving a closed icosahedron, whereas hexamers give a 2-dimensional curvature (or tubular structure). Thus, the conceptual basis of quasiequivalence is the interchangeable formation of hexamers and pentamers by the same protein subunit.

V. SMALL ICOSAHEDRAL VIRUSES

A. Subunit Structure

The coat protein subunits of most small icosahedral viruses are in the range of 20–40 kDa; some are larger but fold to give effective "pseudomolecules" within this range. In contrast to rod-shaped viruses, the subunits of most small icosahedral viruses have a relatively high proportion of ß-sheet structure and a low proportion of a-helix and most have the same basic structure. This comprises an eight-stranded antiparallel ß sandwich, often termed a ß-barrel, which is shown schematically in Figure 5.7A.

The overall shape is a 3-dimensional wedge with the B-C, H-I, D-E, and F-G turns being at the narrow (interior) end. Most variation between the subunit sizes occurs at the N- and C-termini and between the strands of ß-sheet at the broad end of the subunit. Figure 5.7B shows the ß-barrel structure for the coat proteins of three viruses that have icosahedral particles.

It is the detailed positioning of the elements of the ß-barrel and of the N- and C-termini that give the flexibility to overcome the quasiequivalence problems. The coat protein subunits comprise one, two, or three structural domains; the S (shell) domain; the R domain (random-binding); and the P (protruding) domain; all viruses have the S domain. The R domain is somewhat of a misnomer but defines an N-terminal region of the polypeptide chain that associates with the viral RNA. As it is random, no structure can be determined by X-ray crystallography. The P domain gives surface protuberances on some viruses.

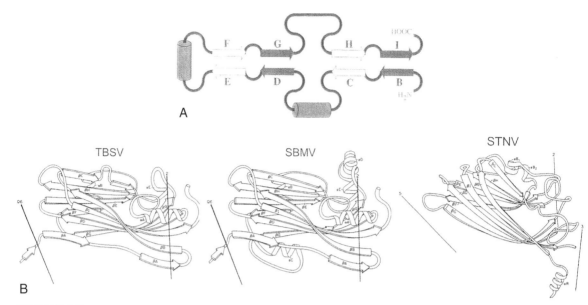

FIGURE 5.7   Structure of protein subunits of icosahedral viruses. A. Basic structure of the ß-barrel with the 8 ß-sheets labelled B-I; the N- and C-termini are shown; the cylinders represent conserved helices. [Reprinted from *Encyclopedia of virology*, Vol. 5, Johnson and Spier, pp. 393–400, Copyright Elsevier (2008).] B. Diagrammatic representation of the backbone folding of TBSV, SBMV, and STNV. [Reprinted from *J. Molec. Biol.*, **165**, M.G., C. Abad-Zapatero, M.R.N. Murty, L. Liljas, A. Jones, and B. Strandberg, Structural comparisons of some small spherical plant viruses, pp. 711–736, Copyright (1983), with permission from Elsevier.]

B. Virion Structure

At present, we can distinguish seven kinds of structure among the protein shells of small icosahedral- or icosahedral-based plant viruses whose architecture has been studied in sufficient detail. These are T = 1 particles, bacilliform particles based on T = 1, geminate particles based on T = 1, T = 3 particles, bacilliform particles based on T = 3, pseudo T = 3 particles, and T = 7 particles.

1. T = 1 Particles

The satellite viruses are the smallest known plant viruses, with a particle diameter of about 17 nm and capsids made up of 17–21 kDa polypeptides. The structure of *Satellite tobacco necrosis virus* (STNV) was the first to be solved and shown to be made of 60 protein subunits of 21.3 kDa arranged in a T = 1 icosahedral surface lattice. The general topology of the polypeptide chain of the coat protein (Figure 5.7C) shows an S

domain but no P domain. Three different sets of metal ion binding sites (probably Ca^{2+}) have been located. These link the protein subunits together.

2. Other Particles Based on T = 1 Symmetry

a. Bacilliform Particles Based on T = 1 Symmetry.
Some virus particles—for instance, those of *Alfalfa mosaic virus* (AMV; see Profile 3 in the Appendix)—are bacilliform with rounded ends separated by a tubular section. The structure of these particles is based on icosahedral symmetry. The rounded ends would have constraints of icosahedra with the three-dimensional curvature determined by 12 pentamers, 6 at each end. The two-dimensional structure of the tubular section would be made up of hexamers. Thus, in these virus particles the hexamers are not interspersed between the pentamers.

Purified preparations of AMV contain four nucleoprotein components present in major

FIGURE 5.8   Structure of AMV particles. A. Sizes (in nm) of the four main classes (components) of particles. [From Hull
(2002).] B. Geodestix models showing the
proposed structures of the components of
AMV shown in panel A. [This article was
published in *Virology*, **37**, R. Hull, G.J.
Mills, and R. Markham, Studies on alfalfa
mosaic virus. II. The structure of the virus
components, pp. 416–428, Copyright
Elsevier (1969).]

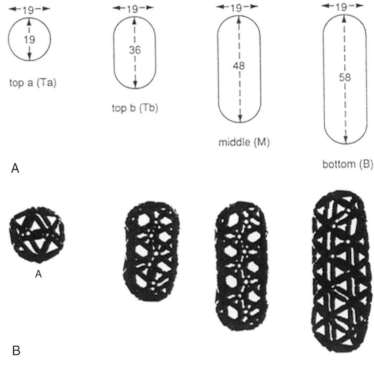

A

B

amounts (bottom, B; middle, M; top b, Tb; and top
a, Ta) that each contain an RNA species of definite
length. Three of the four major components are
bacilliform particles, 19 nm in diameter and differing lengths. The fourth (Ta) is normally spheroidal with a diameter of 19 nm (Figure 5.8A).

It is suggested that the tubular structure of
the bacilliform particles is based on a T = 1 icosahedron cut across its 3-fold axis with rings of
three hexamers (18 coat protein monomers)
forming the tubular portion (Figure 5.8B).

b. Geminiviruses. Geminiviruses   contain
ssDNA and one type of coat polypeptide. Particles in purified preparations consist of twinned
or geminate icosahedra (see Profile 7). The fine
structure of *Maize streak virus* (MSV) particles
was determined by cryoelectron microscopy,
and three-dimensional image reconstruction
shows that they consist of two T = 1 icosahedra
joined together at a site where one morphological

subunit is missing from each, giving a total of 22
morphological subunits in the geminate particle.

3. T = 3 Particles

The isometric particles of many plant viruses
have T = 3 structure. The major structural
features found in many of these viruses are exemplified in *Tomato bushy stunt virus* (TBSV; Profile
16) (Figure 5.7B) and *Turnip yellow mosaic virus*
(Profile 18). The structure of TBSV has been determined crystallographically to 2.9 Å resolution and
the virus shown to contain 180 protein subunits
arranged to form a T = 3 icosahedral surface lattice, with prominent dimer clustering at the outside of the particle, the clusters extending to a
radius of about 17 nm. The detailed structure
shows that coat protein has two different large-
scale conformational states in which the angle
between the P and S domains differ by about
20°; the conformation taken up depends on
whether the subunit is near a quasi dyad (2-fold

axis) or a true dyad in the T = 3 surface lattice. Further into the particle the R domain forms a flexible arm that winds with the equivalent domain of adjacent subunits to form an interlocking network. Thus, the detailed external and internal structures differ.

a. Bacilliform Particles Based on T = 3 Symmetry.

As noted in the preceding section on bacilliform particles, it has been suggested that the structure of bacilliform particles is based on icosahedral symmetry. *Rice tungro bacilliform virus* (Profile 4) has bacilliform particles of 130 × 30 nm, which are thought to be based on a T = 3 icosahedral symmetry cut across its 3-fold axis.

b. Pseudo T = 3 Symmetry.

Comoviruses (Profile 6), and most likely nepoviruses, fabaviruses, and sequiviruses together with picornaviruses, have icosahedral particles formed of one, two, or three coat protein species. If all the coat protein species are considered as one protein, the symmetry would appear to be T = 1. However, the larger polypeptides of viruses with one or two protein species form to give "pseudomolecules," and for each virus the structure can be considered to be made up of effectively three "species." This then gives a pseudo T = 3 symmetry.

4. T = 7 Particles

Cauliflower mosaic virus (CaMV) has a very stable isometric particle about 50 nm in diameter (Profile 4) containing dsDNA. The circular dsDNA is encapsidated in subunits of a protein processed from a 58 kDa precursor to several products, the major ones being approximately 37 and 42 kDa. Electron microscopy shows a relatively smooth protein shell with no structural features. Calculations from MW of the coat protein and the amount of protein in the virus suggested that it may have a T = 7 icosahedral structure, which was supported by cryoelectron microscopy to a resolution of about 3 nm.

C. The Arrangement of Nucleic Acid Within Icosahedral Viruses

For most icosahedral viruses, it is not possible to gain a detailed picture of the arrangement of the nucleic acid within the particles, as it does not form an ordered structure that can be identified by current techniques. In some cases, there is some ordered structure. For example, nearly 20 percent of the RNA in the comovirus *Bean pod mottle virus* particles bind to the interior of the protein shell in a manner displaying icosahedral symmetry. The RNA that binds is single-stranded, and interactions with the protein are dominated by nonbonding forces with few specific contacts. Following are some general points about the arrangement of nucleic acid within these particles.

1. RNA Structure

It is probable that the RNA inside many small icosahedral viruses has some double-helical structure. For example, about 75 percent of the RNA of *Turnip yellow mosaic virus* is in α-helical form, and about 95 percent of *Cowpea chlorotic mottle virus* has an ordered secondary structure. Computer predictions of secondary structure for 20 viral RNAs indicate that the genomes of icosahedral viruses had higher folding probabilities than those of helical viruses. When the folding probability of a viral sequence is compared with that of a random sequence of the same base composition, the viral sequences are more folded. These results suggest that base sequence plays some part in the way in which ssRNA genomes fold within the virus.

2. Interactions Between RNA and Protein in Small Isometric Viruses

Two types of RNA-protein interaction in the small isometric viruses can occur, depending on whether basic amino acid side chains, usually in the mobile R domain, or polyamines neutralize charged phosphates of the RNA.

D. Stabilisation of Small Isometric Particles

Small isometric virus particles are stabilised in three ways: by protein-RNA interactions, by protein-protein interactions, and by a combination of the two.

1. Protein-RNA Stabilisation

The particles of alfamo- and cucumo-viruses are stabilised by electrostatic interactions between basic amino acid side chains and RNA phosphate groups. These viruses are unstable in high salt concentrations dissociating to protein and RNA. One feature is that these viruses are unstable in some negative stains used for electron microscopy.

2. Protein-Protein Stabilisation

The particles of some viruses, such as como-viruses, are very stable, and virus preparations have "empty" particles that do not contain RNA. These particles are stabilised by strong interactions between protein subunits with little or no involvement of the RNA.

3. Protein-Protein + Protein-RNA Stabilisation

The particles of viruses, such as bromoviruses, and sobemoviruses, are relatively stable at low pHs but become unstable above pH 7. As the pH is raised between 6.0 and 7.0, the particles of *Brome mosaic virus* swell from a radius of 13.5 nm to more than 15 nm. This is due to the proton-ation of carboxyl-carboxylate groups on adjacent protein subunits, which interact at low pH but repel each other at higher pH. The swollen particle is then just stabilised by protein-RNA interactions and becomes salt sensitive. A similar situation is found in *Southern bean mosaic virus* (SBMV) with the addition of divalent cation, usually Ca^{2+}, links between adjacent subunits. For swelling of the particles to occur the pH has to be raised, and there must be a divalent cation chelator, such as EDTA, present.

VI. MORE COMPLEX ISOMETRIC VIRUSES

Phytoreoviruses have distinctly angular particles about 65–70 nm diameter that contain 12 pieces of dsRNA. Six different proteins are present in the particle of *Rice ragged stunt virus* (Profile 11; see Appendix), and seven are present in the capsid of *Rice dwarf virus* (RDV). Unlike most animal Reoviruses, which have triple-shelled capsids, the particles of phytoreo-viruses consist of an outer shell of protein and an inner core containing protein. However, there is no protein in close association with the RNA. The particles are readily disrupted during isolation by various agents. Under suitable conditions, subviral particles can be produced, which lack the outer shell and reveal the presence of 12 projections at the fivefold vertices of an icosahedron.

The structure of RDV has been derived to 25 Å resolution using cryoelectron micros-copy and image reconstruction. This revealed two distinct icosahedral shells: a T = 13 outer shell, 700 Å diameter, composed of 260 trimeric clusters of P8 (46 kDa), and an inner T = 1 shell, 567 Å diameter and 25 Å thick, made up of 60 dimers of P3 (114 kDa). The T = 1 core plays a critical role in the organisation of the quasiequivalence of the T = 13 outer capsid and probably guides the assembly of the outer capsid.

VII. ENVELOPED VIRUSES

Two families of plant viruses with envel-oped particles exist. *Rhabdoviridae* is a family of viruses whose members infect vertebrates, invertebrates, and plants. The virus particles (see Profile 13 in Appendix and Figure 2.5C) have a complex structure. Rhabdoviruses from widely differing organisms are constructed on the basic plan shown in Figure 5.9A.

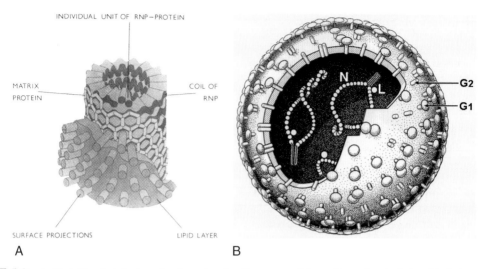

FIGURE 5.9 A. Model for rhabdovirus structure showing the proposed three-dimensional relationship between the viral proteins and the lipid layer. [From Cartwright *et al.* (1972; *J. Virol*. **10**, 256–260; DOI 10.1145/567752.567759, reproduced with permission from American Society for Microbiology).] B. Model showing structure of TSWV particle. The three nucleoprotein components are shown inside the particle with glycoprotein spikes (G1 and G2) extending through the lipid envelope. [From Kormelink (2005; Association of Applied Biologists, Descriptions of Plant Viruses, No. 412).]

Some animal rhabdoviruses may be bullet-shaped, but most, and perhaps all plant members, are rounded at both ends to give a bacilliform shape. The particles comprise a helical ribonucleoprotein core made up of the viral genome and the N protein and the viral polymerase surrounded by the matrix (M) protein. This is enveloped in a membrane through which glycoprotein (G) spikes protrude as surface projections. It is suggested that the structure of the M protein layer is based on half icosahedral rounded ends as in other bacilliform particles and that the G proteins relate to this structure.

Tospovirus particles (Profile 17; see Appendix) are spherical with a diameter of 80–110 nm and comprise a lipid envelope encompassing the genomic RNAs, which are associated with the N protein as a nucleoprotein complex. The viral polymerase protein is also contained within the particle. The lipid envelope contains two types of glycoproteins (Figure 5.9B).

VIII. ASSEMBLY OF ICOSAHEDRAL VIRUSES

A. Bromoviruses

The protein subunits of several bromoviruses can be reassembled *in vitro* to give a variety of structures. In the presence of viral RNA, the protein subunits reassemble to form particles indistinguishable from native virus. However, the conditions used are nonphysiological in several respects. The assembly mechanism of bromoviruses is thought to involve the carboxyl-carboxylate pairs just mentioned.

B. RNA Selection During Assembly of Plant Reoviruses

Every plant reovirus particle appears to contain one copy of each genome segment. Thus, there is a significant problem as to the macromolecular recognition signals that allow one, and one only, of each of 10 or 12 genome segments to be incorporated into each particle during virus assembly. For example, the packaging of the 12 segments of RDV presumably involves 12 different and specific protein-RNA and/or RNA-RNA interactions. Thus, there must be two recognition signals: one that specifies a genome segment as viral rather than host and one that specifies each of the 12 segments. The 5'- and 3'-terminal domains are fully conserved between the genome segments and it is suggested that this determines the recognition of viral rather than host RNA.

In *Reovirus* the (−)-sense strands are synthesised by the viral replicase on a (+)-sense template that is associated with a particulate fraction. These and related results led to the proposal that dsRNA is formed within the nascent cores of developing virus particles and that the dsRNA remains within these particles. If true, this mechanism almost certainly applies to the plant reoviruses. It implies that the mechanism that leads to selection of a correct set of 12 genomic RNAs involves the ss plus strand. It is thought that selection is directed by base-paired inverted repeats found in the genomic RNAs. It may be that other virus-coded "scaffold" proteins transiently present in the developing core are involved in RNA recognition rather than, or as well as, the three proteins found in mature particles.

IX. GENERAL CONSIDERATIONS

Plant viruses, because they occur at much higher concentrations in infected plants and can be relatively easily purified, have provided much of the information on basic virus structure. The basic concepts are not limited to plant viruses but apply equally to viruses from other kingdoms.

The virus particle protects the viral genome outside the cell and is often involved in transmission (see chapter 12). It is also likely that the formation of particles soon after viral replication sequesters the viral nucleic acid, thus controlling the impact of its expression on the host cell.

The idea of quasiequivalence in the bonding between subunits in icosahedral protein shells with $T > 1$ as put forward by Caspar and Klug in 1962 is still useful but has required modification in the light of later developments. Viruses have evolved at least two methods by which a substantial proportion of the potential nonequivalence in bonding between subunits in different symmetry related environment can be avoided: (1) by having quite different proteins in different symmetry-related positions as in the reoviruses and the comoviruses, and (2) by developing a protein subunit with two or more domains that can adjust flexibly in different symmetry positions, as in TBSV and SBMV.

The high-resolution analysis of the structures of isometric viruses is showing that, although the outer surface of the shell shows quasiequivalent icosahedral symmetry, this may not extend into the inner parts. There is frequently interweaving, especially of the C-terminal regions of the polypeptide chain, which forms an internal network.

The many negatively charged phosphate groups on the nucleic acid within a virus are mutually repelling. To produce a sufficiently stable virus particle, these charges, or at least most of them, need to be neutralized. Structural studies to atomic resolution have revealed three solutions to this problem:

1. In TMV, the RNA is closely confined within a helical array of protein subunits. Two of the phosphates associated with each protein subunit are close to arginine residues. However, the electrostatic interactions between protein and RNA are best considered as complementarity between two electrostatic surfaces.

2. In TBSV and a number of other icosahedral viruses, a flexible basic amino-terminal arm with a histone-like composition projects into the interior of the virus interacting with RNA phosphates. Additional phosphates are neutralized by divalent metals, especially Ca^{2+}.

3. In TYMV, where there is little interpenetration of RNA and protein, the RNA phosphates are neutralized by polyamines and Ca^{2+} ions.

In the more complex particles with structures comprising several layers of proteins, there is increasing evidence for interactions between the layers. It seems likely that these interactions drive the structural arrangements of protein subunits in adjacent layers.

X. VIRUSES OF OTHER KINGDOMS

Many of the basic principles of virus structure were determined for plant viruses but are also applicable to viruses of all kingdoms.

XI. SUMMARY

- The three basic virus particle shapes are rod-shape, isometric (spherical), and complex.
- Rod-shaped particles have helical symmetry, and most isometric particles have icosahedral symmetry.
- The assembly of the rod-shaped particles of TMV starts with the interaction of the origin of assembly (OAS) with a double disk of coat protein converting the double disk into a "lock washer," continues with the further addition of double disks ("lock washers") to the 5′ side of the OAS, and is completed by the addition of coat protein subunits to the 3′ side of the OAS.
- Considerations of icosahedral symmetry put constraints on the structural possibilities of icosahedral viruses.

- The bonding between structural subunits of icosahedra larger than the basic icosahedron involves quasiequivalent interactions.
- Variations on the basic icosahedral symmetry lead to bacilliform and geminate particles.
- Isometric virus particles are stabilised by protein-RNA interactions, protein-protein interaction, and divalent cation binding; viruses differ in the forms of particle stabilisation.
- Complex virus structures involve both helical and icosahedral symmetry, two or more shells of icosahedral symmetry, and enveloping in cellular membranes.

References

Caspar, D.L.D. and Klug, A. (1962). Physical principles in the construction of regular viruses. *Cold Spring Harbor Symp. Quant. Biol.* **27**, 1–24.

Crick, F.H.C. and Watson, J.D. (1956). Structure of small viruses. *Nature (London)* **177**, 473–475.

Fraenkel-Conrat, H. and Williams, R.C. (1955). Reconstitution of active tobacco mosaic virus from its inactive protein and nucleic acid components. *Proc. Natl. Acad. Sci. USA.* **41**, 690–698.

Schachman, H.K. (1959). *Ultracentrifugation in biochemistry.* Academic Press, New York.

Further Reading

Chiu, W., Chang, J.T., and Rixon, F.J. (2008). Cryoelectron microscopy. *Encyclopedia of Virology*, Vol. 1, 603–613.

Hull, R. (2002). *Matthew's plant virology.* Academic Press, San Diego.

Johnson, J.E. and Speir, J.A. (2008). Principles [of virus structure]. *Encyclopedia of Virology*, Vol. 5, 393–400.

Luo, M. (2008). Nonenveloped [viruses]. *Encyclopedia of Virology*, Vol. 1, 200–203.

Navaratnarajah, C.K., Warrier, R., and Kuhn, R.J. (2008). Enveloped [viruses]. *Encyclopedia of Virology*, Vol. 1, 193–199.

Peremyslov, V.V., Andreev, I.A., Prokhnevsky, A.I., Duncan, G.H., Taliansky, M.E., and Dolja, V.V. (2004). Complex molecular architecture of *beet yellow virus* particles. *Proc. Natl. Acad. Sci. USA* **101**, 5030–5035.

Speir, J.A. and Johnson, J.E. (2008). Non-enveloped virus structure. *Encyclopedia of Virology*, Vol. 5, 380–392.

Plant Viral Genomes

Viral genomes contain the information for the replication and expression of genes necessary for the functioning of the virus at the right time and in the right place.

I. INTRODUCTION

One of the main features of the last two or three decades has been the great explosion of sequence data on virus genomes. The first plant viral genome to be sequenced was the DNA of CaMV in 1980, followed by the RNA of TMV in 1982. By 2007 full genomic sequences were available for at least one member of the 80 genera of plant viruses and of many species within most of the genera. This plethora of data has led to much comparison between sequences often going into fine detail beyond the scope of this book. However, the comparisons, coupled with the use of infectious clones of viruses, have given valuable

information for use in mutagenesis experiments to elucidate the functions of various gene products and noncoding regions. In this chapter, we examine the various gene products that viruses encode and, in general terms, the genome organisations from which they are expressed.

II. GENERAL PROPERTIES OF PLANT VIRAL GENOMES

Basically, the viral genome, regardless of what kingdom it infects, comprises coding regions that express the proteins required for the viral infection cycle (initial infection, movement through the host, interactions with the host, and movement between hosts) and noncoding regions that control the expression and replication of the genome; control sequences can also be found in the coding regions.

A. Information Content

In theory, the same nucleotide sequence in a viral genome could code for up to 12 polypeptides. There could be an open reading frame (ORF) in each of the three reading frames of both the positive (+)- and negative (−)-sense strands, giving six polypeptides. Usually an ORF is defined as a sequence commencing with an AUG initiation codon and capable of expressing a protein of 10 kDa or more. If each of these ORFs had a leaky termination signal, they could give rise to a second read-through polypeptide. However, in nature, there must be severe evolutionary constraints on such multiple use of a nucleotide sequence, because even a single base change could have consequences for several gene products. However, two overlapping genes in different reading frames do occasionally occur, as do genes on both (+)- and (−)-sense strands. Read-through and frameshift proteins are quite common and are described in Chapter 7.

The number of genes found in plant viruses ranges from 1 for the satellite virus, STNV, to 12 for some closteroviruses and some reoviruses. Most of the single-stranded (ss) (+)-sense RNA genomes code for about four to seven proteins. In addition to coding regions for proteins, genomic nucleic acids contain nucleotide sequences with recognition and control functions that are important for virus replication and expression.

B. Economy in the Use of Genomic Nucleic Acids

Viruses make very efficient use of the limited amount of genomic nucleic acids they possess. Eukaryote genomes may have a content of introns that is 10–30 times larger than that of the coding sequences. Like prokaryote cells, most plant viruses lack introns, but some do not (see Chapter 7). Plant viruses share with viruses of other host kingdoms several other features that indicate very efficient use of the genomic nucleic acids:

- Coding sequences are usually very closely packed, with a rather small number of noncoding nucleotides between genes.
- Coding regions for two different genes may overlap in different reading frames [e.g., in *Potato virus X* (Profile 9; see Appendix) and *Turnip yellow mosaic virus* (Profile 18)], or one gene may be contained entirely within another in a different reading frame [e.g., the Luteovirus genome (Profile 8) and *Tomato bushy stunt virus* (Profile 16)].
- Read-through of a "leaky" termination codon may give rise to a second, longer read-through polypeptide that is coterminal at the amino end with the shorter protein. This is quite common among the virus groups with ss (+)-sense RNA genomes. Frameshift proteins in which the ribosome avoids a stop codon by switching to another reading frame have a result that is similar to a "leaky" termination signal; these are described in Chapter 7.
- A single gene product may have more than one function. For example, the coat protein of *Maize streak virus* (MSV) has a protective function and is involved in insect vector specificity, cell-to-cell transport of the virus, nuclear transport of the viral DNA, and possibly symptom expression and control of replication.
- Functional introns have been found in several geminiviruses and in *Rice tungro bacilliform virus* (RTBV; Profile 4). Thus, mRNA splicing, a process that can increase the diversity of mRNA transcripts available and therefore the number of gene products, may be a feature common in viruses with a DNA genome.

- A functional viral enzyme may use a host-coded protein in combination with a virus-coded polypeptide (see Chapter 8).
- Regulatory functions in the nucleotide sequence may overlap with coding sequences (e.g., the signals for subgenomic RNA synthesis in TMV).
- In the 5′ and 3′ noncoding sequences of the ssRNA viruses, a given sequence of nucleotides may be involved in more than one function. For example, in genomic RNA, the 5′-terminal noncoding sequences may provide a ribosome recognition site and at the same time contain the complementary sequence for a replicase recognition site in the 3′ region of the (−) strand.

C. The Functions of Viral Gene Products

There are two types of viral gene products: structural and functional. Structural gene products comprise the coat proteins described in Chapter 5 and, for complex viruses, scaffold proteins that direct the formation of the viral coat. Few, if any, examples of scaffold proteins in plant viruses exist.

1. Functional Proteins

Table 6.1 lists the presence of functional proteins for viruses of the various kingdoms (see also Box 4.2).

a. Proteins That Initiate Infection. Plants, animals, and bacteria differ in the surfaces that they present to an incoming virus (see Box 2.2). The cells that are first infected in bacteria and animals usually exist in a liquid medium. Thus, viruses reach the cell by simple diffusion, and the initial interaction with the virus is at the cell membrane surface. Entry into these cells is usually mediated by cell surface receptors and, in the case of some bacteriophages, by specialised virus mechanisms. In contrast, the plant surface is a waxy cuticle, and each cell is surrounded by a cellulose cell wall. Thus, the incoming virus has to pass these two barriers by various forms of mechanical damage, and there are no specialised cell surface receptors. There are also differences in bacteria, animals, and plants in the ways that viruses spread from the initially infected cell(s) to surrounding cells. Bacteria are unicellular, and cells in an animal body usually have no or few cytoplasmic connections between them. Thus, the initially infected bacterial or animal cells normally release virus from infected cells and infect further cells by interacting with surface recognition sites. In contrast, although plant cells have a cellulose cell wall, there are cytoplasmic connections (plasmodesmata) between the cells. A plant virus moves from cell to cell via plasmodesmata and vascular tissue throughout almost the entire plant (see Chapter 9). Thus, as far as a virus is

TABLE 6.1 Functional Viral Proteins

Protein	Virus				
	Plant	Vertebrate	Invertebrate	Fungus	Prokaryote
Initiate infection	−/+[a]	+	+	+	+
Replicate viral genome	+	+	+	+	+
Process gene product	+	+	+	+	+
Movement through host	+	−	−	Check	Check
Movement host to host	+	−	−	Check	Check

[a]See text.

concerned, one can consider a bacterial colony or an animal as a collection of cells in a liquid medium from which a virus has to exit through a cell membrane and enter a new cell through the membrane. As far as a virus is concerned, a plant is a single cell. However, plant viruses that circulate in their insect vectors have to cross various barriers, such as the gut and salivary gland accessory cell walls. These barriers are crossed by receptor-mediated mechanisms (see Chapter 12).

Within plant cells, the situation may be quite different. There is increasing evidence for the involvement of membranes in virus replication and in cell-to-cell movement. Thus, it is likely that there are receptor sites on intracellular membranes that are targeted by viral proteins rather than virus particles. Another aspect of initiating and establishing virus infection is overcoming host defences. Various viral gene products are involved in this and are described in Chapters 10 and 11.

b. Proteins That Replicate the Viral Genome. It is now generally accepted that all viruses, except satellite viruses (see Chapter 3), code for one or more proteins that have an enzymatic function in nucleic acid synthesis, either genomic nucleic acid or mRNAs or both. The general term for these enzymes is *polymerase*. As there is some inconsistency in the literature in relation to the terms used for different polymerases the terms used in this book are outlined in Box 6.1.

Some viral gene products are involved in genome replication in a nonenzymatic way. For instance, the 5′ VPg protein (see later in this chapter) found in some virus genera is thought to act as a primer in RNA synthesis, thus having a nonenzymatic role in RNA synthesis. *Alfalfa mosaic virus* (AMV) coat protein has an essential role in the initiation of infection by the viral RNA, possibly by priming (−)-strand synthesis. This protein is discussed in more detail in Box 8.5.

c. Proteins That Process Viral Gene Products. Viruses in which the whole genome or a segment of the genome is first transcribed into a single polyprotein usually encode one

---

BOX 6.1

TERMS USED TO DESCRIBE POLYMERASES

RNA-dependent RNA polymerase (RdRp). Catalyzes transcription of RNA from an RNA template.

RNA-dependent DNA polymerase (reverse transcriptase: RT). The enzyme coded, for example, by members of the *Retroviridae* and *Caulimoviridae*, which copies a full-length viral RNA into genomic DNA.

Replicase. The enzyme complex that makes copies of an entire RNA genome and the subgenomic mRNAs. Replicase enzymes often have either various functional domains or are made up of virus-encoded subunits with different functions. For instance, RNA → RNA replication can involve methyl transferase and helicase activities, as well as the actual polymerase itself.

Transcriptase. RdRp found as a functional part of the virus particle as in the *Rhabdoviridae* and *Reoviridae*.

There are no examples of viruses that replicate DNA directly from DNA using DNA-dependent DNA polymerase (DdDp) among plant viruses. In the *Geminiviridae* the viral gene product(s) associate(s) with the host DdDp.

or more proteinases that process the polyprotein into functional proteins. These are described in more detail in Box 4.3.

d. Proteins That Facilitate Viral Movement Through the Host.

As noted previously, plant viruses differ from those of other kingdoms in that the virus moves from cell to cell though cytoplasmic connections (plasmodesmata). For many plant viruses, one or more specific virus-coded protein(s) is required for this cell-to-cell movement and for systemic movement within the host plant (see Chapter 9).

e. Overcoming Host Defence Systems.

As we will see in Chapter 11, plants have a nucleic acid that targets the defence system. Successful viruses encode one or more protein(s) that suppress this defence system. Similarly, animal and bacterial viruses have ways of overcoming the cognate host defence mechanism.

f. Proteins That Facilitate the Host to Host Movement of Viruses.

Gene products have been identified as essential for successful transmission from plant to plant by invertebrate vectors and possibly by fungal vectors (see Chapter 12).

D. Nucleic Acids

As shown in Table 1.5, plant viral genomes can comprise dsDNA, ssDNA, dsRNA, and ss (−)-sense RNA, but most commonly ss (+)-sense RNA. In this section we examine some of the properties of the ss RNA genomes.

1. Multipartite Genomes

Of the 80 genera of plant viruses, 33 have their genome in two or more separate pieces of different sizes, called multipartite genomes. In most genera the individual pieces are encapsidated in separate particles (multicomponent viruses; Box 5.1). (See *Brome mosaic virus* (BMV) in Profile 3.)

2. Nucleic Acid Structures

In the intact virus particle, the three-dimensional arrangement of the RNA is partly or entirely determined by its association with the virus protein or proteins (Chapter 5). In solution, dsDNA has a well-defined secondary structure imposed by base-pairing and base-stacking in the double helix; viral ssRNAs have no such regular structure but under appropriate conditions contain numerous short helical regions of intrastrand hydrogen-bonded base-pairing interspersed with ss regions. However, as will be discussed in Chapter 11, one of the host defence systems targets dsRNA above a certain length, so there would be selection against extensive secondary structure of free RNA *in vivo*.

RNA molecules can fold into complex three-dimensional shapes and structures to perform their diverse biological functions. The most prevalent of these forms is the pseudoknot, which, in its simplest manifestation, involves the loop of a stem-loop structure base-pairing with a sequence some distance away (Figure 6.1). The involvement of pseudoknots has been recognised in a great variety of functions of viral RNAs, including control of translation by −1 frameshifting (see Chapter 7), by read-through of stop codons, by internal ribosome entry sites (see Chapter 7), and by translational enhancers.

3. Noncoding Regions

a. End-Group Structures.

Many plant viral ssRNA genomes contain specialised structures at their 5′ and 3′ termini. These are shown in the genome maps in the Profiles.

The 5′ Cap

Many mammalian cellular messenger RNAs and animal virus messenger RNAs have a methylated blocked 5′-terminal group of the form:

$$m^7G^{5N}ppp^{5N}X^{(m)}pY^{(m)}p\ldots$$

where $X^{(m)}$ and $Y^{(m)}$ are two methylated bases.

Some plant viral RNAs have this type of 5′ end, known as a "cap," but in the known

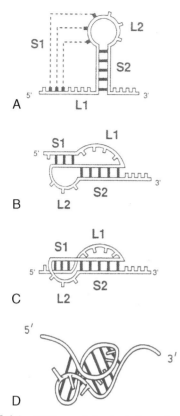

FIGURE 6.1   RNA pseudoknots. A. Secondary structure. The dotted lines indicate the base-pair formation of the nucleotides from the hairpin loop with the complementary region at the 5′ side of the hairpin. B. and C. Schematic folding. D. Three-dimensional folding, showing the quasi-continuous double-stranded helix. The stem regions (S1 and S2) and the loop regions (L1 and L2) are indicated; L1 crosses a deep groove and L2 a shallow groove. [Modified from Deiman and Pleig (1997; *Semin. Virol.* **8**, 166–175).]

plant viral RNAs the bases X and Y are not methylated. The capping activity is virus coded and differs from the host capping activity. Capping activity has been identified in several plant viruses including the 126 kDa TMV protein and in protein 1a encoded by RNA-1 of BMV. These activities methylate GTP using S-adenosyl-methionine (AdoMet) as the methyl donor, the guanyl transferase and transferase activity being specific for guanine position 7, forming a covalent complex with m^7GTP.

5′ Linked Protein

Members of several plant virus groups have a relatively small protein (c 3.5–24 kDa) covalently linked to the 5′ end of the genome RNA. These are known as VPg's (short for *virus protein, genome linked*). All VPg's are coded for by the virus concerned, and for most viruses the viral gene coding for the VPg has been identified. If a multipartite RNA genome possesses a VPg, all the genomic RNAs will have the same protein attached. The VPg is attached to the genomic RNAs by a phosphodiester bond between the ß-OH group of a serine or tyrosine residue located at the NH_2 terminus of the VPg and the 5′-terminal uridine residue of the genomic RNA(s). VPgs are involved in virus replication (see Box 8.6).

3′ Poly(A) Tracts

Polyadenylate sequences have been identified at the 3′ terminus of the messenger RNAs of a variety of eukaryotes. Such sequences have been found at the 3′ terminus of several viral RNAs that can act as messengers. The length of the poly(A) tract may vary for different RNA molecules in the same preparation, and such variation appears to be a general phenomenon. Internal poly(A) tracts are found in bromoviruses and hordeiviruses.

3′ tRNA-Like Structures

The 3′ termini of the RNA genomes of several plant viruses have the property of accepting and binding specific host tRNAs through an ester linkage. The accepting activity is also present in the ds replicative form of the viral RNA and in this state is resistant to RNase attack, thus demonstrating that the amino acid accepting activity is an integral part of the viral (+)-sense ssRNA.

The tRNA-like structures at the 3′ termini of a variety of plant viral RNAs form pseudoknots (Figure 6.2). These structures are involved in the regulation of RNA replication (see Boxes 8.4 and 8.5).

Barley stripe mosaic virus (BSMV) RNAs are unusual in that they have an internal poly(A) sequence between the end of the coding region

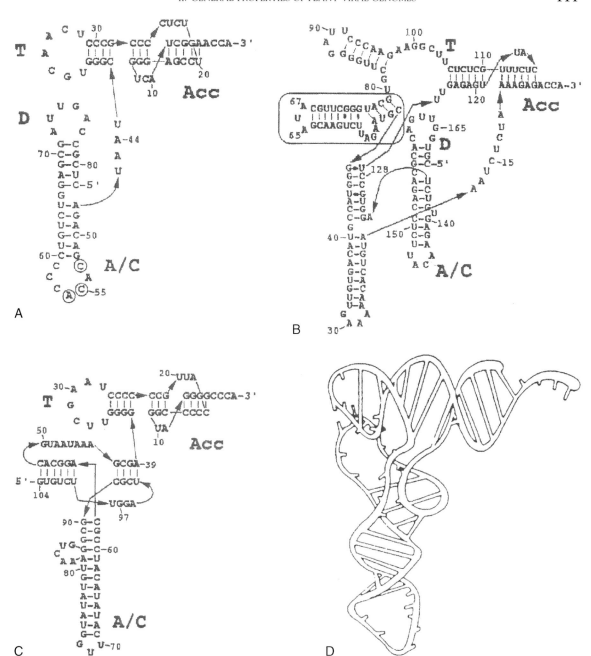

FIGURE 6.2 tRNA-like structures at the 3' end of plant viral RNAs. A. TYMV RNA, valine aminoacylation; B. BMV RNA, tyrosine aminoacylation; C. TMV, histidine aminoacylation. Acc, aminoacyl acceptor stem; T, Tψ-arm; D, D-arm; A/C, anticodon arm. [From Dreher (1999; *Annu. Rev. Phytopathol.* **37**, 151–174), reprinted, with permission, from the *Annual Review of Phytopathology*, Volume 37 © 1999 by Annual Reviews. www.annualreviews.org.] D. Three-dimensional model of canonical tRNA. [From Dumas *et al.* (1987; *J. Biomol. Struct. Dyn.* **4**, 707–728).]

and the 3′ tRNA-like sequence that accepts tyrosine.

Complementary 5′ and 3′ Sequences

The RNA genome segments of members of the tospoviruses and tenuiviruses have complementary sequences at the 5′ and 3′ ends that enable the termini of the RNAs to anneal to form "panhandle structures." These sequences are conserved across the genome segments of members of each genus. Similar structures are found in the genomes of vertebrate- and invertebrate-infecting Bunyaviridae.

b. 5′ and 3′ Noncoding Regions.   The 5′ and 3′ noncoding regions control both translation and replication. As we will see in Chapter 7, these two regions interact in the initiation of translation of, at least, the 5′ open reading frames (ORFs). The 3′ noncoding region is the site of initiation of (−)-strand RNA synthesis and the 5′ noncoding region (the 3′ end of (−)-strand RNA) is the site of initiation of (+)-strand synthesis (see Chapter 8).

c. Intergenic Regions.   Sequences in intergenic regions are also involved in both RNA synthesis and the translation of downstream ORFs. The initiation of synthesis of subgenomic RNAs is often in these regions and these RNAs are the messengers for translation of non-5′ ORFs in many viruses, as we will see in Chapter 7.

An increasing number of interactions are occurring between terminal and internal sequence regions in the control of expression of the genomic information from (+)-strand RNAs. It is likely that similar interactions will be found that control the expression of (−)-strand and dsRNA genomes and the genomes of ss and ds DNA viruses.

III. PLANT VIRAL GENOME ORGANISATION

The genome organisations of representative plant viruses are shown in the Profiles in the Appendix.

A. Structure of the Genome

Several questions must be answered to determine the structure of a viral genome:

1. Is the genomic nucleic acid single- or double-stranded, DNA or RNA, linear or circular?
2. How many pieces of nucleic acid make up the basic infectious genome? How many pieces make up the genome of a virus functioning in the natural situation?
3. What is the full nucleic acid sequence of the infectious genome?
4. What are the structures of the 5′ and 3′ termini of a linear nucleic acid?
5. How many ORFs does the genome contain? The answer to this question raises two more questions:
   a. What is an ORF? The simple definition is that it is a piece of the (+)-sense mRNA coding for a protein of more than 10 kDa with an AUG translation start codon and an appropriate stop codon. However, there are an increasing number of proteins of less than 10 kDa that are being shown to be functional. Factors to be considered with such proteins include whether they have highly improbable amino acid composition or whether they have sequence similarity between small ORFs in several viruses which may indicate that they are functional (e.g., the 6 kDa ORF3 of *Citrus tristeza virus;* Profile 5).

   ORFs of significant size representing possible proteins of 100 amino acids or more occur in the (−)-sense strands of several viruses that are normally regarded as being (+) stranded (e.g., TMV, AMV RNA1, and RNA2). There is no evidence that any of these have functional significance. However, there is no reason, in principle, why functional ORFs should not occur in the (−)-sense strand. Such ORFs are found in the geminiviruses, tenuiviruses, and tospoviruses (Profiles 7, 12, and 17, respectively; see Appendix).

ORFs do not necessarily start with the conventional AUG start codon. An AUU start codon has been recognised for ORFI of RTBV (see Box 7.5) and a CUG start codon for the capsid protein of *Soil-borne wheat mosaic virus*. This phenomenon raises the question of the definition of an ORF. Conventionally, it starts with an AUG codon and stops with one of the three stop codons. If non-AUG start codons are more widely used than believed, an ORF should be a largish in-frame region without a stop codon.

b. Is the ORF functional? Some of the ORFs revealed by the nucleotide sequence will code for proteins *in vivo*, whereas others may not. The functional ORFs can be unequivocally identified by finding the relevant protein in infected cells or by *in vitro* translation studies using viral mRNAs (Box 6.2).

Virus-coded proteins, other than those found in virus particles, may be difficult to detect *in vivo*, especially if they occur in very low amounts and are only transiently expressed during a particular phase of the virus replication cycle. However, a battery of methods is now available for detecting virus-coded proteins *in vivo* and matching these with the ORFs in a sequenced viral genome. In particular, the nucleotide sequence information gives a precise estimate of the size and amino acid composition of the expected protein. Knowledge of the expected amino acid sequence can be used to identify the *in vivo* product either from a partial amino acid sequence of that product or by reaction with antibodies raised against either a synthetic polypeptide that matches part of the expected amino acid sequence or against the ORF or part thereof expressed in, say, *E. coli*.

6. What are the regulatory and recognition signals for expression of the mRNAs? These are found in various parts of the genome,

---

BOX 6.2

IN VITRO TRANSLATION SYSTEMS

In vitro translation systems are derived from cells that have a high rate of protein synthesis. Here are the most commonly used systems:

- *E. coli* cell-free system: The prokaryotic *in vitro* system derived from *Escherichia coli* translates some monocistronic viral RNAs with fidelity but produces equivocal results with other RNAs.
- The eukaryotic rabbit reticulocyte lysate system.
- The eukaryotic wheat embryo system.

These systems can be used for analysis of cloned genes [using coupled transcription:translation.]

Some *in vivo* systems, such as the Toad oocyte system, have also been used for analysing plant viral RNAs. However, *in vivo* systems can have problems with proteins that are toxic to the cells and when the expressed protein is rapidly processed by intracellular proteases.

Details of *in vitro* systems can be found in Jagus, R. (1987). Translation in cell-free systems. In *Methods in enzymology* (S.L. Berger and A.R. Kimmel, Eds.), Vol. 152, pp. 267–296. Academic Press, San Diego.

particularly the 5′ and 3′ noncoding sequences and between ORFs in polycistronic RNAs. Regulatory sequences may also be found in coding regions.

7. What are the mRNAs for ORFs of polycistronic RNA viral genomes and DNA viruses? It may be difficult to establish whether a viral RNA of subgenomic size is a functional mRNA or merely a partly degraded or partly synthesised piece of genomic RNA.

Quite often, genuine viral subgenomic mRNAs are encapsidated along with the genomic RNAs. These can then be isolated from purified virus preparations and characterised. When the sequence of the genomic nucleic acid is known, two techniques can be used to locate precisely the 5′ terminus of a presumed subgenomic RNA. In the S1 nuclease protection procedure, the mRNA is hybridized with a complementary DNA sequence that covers the 5′ region of the subgenomic RNA. The ss regions of the hybridized molecule are removed with S1 nuclease. The DNA that has been protected by the mRNA is then sequenced. In the second method, primer extension, a suitable ss primer molecule is annealed to the mRNA. Reverse transcriptase is then used to extend the primer as far as the 5′ terminus of the mRNA and the DNA produces is sequenced.

B. Recognising Activities of Viral Genes

Before information on the sequence of nucleotides in viral genomes became available and before the advent of *in vitro* translation systems, there were two ways to determine the activities of viral genes: identification of proteins in the virus particle and classic virus genetics. These approaches are still relevant. Classic genetic studies have identified many biological activities of viral genomes. The discovery of viruses with the genome divided between two or three particles opened up the possibility of locating specific functions on particular RNA species by comparing the physical and biological properties of reassortments of the various components with those of the parent viruses after inoculation to appropriate hosts. This approach has been used on both natural and artificially induced mutants.

The advent of a wide range of technologies—such as sequencing, mutagenesis, and recombinant DNA—has provided many approaches to addressing the question of viral gene function. A few of these give unequivocal proof of function, whereas others are more or less strongly indicative of a particular function. There are two basic groups of methods. In the first, which may not be generally applicable, the natural gene product produced in infected tissue is isolated, and its activity is established by direct methods. Some virus-coded proteins besides coat proteins have functions that can be identified in *in vitro* tests, such as the use of *in vitro* translation products to detect protease activity. Another approach is to express the viral gene in either a prokaryotic system such as *E. coli* or in a eukaryotic system such as baculovirus vectors in insect cells or a yeast vector in yeast. The eukaryotic systems are preferred as they provide posttranslational modification not found in prokaryotic systems. Viral proteins produced in such systems can be used for *in vitro* experiments such as demonstrating the aphid transmission helper component activity of the product of CaMV ORF II (see Chapter 12). The second group of methods involves, directly or indirectly, the use of recombinant DNA technology. This consists of the following approaches.

1. Location of Spontaneous or Artificially Induced Mutations

Knowledge of nucleotide sequences in natural virus variants allows a point mutation to be located in a particular gene, even if the protein product has not been isolated. In this way, the changed or defective function can be allocated to a particular gene. The temperature-sensitive mutant of TMV, known as LS1, serves as an example. At the nonpermissive temperature, it replicates and forms virus particles normally found in protoplasts and infected leaf cells but is unable to move from cell to cell in leaves. A nucleotide comparison of the LS1 mutant and the parent virus showed that the LS1 mutant had a single base change in the 30 kDa protein gene that substituted a serine for a proline. This was a good indication, but not definitive proof, that the 30 kDa protein is involved in cell-to-cell movement.

The genomes of many DNA viruses and cDNAs to many RNA viruses have been cloned and the DNA or transcripts thereof shown to be infectious. There are numerous examples of experiments in which point mutations, deletions, or insertions have been used to elucidate the function(s) of the gene produced by the modified ORF. The introduction of defined changes in particular RNA viral genes to study their biological effects, and thus define gene functions, is commonly known as reverse genetics. This approach has been of major importance in understanding gene functions.

2. Recombinant Viruses

Recombinant DNA technology can be used to construct viable viruses from segments of related virus strains that have differing properties and thus to associate that property with a particular viral gene. For example, various viable recombinants were constructed containing parts of the genome of two strains of TMV, only one of which caused necrotic local lesions on plants such as *Nicotiana sylvestris*, which contain the *N'* gene. Infection with these recombinants indicated that the viral factor responsible for the necrotic response in *N'* plants is coded for in the coat protein gene.

One further application of recombinants is the tagging of gene products with fluorescent or other probes that report where in the plant or protoplast that gene product is being expressed or accumulates. By using video imaging, the sequence of events involved in the functioning of the gene product can be recorded. This approach has been used to study the movement of plant viruses around the cell.

3. Expression of the Gene in a Transgenic Plant

As with mutagenesis of infectious cloned genomes of viruses, the technique of transforming plants with viral (and other) sequences has had a major impact on understanding viral genes and control functions, and there are numerous examples of their expression in transgenic plants. The basic features of the technique are that a construct comprising the gene of interest, a promoter (often the 35S promoter of CaMV), and a transcriptional terminator sequence are introduced into suitable plant material. The gene of interest can be tagged with a marker, frequently a fluorescent dye, to enable its location in the transformed plant.

4. Hybrid Arrest and Hybrid Selection Procedures

Hybrid arrest and hybrid selection procedures can be used to demonstrate that a particular cDNA clone contains the gene for a particular protein. In hybrid arrest, the cloned cDNA is hybridized to mRNAs, and the mRNAs are translated in an *in vitro* system. The hybrid will not be translated. Identification of the missing polypeptide defines the gene on the cDNA.

In the hybrid select procedure, the cDNA-mRNA hybrid is isolated and dissociated. The mRNA is translated *in vitro* to define the encoded protein. In appropriate circumstances, these procedures can be used to identify gene function—for example, identifying the protease gene in *Tobacco vein mottling virus*.

5. Sequence Comparison with Genes of Known Function

As noted earlier in this chapter, sequence comparisons can be used to obtain evidence that a particular ORF may be functional. The same information may also give strong indications as to actual function. For example, using sequence information amino acid sequence similarities were found between the gene for an RNA-dependent RNA polymerase (RdRp) in poliovirus and proteins coded for by several plant viruses. This similarity implied quite strongly that these plant virus-coded proteins also have a polymerase function. The conserved amino acid sequences (motifs) of RdRps and many other viral gene products are described at the appropriate places in this book (e.g., for RdRps see Box 8.2).

6. Functional Regions Within a Gene

Spontaneous mutations and deletions can be used to identify important functional regions within a gene. However, mutants obtained by site-directed mutagenesis and deletions constructed *in vitro* can give similar information in a more systematic and controlled manner. For example, the construction and transcription of cDNA representing various portions of the *Tobacco etch virus* genome, translation *in vitro*, and testing of the polypeptide products showed that the proteolytic activity of the 49 kDa viral proteinase lies in the 3'-terminal region. The amino acid sequence in this region suggested that it is a thiol protease related in mechanism to papain.

However, care must be taken with this approach. Many functions depend on the three-dimensional structure of the protein and mutations, not at the active site, may have a secondary effect on the protein structure.

IV. VIRUSES OF OTHER KINGDOMS

Most of the points made in this chapter apply to viruses of other kingdoms. Differences, such as gene products that control the initiation of infection, have been explained in the text.

V. SUMMARY

- Viruses are very efficient in the use of the limited amount of genomic nucleic acid that they possess.
- Viral genomes contain coding sequences and sequences that control the expression of the viral genome.
- Viral genomes encode proteins required for successful infection, including proteins initiating infection (not most plant viruses) and proteins that replicate the viral genome, process viral gene products, facilitate movement through the host, overcome host defence systems, and facilitate movement between hosts.
- The 5' and 3' regions have structures that are important for genome expression and replication.

Further Reading

Hull, R. (2002). *Matthews' plant virology*. Academic Press, San Diego.

Chung, B.Y., Miller, W.A., Atkins, J.F., and Firth, A.E. (2008). An overlapping essential gene in the Potyviridae. *Proc. Natl. Acad. Sci. USA* **105**, 5897–5902.

Expression of Viral Genomes

Viral genomes are expressed via mRNAs. Eukaryotic cells pose constraints on how the information on an mRNA is expressed. Viruses have a variety of ways of overcoming these constraints.

I. STAGES IN VIRUS INFECTION CYCLE

The infection cycle of a virus, be it of a plant, vertebrate, invertebrate, or bacterium, has seven stages (see Figure 7.1):

1. Virus initial entry into the cell (discussed in Chapter 12).
2. Genome uncoating (discussed in this chapter).

3. Production of mRNAs. As will be shown in this chapter, the route used for the production of mRNAs depends on the nature of the viral genome.
4. Translation of the viral genetic information from the mRNAs. Some of this information is expressed early and some late in the infection cycle, depending on when the product is required.

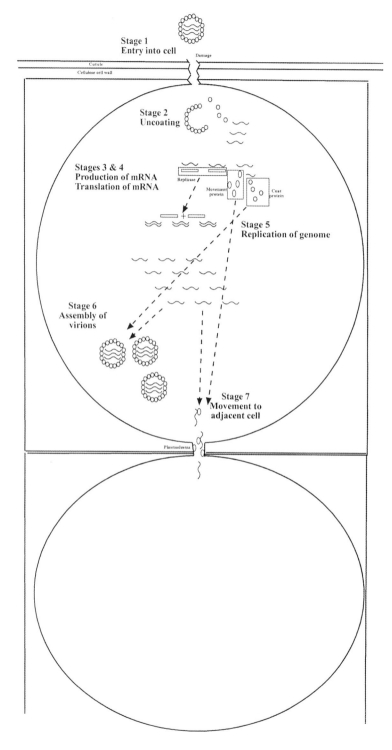

FIGURE 7.1 The seven stages of a virus infection cycle. The square box represents the plant cell wall and the circle within it, the plasma-membrane.

5. Replication of the viral genome using at least some of the factors expressed from the viral genome (discussed in Chapter 8).
6. Assembly of the progeny virions (discussed in Chapter 5).
7. Release of virus from initially infected cell and infection of adjacent cells (discussed in Chapter 9).

However, we must bear in mind that these seven stages are not completely separate and sequential and that they are closely integrated and coordinated.

II. VIRUS ENTRY AND UNCOATING

A. Virus Entry

As we will see in Chapter 12, plant viruses require damage to the cuticle and cell wall to be able to enter a plant cell. There is no evidence for a specific entry mechanism such as plasma membrane receptor sites or endocytotic uptake as with viruses of vertebrates and invertebrates, and it is generally considered that entry is accomplished by "brute force".

B. Uncoating

Once in the initially infected cell, the virus particle has to be uncoated to release the genomic nucleic acid. There is a dichotomy in the structural stabilisation of viruses in that the particles have to be stable enough to protect the viral genome when being transported outside the host but must be able to present the genome to the cellular milieu for the first stages in replication. The uncoating process depends on the chemical bonds that stabilise the virus particles (see Chapter 5). This has been studied for the rod-shaped *Tobacco*

mosaic virus (TMV) and for several isometric viruses.

1. Uncoating of TMV

Using TMV radioactively labeled in the protein or the RNA or in both components, the following has been observed:

1. Within a few minutes of inoculation, about 10 percent of the RNA may be released from the virus retained on the leaf.
2. Much of the RNA is degraded, but some are still full-length RNA molecules.
3. The early stages of uncoating of the particles do not appear to depend on preexisting or induced enzymes.
4. The process is not host specific, at least in the early stages.

However, there is a fundamental difficulty with all such experiments. Concentrated inocula must be used to provide sufficient virus for analysis, but this means that large numbers of virus particles enter cells rapidly. Thus, it is impossible to know which among these particles actually establish an infection.

To initiate infection, TMV RNA must be uncoated, at least to the extent of allowing the first ORF to be translated. By various *in vitro* experiments, it was shown that TMV is uncoated by a process termed *cotranslational disassembly* (Box 7.1).

2. Uncoating of Brome Mosaic Virus and Southern Bean Mosaic Virus

The isometric particles of *Brome mosaic virus* (BMV) and *Southern bean mosaic virus* (SBMV) are stabilised by carboxyl-carboxylate bonds and by protein-RNA interactions; SBMV particles are also stabilised by Ca^{2+} (see Chapter 5). Above pH 7, the carboxyl-carboxylate bonds protonate, and after the removal of Ca^{2+} from SBMV, the particles of both viruses swell, being just stabilised by protein-RNA interactions. These swollen particles can be uncoated by

BOX 7.1

COTRANSLATIONAL DISASSEMBLY OF TMV

The first event in cotranslational disassembly of TMV is that the structure of the virion relaxes so the 5′ terminus of the RNA is accessible to a ribosome. The 68 nucleotide 5′ leader sequence, which lacks G residues, interacts more weakly with coat protein subunits than do other regions of the genome. As discussed in Chapter 5, TMV particles are stabilised by carboxylate interactions, which become protonated at slightly alkaline pHs. This allows a ribosome to attach to the 5′ leader sequence and then to move down the RNA, displacing coat protein subunits as it moves. The ribosome-partially-stripped-rod complexes are termed *striposomes* (Figure A), which have been found both *in vivo* and *in vitro*.

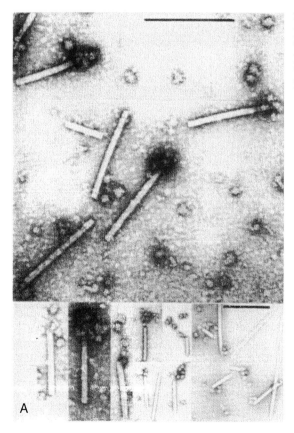

Fig.A. Electron micrographs of *striposome* complexes showing ribosomes clustered around one end (the 5′ end) of TMV rods. [This article was published in *Virology*, **137,** T.M. Wilson, Cotranslational disassembly of *tobacco mosaic virus in vitro*, pp. 255–265, Copyright Elsevier (1984).]

Having initiated translation, ribosomes proceed along TMV RNA, translating the 5′ ORF and the 126/183 kDa replicase protein (see Profile 14 in the Appendix for TMV genome map) and displacing

(continued)

coat protein subunits. When the ribosomes reach the stop codon of the 126/183 kDa ORF, they disengage. This raises the question of how the 3′ quarter of the particle is disassembled. It is considered that the replicase performs this task in a 3′→5′ direction in synthesising the (−)-strand replication intermediate in a process termed *coreplicational disassembly* from the 3′ end.

Thus, TMV is uncoated in a bidirectional manner, using the cotranslational mechanism for the 5′→3′ direction, yielding the replicase that disassembles the rest of the particle in the 3′→5′ direction, showing that disassembly and replication are coupled processes. The process happens rapidly with the whole capsid uncoated within about 30 minutes (Figure B).

Fig.B. Time scale of bidirectional disassembly and assembly of TMV particle *in vivo*. [From Wu and Shaw (1996), *Proc. Natl. Acad. Sci. USA*, **93**, 2981–2984, Copyright (1996) National Academy of Sciences, U.S.A.]

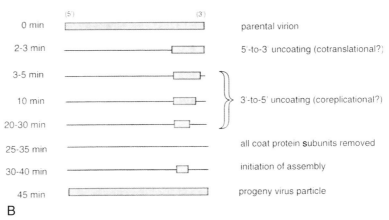

B

cotranslational disassembly *in vitro* translation. However, mutants of another bromovirus that did not swell under alkaline conditions showed that swelling was not necessarily required for cotranslational disassembly, leading to the suggestion that there is a pH-dependent structural transition in the virion, other than swelling, which enables the RNA to be accessible to the translation system. The proposed model, which is similar to those of some vertebrate and insect viruses, postulates that the N termini of the five subunits in the pentameric capsomere undergo a major structural transition from the interior to the exterior of the virion. This provides a channel through which the RNA passes to be accessible for translation. However, the 5′ end of the RNA must be released, which suggests that it is located in association with a pentameric capsomere.

3. Uncoating of Turnip Yellow Mosaic Virus

The isometric particles of *Turnip yellow mosaic virus* (TYMV) do not cotranslationally disassemble in the *in vitro* translation system just described for BMV and SBMV. *In vitro* studies show that under various nonphysiological conditions, such as pH 11.5 or freezing and thawing, the RNA can escape from TYMV particles without disintegration of the protein shell. When Chinese cabbage leaves are inoculated with TYMV, a significant proportion of the inoculum is uncoated after two minutes, with empty virus particles and low-MW protein being formed following RNA release. At least 80 to 90 percent of this uncoating takes place in the epidermis. This uncoating process is not confined to known hosts of TYMV.

4. Uncoating Other Plant Viruses

The requirements in the first stages of infection with other types of genomes are different to those of the (+)-sense ssRNA viruses. Viruses with dsRNA or (−)-sense ssRNA have to transcribe their genome to give mRNA. These viruses carry the viral RNA-dependent RNA polymerase (RdRp) in the virus particle and, presumably, transcription is an early event. It is not known if this occurs within the virus particle, possibly in a relaxed structure, or if the viral genome is released into the cell. However, it is most likely that this process takes place in an environment that is protected from cellular nucleases and that it is coupled to translation of the mRNA.

The dsDNA genomes of members of the *Caulimoviridae* must be transported to the nucleus, where they are transcribed to mRNA by the host RNA-dependent RNA polymerase (see later in this chapter). The coat protein of *Cauliflower mosaic virus* (CaMV) has a nuclear localisation signal that will presumably target the particle into the nucleus. Particles of some caulimoviruses and badnaviruses are particularly stable, capable of resisting phenol, and nothing is known about how they disassemble.

The ssDNA genomes of members of the *Geminiviridae* also have to be transported to the nucleus so they can be replicated before being transcribed to give mRNAs. Nuclear

localisation signals have been recognised in some geminiviral proteins, but nothing is known about how the particles uncoat.

III. INITIAL TRANSLATION OF VIRAL GENOME

Viral genomes are expressed from mRNAs that are either the nucleic acid (+)-sense ssRNA viruses or transcripts from the (−)-sense or dsRNA, or from ds or ss DNA viruses. Baltimore (1971) pointed out that the expression of all viral genomes, be they RNA or DNA, ss or ds, (+)- or (−)-sense, converges on the mRNA stage (Figure 7.2). As we will see later in this chapter, expression of the viral mRNA faces various constraints imposed by the eukaryotic translation system.

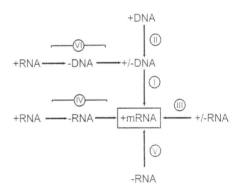

FIGURE 7.2 Routing of viral genome expression through mRNA. Route I is transcription of dsDNA usually by the host DNA-dependent RNA polymerase; route II is transcription of ssDNA to give a dsDNA template for I (e.g., geminiviruses); route III is transcription of dsRNA, usually by virus-coded RdRp (e.g., reoviruses); route IV is replication of (+)-strand RNA via a (−)-strand template by virus-coded RdRp—the viral (+) strand is often the template for early translation (the (+)-strand RNA viruses); route V is transcription of (−)-strand RNA viral genome by virus-coded RdRp (e.g., tospoviruses); route VI is reverse transcription of the RNA stage of retro- and pararetroviruses leading to a dsDNA template for mRNA transcription (for pararetroviruses the input viral dsDNA can be the template. [From Baltimore (1971).]

IV. SYNTHESIS OF mRNAs

The (+)-sense ssRNAs of many genera of plant viruses can act as mRNAs directly on entry into the host cell. For viruses with other types of genome, mRNAs have to be synthesised at some stage of the infection cycle.

A. Negative-Sense Single-Stranded RNA Viruses

All viruses with a (–)-sense ssRNA genome carry the viral RdRp in their virus particles. Thus, one of the early events on entry into a host cell is the transcription of the viral genome to (+)-sense RNA required for both translation of the viral genetic information and as an intermediate for replication.

Plant rhabdoviruses, like those infecting vertebrates, possess a genome consisting of a single piece of (–)-sense ssRNA, with a length in the range of 11 to 13 kb and encoding six proteins, one more than animal rhabdoviruses (for genome organisation of plant rhabdoviruses, see Profile 13 in the Appendix). The plant rhabdoviruses appear to be expressed in a manner similar to animal rhabdoviruses such as *Vesicular stomatitis virus* (VSV), which has been studied much more extensively. For VSV, the active transcribing complex consists of the RNA genome tightly associated with the N protein, and the polymerase made up of the phospho-protein (P) and the large (L) protein. This complex starts transcribing (+)-sense RNA at a single entry site at the 3′ end of the genome and transcribes the leader RNA that is transported to the nucleus, where it inhibits host cell transcription. The complex then transcribes the mRNA for the N protein, which is capped during synthesis by the polymerase. At the end of the N gene, and of all genes, is the sequence 5′-AGUUUUUUU-3′ (element I), which signals termination and polyadenylation of the mRNA. This intergenic sequence also comprises a short untranscribed sequence (element II) and the start site for transcription of the next mRNA (element III). Similar sequences are found in plant rhabdoviruses.

Thus, the viral genes are transcribed separately from the 3′ end and are transcribed in decreasing amounts (N > P > sc4 > M > G > L). This is an efficient way of regulating gene expression, as the genes that are located at the 3′ end are those that are required in greatest amounts.

The genomes of tospoviruses comprise three ssRNA segments (see Profile 17). L RNA is (–)-sense and monocistronic encoding the viral RdRp. The mRNA is transcribed from the virion RNA by the virion-associated polymerase. The other two RNAs have an ambisense gene arrangement with one ORF in the viral strand and one in the complementary strand. The two ORFs are separated by an AU-rich intergenic region of variable length. For both RNAs the virion-sense ORF is expressed from a subgenomic (sg) RNA transcribed from the complementary RNA and the complementary-sense ORF from an sgRNA transcribed from the virion RNA.

The intergenic region between the ambisense ORFs is predicted to form stable hairpin structures that are suggested to control the termination of transcription of sgRNAs. However, as noted in the following section, this should be considered with circumspection. The formation of tospovirus mRNAs involves cap snatching (see later in this chapter).

B. Double-Stranded RNA Viruses

Plant members of the *Reoviridae* family are placed in three genera: *Phytoreovirus* with 12 dsRNA genome segments and *Fijivirus* and *Oryzavirus* each with 10 dsRNA genome segments. The genome organisation of the Oryzavirus *Rice ragged stunt virus* is described in Profile 11. Most of the dsRNA segments are monocistronic, but segment 4 is bicistronic and segment 8 has three ORFs.

The plant reoviruses, like their counterparts that infect vertebrates and insects, contain a transcriptase that uses the RNA in the particle as template to produce ssRNA copies. In animal reoviruses, this occurs in subviral particles comprising part of the capsid, the polymerase, and the dsRNAs. Early in infection only (+)-sense ssRNAs are synthesised, which act as mRNAs. Later, (–)-sense strands are synthesised leading to viral replication (see Chapter 8). It is likely that a similar series of events occurs in the plant reoviruses, especially when they multiply in their insect vectors.

C. DNA Viruses

The synthesis of mRNAs from either the dsDNA members of the *Caulimoviridae* or the ssDNA members of the *Geminiviridae* and the nanoviruses does not involve a virus-coded enzyme but is performed by the host DNA-dependent RNA polymerase II located in the nucleus. This synthesis is initiated by a viral promoter.

1. Caulimoviridae

The genome organisations of the *Caulimoviridae* genera are described in Profile 4. Most of the detailed studies have been performed on *Cauliflower mosaic virus* (CaMV).

As we will see in Chapter 8, there are two phases in the nucleic acid replication cycle of CaMV: the nuclear phase of transcription and the cytoplasmic phase of gene expression and reverse transcription. In the first, the dsDNA of the infecting particle moves to the cell nucleus, where the overlapping nucleotides at the gaps are removed, and the gaps are covalently closed to form a fully dsDNA. These minichromosomes form the template used by the host DNA-dependent RNA polymerase to transcribe two RNAs of 35S and 19S, as indicated in Profile 4. As well as promoters for these two mRNAs, the viral DNA also has signals for the polyadenylated termination of transcription. The 35S promoter comprises the core promoter upstream of the transcription start site and various control elements both upstream and downstream of the start site (Box 7.2).

Other caulimovirus promoters also show similar modular structures. The promoters of some viruses require sequences downstream of the transcription start site for maximum expression. An example of this is the *Rice tungro bacilliform virus* (RTBV) promoter, which requires an enhancer located in the first 90 nucleotides of the transcript for efficient transcription. There are two subelements in this enhancer region, one being position and orientation independent and the other being position dependent.

Some promoters are specific to the vascular tissue. In that of RTBV the region between 164 and 100 nucleotides upstream of the transcription start site is essential for vascular tissue expression, and deletion leads to specificity in the epidermis. This tissue specificity is not surprising for RTBV, as the virus itself is phloem-limited.

Most of the caulimovirus promoters act in both monocot and dicot plant species even though the parent virus is restricted in host range. The CaMV 35S promoter has been used for the expression of transgenes in many dicot and monocot plant species and is considered a good, strong constitutive promoter. It has also been shown to be active in bacteria, yeast, animal HeLa cells, and Xenopus oocytes.

The caulimovirus 19S promoter has been much less studied. Only that of CaMV has been analysed and shown to be weaker than the 35S promoter when tested in transgenic constructs. This is in contrast to virus infections leading to comparable levels of the 35S and 19S RNAs and the product of the gene encoded by the 19S RNA, the ORFVI product, being the most abundant viral protein. The core 19S promoter can be strongly activated by the 35S promoter enhancer elements, but no enhancer elements have been detected for the promoter itself.

The caulimovirus 35S and 19S RNAs are 3′ coterminal and share a polyadenylation signal. The signal motif, AAUAAA, is found upstream

---

BOX 7.2

THE CAULIFLOWER MOSAIC VIRUS 35S PROMOTER

The 35S RNA is the major transcript from CaMV (see Profile 4). Its expression is controlled by a promoter that is a sequence of the viral genomic DNA.

The core promoter is characterised by what is termed a "TATA box," about 25 nucleotides upstream of the transcription start site. A detailed analysis of the CaMV promoter revealed that it has a modular nature with subdomains conferring patterns of tissue-specific expression. The two major domains are domain A (-90 to $+8$; numbering relative to transcription start site at $+1$), which is important for root-specific expression, and domain B (-343 to -90), which is mainly involved in expression in the aerial parts of the plant. The region of the A domain between -83 and -63 contains an as-1 (activation sequence-1)-like element that is important for the root-specific expression. The as-1 element is present in several nonviral promoters and can be recognised in many of the caulimovirus promoters, where it is important. The B domain comprises five subdomains, B1 to B5, each conferring specific expressions patterns in developing and mature leaves. A 60-nucleotide region downstream of the transcription start site also enhances gene expression. Thus, in the full promoter these domains and subdomains act coordinately and synergistically to give the constitutive expression of the CaMV 35S promoter. Plant nuclear factors have been identified that bind to various regulatory regions in this promoter.

---

of the transcription termination or cleavage site but downstream of the transcription initiation site.

2. Geminiviridae

The circular ssDNA genomes of members of the *Geminiviridae* have ORFs both in the virion-sense and complementary-sense orientations (see Profile 7). All geminiviruses employ the same basic strategy to transcribe their genomes in that it is bidirectional from the long intergenic region and terminating diametrically opposite in the short intergenic region. However, there are differences between the genera in the details of transcription.

The long intergenic region (which also contains the common region involved in replication) is about 300 nucleotides. Each of the virion-sense and complementary-sense transcripts has a characteristic eukaryotic RNA polymerase II promoter sequence, with transcription being initiated 20–30 bp downstream of the TATA box motifs. For at least some of the promoters, there are upstream sequences that enhance promoter activity. The C1 promoter for the Rep protein (see Chapter 8 for Rep protein) lies in the common region and overlaps the origin of (+)-strand DNA synthesis; this illustrates the close interactions between transcription and replication. The short AT-rich intergenic region contains the polyadenylation sites for the virion- and complementary-sense RNAs, which overlap so that they share a few 3′ nucleotides.

V. PLANT VIRAL GENOME STRATEGIES

A. The Eukaryotic Translation System Constraints

It is generally accepted that the eukaryotic protein-synthesising system translates the information from viral mRNAs. This translation system has various features and controls all of the following:

- Plant cellular mRNAs have a cap—an inverted and methylated GTP at the $5'$ terminus [$m^7G(5')ppp(5')N$] and a poly(A) tail at the $3'$ terminus.
- In most circumstances, mRNAs contain a single open reading frame (ORF).
- Translation is initiated at an AUG start codon, the context of which controls the efficiency of initiation.
- The cap, $5'$ untranslated region, the coding region, the $3'$ untranslated region, and the poly(A) tail all have potential to influence translational efficiency and mRNA stability.

However, the feature of the single ORF (bullet two above) presents major difficulties for viruses (Box 7.3).

B. Virus Strategies to Overcome Eukaryotic Translation Constraints

On current knowledge, there are at least 12 strategies by which RNA viral genomes and transcripts from DNA viruses ensure that all their genes are accessible to the eukaryotic protein-synthesising system and overcome the problem outlined in Box 7.3. The strategies fall into three groups (Figure 7.3):

- *Making the viral genomic RNA or segment thereof effectively monocistronic by bringing any downstream AUG to the $5'$ end.* This is done by either having a single ORF expressing a polyprotein that is subsequently cleaved to give the functional proteins (strategy 1) or by dividing up the viral genome to give monocistronic RNAs either during expression (strategies 2, 4, and 5) or permanently (strategy 3);
- *Avoiding the constraints of the $5'$ AUG.* This can be done in several ways (strategies 6 to 10).
- *Maximising the information expressed from a viral RNA by bringing together two adjacent ORFs to give two proteins, one from the $5'$ ORF and the other from both ORFs.* Thus, the second protein comprises the upstream protein in its N-terminal region, the C-terminal region being from the downstream ORF (strategies 11 and 12).

1. Strategy 1. Polyproteins

In this strategy the coding capacity of the RNA for more than one protein, and sometimes for the whole genome, is translated from a single ORF. The polyprotein is then cleaved at specific sites by one or more virus-coded

---

BOX 7.3

VIRUS mRNA TRANSLATION PROBLEM

In the scanning model for translation, the 40S ribosomal subunit binds to the $5'$ cap and translocates to the first AUG in a suitable context, where it forms the 80S ribosome that translates only that ORF immediately downstream from the $5'$ region of an mRNA; at the stop codon of this ORF, the ribosomes dissociate. Thus, ORFs beyond this point normally remain untranslated.

Viral genomes, except those of satellite viruses, encode two or more proteins and therefore are presented with a problem of how to express their downstream proteins in the eukaryotic system. Much of the variation in the way gene products are translated from viral RNA genomes appears to have evolved to meet this constraint.

FIGURE 7.3 Diagram illustrating the 12 strategies that viruses have for overcoming the constraints of the eukaryotic translation system. A. Diagrammatic viral genome with three open reading frames; B. The 12 strategies described in the text.

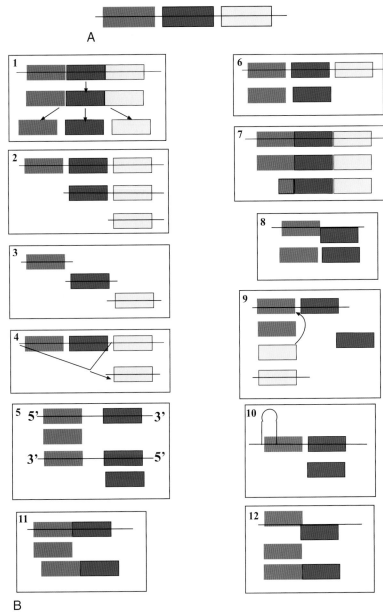

proteinases (see Box 4.3) to give the final gene products. Some viruses use one type of proteinase, others two or three.

The use of the polyprotein strategy is exemplified by the potyviruses, which have genomes of approximately 10 kb that contain a single ORF for a polyprotein of about 3,000 to 3,300 amino acids (see Profile 10). The polyprotein is cleaved to give 10 proteins (Figure 7.4), using three virus-coded proteases.

The 35 kDa P1 is a serine protease that cleaves itself from the polyprotein at Phe-Ser. The 52 kDa HC-Pro is a papain-like cysteine protease that cleaves at its C-terminal Gly-Gly.

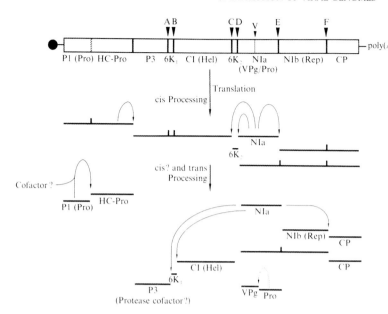

FIGURE 7.4 Schematic for the processing of the potyvirus polyprotein. The top part shows the potyvirus genome with 5′ VPg and 3′ poly(A) sequence and the gene products within the polyprotein. Below this is shown the cascade of processing of the polyprotein. The primary events are probably cotranslational and autocatalytic, yielding precursors and mature products. For gene products, see Profile 10. [From Riechmann *et al.* (*Journal of General Virology* **73**, 1–16, 1992).]

The 27 kDa protease domain of the NIa region is a serine protease responsible at most, if not all, for the other cleavages at Gln-(Ser/ Gly). Thus, the P1 and HC-Pro proteases act autocatalytically and the NIa protease acts both trans- and autocatalytically. The cleavage events take place at different rates yielding intermediate products that have different properties to the final products.

There are both advantages and disadvantages to the polyprotein processing strategy. Apart from overcoming the non-5′ start codon problem, there are advantages both in the fact that several functional proteins are produced from a minimum of genetic information and also in the potential for regulating the processing pathway. This is shown in the differences in the cleavage sites for the NIa protease and the influence that the surrounding residues can have on the rates of cleavage. Similarly, the requirement for the processing at the C-terminus of the HC-Pro before P1 is cleaved from the product most probably represents a control mechanism.

The major disadvantage is that it is difficult to visualise how the polyprotein strategy of the potyviruses can be efficient. The coat protein gene is at the 3′ end of the genome (see Figure 7.4). Thus, for every molecule of the coat protein produced, a molecule of all the other gene products must be made. Since about 2,000 molecules of coat protein are needed to encapsidate each virus particle but probably only one replicase molecule to produce it, this appears to be a very inefficient procedure. Indeed, large quantities of several gene products, apparently in a nonfunctional state, accumulate in infected cells (see Box 2.4). Nevertheless, the potyviruses are a very successful group. There are many member viruses, and they infect a wide range of host plants. Other viruses using polyprotein processing have additional devices that can avoid this problem. Comoviruses have their two coat proteins on a separate genome segment (see Profile 6). There does not appear to be any massive accumulation of noncoat gene products in cells infected with these viruses.

2. Strategy 2. Subgenomic RNAs

Subgenomic RNAs (sgRNA) are synthesised during viral replication from a genomic RNA that contains more than one ORF, giving 5′-truncated, 3′ coterminal versions of the genome. This then places the ORFs that were originally downstream at the 5′ end of the mRNA, as shown in the expression of TMV (Figure 7.5A).

FIGURE 7.5 Subgenomic RNAs. A. sgRNAs in the expression of TMV. The top line shows the genome organisation of TMV (see Profile 14); below is the expression of the four ORFs, the 126K protein being translated directly from the genomic RNA and read through to give the 183K protein; the 32k movement protein ORF3 and the 17K coat protein ORF4 are translated from sgRNAs. [From Hull (2002).] B. Expression of CTV. The top line is the genome organisation (see Profile 5). Below are the ways that the gene products are expressed by translation of the 5′ ORF directly from the genomic RNA, frameshift with the 5′ ORF, proteolytic processing of these products and expression of ORFs 2–11 from sgRNAs. [From Karasev and Hilf (1997; in *Filamentous viruses of woody plants* (P.L. Monette, Ed.), pp. 121–131, Research Signpost, Trivandrum).]

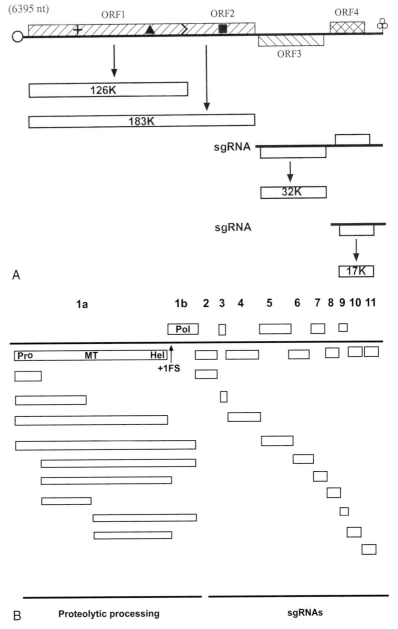

When several genes are present at the 3′ end of the genomic RNA, a family of 3′ collinear sgRNAs may be produced; *Citrus tristeza virus* (CTV) has a nested set of at least 9 sgRNAs (see Figure 7.5B). sgRNAs may be encapsidated (e.g., BMV; see Profile 3) and can cause uncertainty as to what comprises an infectious genome.

At least four models have been proposed for the synthesis of sgRNA from the genomic RNA (Box 7.4). The first and second mechanisms have been proposed for plant viruses.

The use of sgRNAs is widespread in plant viruses as a strategy to obviate the limitations of eukaryotic translation. The synthesis of these

---

BOX 7.4

SUBGENOMIC RNAs AND THEIR PROMOTERS

A subgenomic RNA (sgRNA) is a 5′ truncated version of the genomic RNA that places an ORF that was originally downstream of the 5′ ORF at the 5′ end (see Figure 7.2B, panel 2). This enables the downstream ORF to be translated. sgRNAs are produced during the replication of the genomic RNA, and at least four models have been suggested for their synthesis:

1. De novo internal initiation on the full-length (–)-strand of the genome during (+)-strand synthesis.
2. Premature termination during (–)-strand synthesis of the genome, followed by the use of the truncated nascent RNA as a template for sgRNA synthesis.
3. Initiation on the full-length (–) strand primed by a short leader from the 5′ end of the genomic DNA during (+)-strand synthesis. This has been found for coronaviruses.
4. Intramolecular recombination during (–)-strand synthesis, in which the replicase jumps from the subgenomic RNA start site on the full-length (+) strand and reinitiates near the 5′ end of the genome. This also has been found for coronaviruses.

The simplest model for de novo internal initiation of sgRNAs necessitates the replicase recognising a sequence upstream of the sgRNA 5′ end. This is termed the subgenomic promoter.

Brome mosaic virus (BMV) has a tripartite genome (Profile 3). The coat protein ORF is downstream on the bicistronic RNA-3 and is expressed from sgRNA-4. A subgenomic promoter for RNA-4 is in an intergenic region upstream of its 5′ end. The subgenomic promoter of BMV comprises two major parts. A "core" promoter of 20 nucleotides (–20 to +1) upstream of the subgenomic initiation site (the subgenomic initiation site is designated as + 1 and – numbers are 5′ of that site as read on the positive-strand RNA) is the smallest region capable of promoting sgRNA synthesis with low accuracy at a basal level; the core promoter is predicted to form a hairpin structure. Three "enhancer" regions, one downstream of the start site and two upstream, provide accuracy of replication initiation and control yields of sgRNAs. The fully functional subgenomic promoter encompasses about 150 nucleotides.

Premature termination of (–)-strand synthesis is effected by either *cis* or *trans* long-distance interactions between a region just upstream of the subgenomic promoter and another region of the viral nucleic acid. This can either be on the same nucleic acid molecule as that giving rise to the sgRNA (*cis* interaction) as exemplified by *Tomato bushy stunt virus* or on another genomic fragment of a split genome virus (*trans* interaction; e.g., *Red clover necrotic mosaic virus*).

RNAs involves close interlinks with viral replication and having strong controlling systems. The subgenomic promoter may having elements in both intergenic and coding regions, the latter suggesting that the position may control expression of the promoter.

3. Strategy 3. Multipartite Genomes

Viruses with multipartite genomes have the information required for the virus infection cycle divided between two or more nucleic acid segments. This is found for both DNA and RNA plant viruses. For the (+)-sense ssRNA viruses, this strategy places the gene at the 5' end of each RNA segment, and thus it is open to translation (e.g., RNAs 1 and 2 of BMV; see Profile 3).

Of the 80 plant virus genera, 33 have multipartite genomes. In most of these, the genome segments are encapsidated in separate particles, such viruses being termed multicomponent. (see Box 5.1). Members of the *Reoviridae,* and possibly the *Partitiviridae,* have all their multipartite genome segments in one particle.

4. Strategy 4. Splicing

The production of mRNAs from DNA in eukaryotes involves splicing, which removes internal noncoding sequences and can give various versions of an mRNA. Two of the families of plant viruses with DNA genomes, the *Caulimoviridae* and *Geminiviridae,* use splicing in the production of mRNAs, a process that, at least in the caulimoviruses, opens up downstream ORFs.

ORF4 of *Rice tungro bacilliform virus* (RTBV) is expressed from an mRNA spliced from the 35S RNA (see Profile 4). The splice removes an intron of about 6.3 kb and brings a short ORF (sORF) into frame with ORF4. The splice donor and acceptor sequences correspond to plant splice consensus sequences.

The circular ssDNA genome of *Maize streak virus* has four ORFs, two being expressed from transcripts, V1 and V2, in the virion sense, and

two, C1 and C2, from transcripts in the complementary sense (see Profile 7 for genome organisation). The C transcripts are of low abundance, and a splicing event fuses ORF C1 to C2.

5. Strategy 5. Translation for Both Viral and Complementary Strands (Ambisense)

Some of the genome segments of the tospoviruses and tenuiviruses encode two proteins, with one ORF in the virion sense and the other in the complementary sense (see *Rice stripe virus* and *Tomato spotted wilt virus* (TSWV) in Profiles 12 and 17, respectively). Thus, one of the proteins is expressed from complementary-sense RNA. This, called the ambisense expression strategy, was discussed previously in this chapter and is another means by which viruses overcome the eukaryotic translation constraints.

6. Strategy 6. Internal Initiation

As we noted previously, most eukaryotic mRNAs are 5'-capped and translation starts at the 5' end. However, in some cases, the 5' nontranslated regions or other regions of uncapped viral (and host) mRNAs contain internal ribosome entry sites (IRES), allowing eukaryotic ribosomes to be loaded onto the mRNA substantially downstream of the 5' end. It is suggested that the IRES forms a complex secondary/tertiary structure to which ribosomes and transacting factors bind. Although this internal initiation strategy was first demonstrated for animal viruses, an increasing number of plant viruses (e.g., a crucifer-infecting tobamovirus, a luteovirus, and a nepovirus) are also being shown to employ it.

It is considered that the IRES strategy enables a potentially inefficient mRNA (no cap and long 5' UTR) to be translated efficiently and might also provide translational control so gene products can be expressed at the appropriate time.

BOX 7.5

EXPRESSION OF RICE TUNGRO BACILLIFORM VIRUS (RTBV)

The genome of RTBV contains four ORFs (see Profile 4). The dsDNA genome is transcribed to give a 35S RNA, which is spliced to form the mRNA for ORF IV. Thus, the expression of ORFs I–III faces the problems of translation in a eukaryotic system. RTBV has a long leader sequence (more than 600 nucleotides) that contains 12 short ORFs. Long ORF I has an AUU start codon, and the next approximately 1,000 nucleotides have only two AUG codons in any reading frame: those for ORFs II and III. ORFs I, II, and III each overlap the next by one nucleotide having a "stop/start" signal of AUGA.

This has led to the development of the following model for the expression of RTBV ORFs I–III. (Fig.)

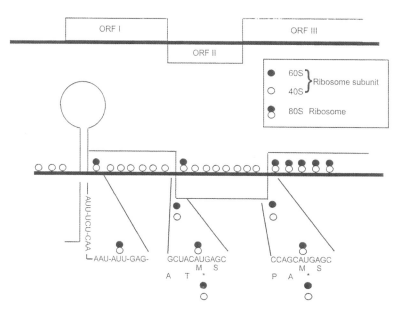

Fig. Expression of the first three ORFs of RTBV. Top line shows genome organisation of first three ORFs. Lower panel shows translation shunt with much pf the leader sequence folded into a hairpin followed by the non-AUG start codon for ORF I and "weak" start for ORF II. The translation process is described below. [From Hull (2002).]

The long leader sequence of RTBV is bypassed by a ribosome shunt mechanism by forming an elongate hairpin conformation (Fig.). In the shunt process the ribosome enters at the 5′ cap, which is essential; translocates to sORF A; and shunts across the stable hairpin from the shunt "donor" site just downstream of sORF A to the shunt "landing" site just upstream of ORF I. This places the 40S ribosome subunits at the AUU start codon of ORF I, a start codon that is about 10 percent as efficient as an AUG codon. It is suggested that only some of the 40S ribosome subunits associate with 60S ribosomes and initiate translation at the ORF I start codon, the rest translocating to the next AUG, which is the start codon for ORF II but is in a poor context. The model suggests that only some of these 40S ribosome subunits initiate here, the remainder passing to the next AUG, which is at the start of ORF III and in a good context.

7. Strategy 7. Leaky Scanning

Leaky scanning is where the 40S ribosomal subunits start scanning from the 5′ end of the RNA but do not all start translation at the first AUG. Some, or all, pass the first AUG and start translation at downstream ORFs. In some cases, the 40S subunits of the 80S ribosomes fail to disengage at a stop codon, and they reinitiate at a downstream start codon. There are three forms of leaky scanning:

a. Two Initiation Sites on One ORF (Two Start).

The genome of *Cowpea mosaic virus* (CPMV) is divided between two RNA species, the shorter of which, M-RNA, codes for two C-terminal collinear polyproteins of 105 and 95 kDa initiated from two in-frame AUG codons (see Profile 6 for CPMV genome organisation). The AUG for the 95 kDa protein is in a more favourable context (G at positions −3 and +4) than is that for the 105 kDa protein (A at −3 and U at +4).

b. Overlapping ORFs.

The genomes of members of the *Luteovirus* and *Polerovirus* genera contain six ORFs, the 3′ ones of which are expressed from sgRNAs (see Profile 8 for genome organisations). ORF4 (17 kDa) is contained in a different reading frame within ORF3 (coat protein). The translation of ORF4 fits very well with leaky scanning from the translation initiation of ORF3, as the context of the ORF3 AUG (U at position −3 and A at +4) is unfavourable.

c. Two or More Consecutive ORFs.

The expression of ORFs I, II, and III of RTBV is another example of leaky scanning due to the lack of AUG codons in upstream regions (see Box 7.5).

The translation of non-5′ ORFs by leaky scanning usually results from the AUG start codon in the upstream ORF being in a poor context. A lack or dearth of AUG codons in the upstream region can enhance the effect.

However, care must be taken in interpreting *in vitro* translation information as evidence for leaky scanning. Parameters such as translation system and conditions, especially the presence of divalent cations, can affect the expression from non-5′ ORFS.

8. Strategy 8. Non-AUG Start Codon

Some viral ORFs appear to start with a codon that is not the conventional AUG start codon; the initiation at these ORFs is inefficient. ORF1 of RTBV is AUU (Box 7.5).

9. Strategy 9. Transactivation

The dsDNA genome of *Cauliflower mosaic virus* (CaMV) has six closely spaced functional ORFs (I–VI; see Profile 4 for genome organisation) and is transcribed to give a more-than-genome-length RNA, the 35S RNA, and an mRNA (19S) for ORF VI. Although there is some evidence for the use of spliced RNAs for expressing some of the ORFs, it appears that some of the ORFs are expressed from the long RNA. Most of the downstream ORFs are not, or are poorly, expressed in protoplasts or transgenic plants unless the product of ORF VI is present. This gene product is termed a transactivator (TAV) and is thought to facilitate internal initiation.

For reinitiation at a downstream ORF, there should not be overlapping of long ORFs, and it is particularly efficient when the first ORF is about 30 codons long. The polar effects of the insertion of stem-loop structures, which would inhibit the translocation of ribosomes, into polycistronic mRNAs and the specificity for nonoverlapping ORFs indicated that transactivation causes enhanced reinitiation of ribosomes. The optimal 30-codon length for the first ORF is just long enough to emerge from a translating ribosome and suggests that the TAV acts directly or indirectly on the translating or terminating ribosome.

The transactivating function locates to the central one-third of the TAV protein. TAV associates with polysomes, with an 18 kDa ribosomal or ribosome-associated protein from plants and yeast and also with the 60S ribosomal subunit protein L 18 (RPL18) from *Arabidopsis thaliana*.

10. Strategy 10. Translational (Ribosome) Shunt

Short ORFs (sORFs; defined arbitrarily as having fewer than 50 codons and no known function for the product) in a leader sequence can interfere with translation of a downstream ORF. In the translational ribosome shunt mechanism, initially scanning ribosomes are transferred directly from a donor to an acceptor site without linear scanning of the intervening region, thus avoiding sORFs.

The 35S RNAs of members of the *Caulimoviridae* have long and complex leader sequences ranging in length from about 350 to more than 750 nucleotides and contain between 3 and 19 sORFs. They fold to give complex stem-loop structures. Translational shunting has been proposed as the mechanism by which these constraints to translation are bypassed. For translational shunting, there must be shunt donor and acceptor sites and a well-defined structure to bring these together. Most of the studies on these features have been on CaMV and RTBV (see Box 7.5).

11. Strategy 11. Read-Through Proteins

The termination codon of the 5′ gene may be "leaky" and allow a proportion of ribosomes to carry on translation to another stop codon downstream from the first, giving rise to a second, longer, functional polypeptide. This is termed *read-through* or *stop-codon suppression*, resulting in a protein that has the same sequence as the upstream protein in its N-terminal portion and a unique sequence in its C-terminal portion. An example is shown in Figure 7.5A, where the amber stop codon of TMV ORF1 (which encodes a 126 kDa protein) is read-through into ORF2 to give a protein of 183 kDa. The 183 kDa protein has the same N-terminal amino acid sequence as the 126 kDa protein but a unique C-terminal sequence.

The read-through strategy is found in several plant virus genera. The proteins that are produced by read-through are either replicases (e.g., TMV) or extensions to the coat protein (e.g., luteoviruses) thought to be involved in transmission vector interactions (see Chapter 12).

In read-through the stop codon is read by a suppressor tRNA instead of the ribosome being released by the eukaryotic release factor complex. Essentially, competition exists between the release factor complex and the suppressor tRNAs. Two main factors are involved in read-through of stop codons: the context of the stop codon and the suppressor tRNAs involved.

Stop codons have different efficiencies of termination (UAA > UAG > UGA), and also the first, and possibly the second, nucleotide 3′ of the stop codon acts as an important efficiency determinant. In plant viruses either amber (UAG) or opal (UGA) stop codons are read-through; there are no examples of read-through of the natural ochre (UAA) stop codon. Unlike retroviruses, no structural requirements for stop codon suppression in plant systems exist.

The synthesis of a read-through protein depends primarily on the presence of appropriate suppressors tRNAs. Thus, two normal tRNAs^{tyr} from tobacco plants were shown to promote UAG read-through during the translation of TMV RNA *in vitro*. The tRNA^{tyr} must have the appropriate anticodon, shown to be GψA, to allow effective read-through. tRNA^{gln} will also suppress the UAG stop codon.

The proportion of read-through protein produced may be modulated by sequence context of the termination codon, by long-distance effects, and by the availability of the suppressor tRNA. It is reasonable to suggest that about 1 to 10 percent of the times that a ribosome reaches a suppressible stop codon result in read-through.

As well as overcoming the constraints of the eukaryotic translation system, read-through of stop codons provides a mechanism for the control of the expression of gene products. Transmission helper factors have to be incorporated in the virus capsid, but it is probably not necessary to have them on all coat protein subunits. Thus, it could be more efficient for the virus expression system if 1 to 10 percent of the coat protein subunits also contain the transmission factor. Similarly, the viral replicase comprises several functional domains (described in Chapter 8) that are probably required in different amounts and even at different times. The production of two proteins that both contain some of the domains and one, the read-through protein, that also contains the other domains would give control over the availability of these functions.

12. Strategy 12. Frameshift Proteins

Another mechanism by which two proteins may commence at the same 5′ AUG is by a switch of reading frame before the termination codon of the 5′ ORF to give a second, longer, "frameshift" protein. This translational frameshift event allows a ribosome to bypass the stop codon at the 3′ end in one reading frame and switch to another reading frame so translation can continue to the next stop codon in that reading frame. A frameshift is illustrated in Figure 7.6.

At the frameshift site, ribosomes change their reading frame either one nucleotide in the 5′ direction (–1 frameshift) or one nucleotide in the 3′ direction (+1 frameshift). The frameshift process gives two proteins (the frame and frameshift proteins) that are identical from the N-terminus to the frameshift site but different beyond that point. The frame protein is always produced in greater quantity than the frameshift protein. Frameshift obviously occurs where ORFs overlap and may be at any place within that overlap.

The frameshift strategy is found in several plant virus genera and, in all cases that are known, involves the replicase. Most instances of frameshift are in the –1 direction (as shown in Figure 7.6); only those of the *Closteroviridae* are in the +1 direction (Figure 7.5B).

For a –1 frameshift three features are needed: a heptanucleotide sequence, termed the "slippery" or "shifty" sequence at the frameshift site; a strongly structured region downstream of the frameshift site; and a spacer of four to nine nucleotides between the slippery sequence and the structured region.

The slippery sequence comprises two homopolymeric triplets of the type XXX YYY Z (X = A, G, U; Y = A, U; Z = A, C, U). Upon reaching this heptanucleotide sequence, the two ribosome-bound tRNAs that are in one reading frame (X.XXY.YYZ) shift by one nucleotide in the 5′ direction (XXX.YYY.Z), retaining two out of the three base-paired nucleotides with the viral RNA. This mechanism was deduced for retroviruses, but the evidence points to a similar mechanism in plant RNA viruses.

The strongly structured regions are separated by the spacer region from the frameshift point and are either hairpins or pseudoknots (see Chapter 6 for pseudoknots). It is considered that the structure causes the ribosome to pause, thereby initiating frameshift.

Frameshifting in the +1 direction requires a run of slippery bases and a rare or "hungry" codon or termination codon. A downstream structures region is not necessary but may be found, as has been suggested for *Beet yellows virus* (BYV). The following mechanism has been

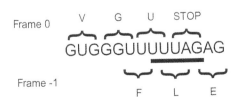

FIGURE 7.6 Translational frameshift. The ribosome bypasses the stop codon in frame 0 by switching back one nucleotide to frame –1 at a UUUAC sequence before continuing to read in frame –1 to give the fusion or frameshift protein. [From Hull (2000).]

suggested for the +1 frameshift between *Lettuce infectious yellows virus* ORF1a and ORF1b:

```
          K (protein 1a)                K (protein 1b)
LIYV RNA 5′...AAAG...3′ slippage 5′...AAAG...3′
             | | |        ---------->        | | -
tRNAˡʸˢ   3′...UUU...5′          3′  ...UUU...5′
```

As with the read-through strategy, most of the studies on the proportion of translation events that result in as read-through protein involve *in vitro* systems. These can give frameshift rates as high as 30 percent; *in vivo* rates of 1 to 5 percent are more likely. Similar to the read-through strategy, frameshift gives control of the production of functional domains in proteins.

C. Control of Translation

Various mechanisms for the control of expression of viral genomes have been described in the preceding section. These relate primarily to mechanisms that viruses have developed to overcome the problem of the limitation of translation of mRNAs in eukaryotic systems to the 5′ ORF. Among other features that control or regulate the translation of eukaryotic host mRNAs are the various noncoding regions (untranslated regions, or UTRs), which include the termini of the RNAs, the 5′ terminus being capped and the 3′ terminus having a poly(A) tail. Also involved in the control and efficiency of translation are the 5′ leader sequence and the 3′ noncoding region. In eukaryotic mRNAs, there is coordinated interaction between the 5′ and 3′ UTRs and even evidence for circularised mRNAs. Only some plant viral RNAs are capped and have poly(A) tails. The majority either have a cap or a poly(A) tail or have neither, although these may be translated very efficiently.

1. Cap but No Poly(A) Tail

The genome of TMV is capped but lacks a poly(A) tail. The structure of the 3′ UTR is complex being composed of five pseudoknots covering a 177 base region. The 3′-terminal two pseudoknots form the tRNA-like structure that is involved in virus replication (see Figure 6.2 and Box 8.4). The remaining three pseudoknots make up the upstream pseudoknot domain that is conserved in the tobamoviruses. This domain appears to functionally substitute for a poly(A) region in promoting interactions between the 5′ and 3′ termini and enhancing translation initiation in a cap-dependent manner. A 102 kDa host protein binds to the pseudoknot domain and also to the 5′ UTR. It is likely that this protein is involved in bringing the 5′ and 3′ ends together in the manner shown for eukaryotic mRNAs.

2. Poly(A) Tail but No Cap

Potyviruses are polyadenylated at their 3′ termini, but the 5′ terminus is attached to a VPg (see Chapter 6 for VPg). The 5′ UTR of *Tobacco etch virus* confers cap-independent enhancement of translation of reporter genes by interactions between the leader and the poly(A) tail. Two centrally located cap-independent regulatory elements that promote cap-independent translation have been identified in the 143-nucleotide leader sequence.

The VPg of the potyvirus *Turnip mosaic virus* (TuMV) interacts with the eukaryotic translational initiation factor eIF(iso)4E of *Arabidopsis thaliana,* and wheat (*Triticum aestivim*). eIF(iso)4E binds to cap structures of mRNAs and plays an important role in regulating the initiation of translation.

3. Neither Cap nor Poly(A) Tail

Many plant viruses have neither a 5′ cap nor a 3′ poly(A) tail, but these mRNAs are expressed very effectively and thus can be considered to have translation enhancement mechanisms.

4. Cap Snatching

All negative-strand RNA viruses with segmented genomes use a mechanism, termed "cap snatching" to initiate transcription of their mRNAs; this applies to viruses both from plants (tospoviruses and tenuiviruses) and from vertebrates and invertebrates (e.g., bunyaviruses and influenzaviruses). In this process, cap structures comprising about 12–20 5' nucleotides are cleaved from host mRNAs by a virus-encoded endonuclease and are then used to prime transcription. Cap snatching has been demonstrated for TSWV and the tenuivirus *Maize stripe virus*, both of which can snatch caps from positive-sense RNA viruses.

5. 5' UTR

As well as being involved in the enhancement of translation initiation, the 5' UTR of TMV also enhances the efficiency of translation. The 67 nucleotide 5' UTR, termed the Ω sequence, dramatically enhances translation of downstream genes in both plant and animal cells; in constructs in transgenic plants it enhanced translation by four- to sixfold. The Ω sequence has reduced secondary structure and a 25 base poly (CAA) region, which mutagenesis indicated is the primary element for *in vivo* translational enhancement. The 36 base 5' leader sequence from AMV RNA 4 also enhances translation, as does the 84-nucleotide leader sequence of PVX genomic RNA.

D. Discussion

The preceding descriptions show the great diversity of mechanisms that viruses use to express the information required for their function from what are often compact genomes. Viral mRNAs have to compete with host mRNAs for the translation machinery without causing significant damage to the normal functioning of the cell. The diversity of mechanisms overcomes constraints imposed on viruses using their host translational machinery. Most viruses use more than one of the strategies outlined in this chapter to express their genetic information, as shown by TMV, CTV, and RTBV.

These mechanisms can be viewed in two ways: specifically overcoming the constraints of the eukaryotic system—say, in the requirement for a cap for translation initiation and the translation of only the 5' ORF—and the control of translation so the right product is in the right place in the right quantity at the right time. The two uses of the mechanisms cannot be separated. For instance, in many cases the frameshift and read-through mechanisms provide different functions of the replication complex, the upstream one from the shorter protein containing the helicase and capping activities and the downstream one the replicase. The regulation is not only by the recognised ORFs but can be by noncoding sequences and possibly by short ORFs that normally might not be considered. There are many examples of the coat protein gene being expressed more efficiently than that for the replicase; more coat protein is required than the replicase.

VI. VIRUSES OF OTHER KINGDOMS

Most of the features in the expression of plant genomes are applicable to viruses of members of other eukaryotic kingdoms. Translation mechanisms in prokaryotes differ from those in eukaryotes, and thus there are different constraints on translation of viral genomes in these hosts.

VII. SUMMARY

- All viral infection cycles must produce mRNA from which to express the viral genome.
- The uncoating of many plant viruses with (+)-strand ssRNA genomes is integrated with the initial translation process.

- There are constraints in the eukaryotic translation system to the expression of polycistronic viral genomes.
- Viruses have various strategies to overcome the translation constraints.
- There are strong controls on translation so the right protein is expressed in the right amount at the right place and at the right time.

Reference

Baltimore, D. (1971). Expression of animal virus genomes. *Bacteriol. Rev.* **35**, 235–241.

Further Reading

Dreher, T.W. and Miller, W.A. (2006). Translational control in positive strand RNA plant viruses. *Virology* **344**, 185–197.

Dolja, V.V., Kreuze, J.F., and Valkonen, J.P.T. (2006). Comparative and functional genomics of closteroviruses. *Virus. Res.* **117**, 38–51.

Hanley-Bowdoin, L., Settlage, S.B., Orozco, B.M., Nagar, S., and Robertson, D. (1999). Geminiviruses: Models for plant DNA replication, transcription and cell cycle regulation. *Crit. Rev. Plant Sci.* **18**, 71–106.

Hull, R. (2000). *Matthew's plant virology.* Academic Press, San Diego.

López-Lastra, M., Rivas, A., and Barria, M.I. (2005). Protein synthesis in eukaryotes: The growing biological relevance of cap-independent translation initiation. *Biol. Res.* **38**, 121–146.

Miller, W.A. and Koev, G. (2000). Synthesis of subgenomic RNAs by positive-strand RNA viruses. *Virology* **273**, 1–8.

Rothnie, H.M., Chen, G., Fütterer, J., and Hohn, T. (2001). Polyadenylation in rice tungro bacilliform virus: cis-acting signals and regulation. *J. Virol.* **75**, 4148–4198.

Ryabova L., Park, H.S., and Hohn, T. (2004). Control of translation reinitiation on the cauliflower mosaic virus (CaMV) polycistronic. *RNA Biochem. Soc. Transact.* **32**, 592–596.

Shaw, J.G. (1999). Tobacco mosaic virus and the study of early events in virus infection. *Phil. Trans. R. Soc. Lond. B.* **354**, 603–611.

Virus Replication

One of the major features of viruses is their ability to replicate their genomic nucleic acid, often to high levels, in cells in which there are normally strict limits on the production of new nucleic acid molecules. Some viruses do this by adapting the existing cellular machinery, and others replicate their nucleic acid by mechanisms not widely used in host cells.

I. HOST FUNCTIONS USED BY PLANT VIRUSES

Like all other viruses, plant viruses are intimately dependent on the activities of the host cell for many aspects of replication, which include the following:

• *Components for virus synthesis.* Viruses use amino acids and nucleotides synthesised by host-cell metabolism to build viral proteins and nucleic acids. Certain other, more specialised, components found in some viruses—for example, polyamines, are also synthesised by the host.

- *Energy.* The energy required for the polymerisation involved in viral protein and RNA synthesis is provided by the host cell, mainly in the form of nucleoside triphosphates.
- *Protein synthesis.* Viruses use the host cell's protein-synthesising system for the synthesis of viral proteins using viral mRNAs. Many viruses also depend on host enzymes for any posttranslational modification of their proteins—for example, glycosylation.
- *Nucleic acid synthesis.* Almost all viruses code for an enzyme or enzymes involved in the synthesis of their nucleic acids, but they may not contribute all the polypeptides involved. For example, in the first phase of the replication of caulimoviruses, the viral DNA enters the host-cell nucleus and is transcribed into RNA form by the host's DNA-dependent RNA polymerase II. In most, if not all, RNA viruses, the replication complex comprises the viral RNA-dependent RNA polymerase (RdRp), several other virus-coded activities, and various host factors. ssDNA viruses alter the cell cycle constraints on the host DNA replication system. These aspects will be developed in greater detail in this chapter.
- *Structural components of the cell.* Structural components of the cell, particularly membranes, are involved in virus replication. For example, viral nucleic acid synthesis usually involves a membrane-bound complex. If we count a membrane as a compartment, eukaryotic cells have at least 20 compartments. In their replication, plant viruses have adapted in a variety of ways to the opportunities provided by this intracellular metabolic diversity.

II. METHODS FOR STUDYING VIRAL REPLICATION

Because of the involvement of host systems and the close integration with other stages of the infection cycle, it is generally accepted that a full picture of viral replication can only be obtained from *in vivo* systems. However, owing to their complexity, *in vivo* systems are extremely difficult to establish, and many of the questions of detailed interactions and functions must be addressed by *in vitro* systems. One of the major constraints to studying the replication of plant viruses is the establishment of a system where the replication events are synchronous. In inoculated whole plants, most cells are infected sequentially, and so at any one time a virus will be at different stages of replication in different cells. The three approaches to studying plant virus replication are *in vivo* systems, manipulation of the viral genome, and *in vitro* systems (Box 8.1).

III. REPLICATION OF PLUS-SENSE SINGLE-STRANDED RNA VIRUSES

The basic mechanism of replication of (+)-sense RNA genomes is that the virus-encoded replicase synthesises a complementary (–) strand, using the (+) strand as a template, and then new (+) strands are synthesised from the (–)-strand template. Synthesis of new RNA is from the 3' to 5' ends of the templates. Replication occurs in a replication complex that comprises the templates, newly synthesised RNA, the replicase, and host factors.

A. Viral Templates

Two kinds of RNA structures have been isolated from viral RNA synthesising systems. One, known as *replicative form* (RF; Figure 8.1A), is a fully base-paired ds structure, whose role is not certain. For example, it may represent RNA molecules that have ceased replicating. The other, called *replicative intermediate* (RI), is only partly ds and contains several ss tails (nascent product strands; Figures 8.1A and B).

The ds nature of RFs and RIs is apparent when these molecules have been isolated from infected

BOX 8.1

METHODS FOR STUDYING PLANT VIRUS REPLICATION

A. *In Vivo* Systems

Plant viruses have been studied using several *in vivo* systems:

1. *Protoplasts.* Protoplasts are isolated plant cells that lack the rigid cellulose walls found in intact tissue. Metabolically active protoplasts isolated from leaf cells have been infected with a range of viruses that replicate in a near-synchronous manner. This enables one-step virus growth experiments to be carried out, an important kind of experiment that has long been available to those studying viruses of bacteria and mammals.

2. *Temperature manipulation.* The synchrony of infection in the young systemically infected leaf can be greatly improved by manipulating the temperature. The lower inoculated leaves of an intact plant are maintained at normal temperatures (~ 25–$30°$), while the upper leaves are kept at 5–12°. Under these conditions, systemic infection of the young leaves occurs, but replication does not. When the upper leaves are shifted to a higher temperature, replication begins in a fairly synchronous fashion.

3. *Yeast.* Yeast, *Saccharomyces cerevisiae*, is a single-cell organism for which there is a considerable resource of classical and molecular genetics. Although yeast is the host for several viruses and virus-like agents, no ssRNA virus is known to naturally infect it. However, yeast has been shown to support the replication of some (+)-strand ssRNA viruses, including *Brome mosaic virus* and *Flockhouse virus,* and has been used to advance the understanding of how RNA viruses replicate.

B. Manipulating Viral Genomes

1. *Virus mutagenesis.* Infectious cDNA or DNA clones are available for many viruses and open the possibility to make specific mutations, the effects of which on virus replication can be studied in intact plants.

2. *Viral reporter systems.* By manipulating cloned viral genomes reporter molecules, usually fluorescent proteins can be attached to specific viral gene products or expressed separately from the viral genome. This enables the virus to be studied in intact plants in real time and for details to be obtained on the exact location of the gene function being studied. Radioactively labelled virus precursors are often used for identifying intermediates of replication.

3. *Metabolic inhibitors.* Inhibitors of certain specific processes in normal cellular metabolism have been widely applied to the study of virus replication. Three have been of particular importance: actinomycin D, which inhibits DNA-dependent RNA synthesis but not RNA-dependent RNA synthesis; cycloheximide, which is used as a specific inhibitor of protein synthesis on 80S cytoplasmic ribosomes; and chloramphenicol, which inhibits protein synthesis on 70S ribosomes (e.g., in chloroplasts and bacteria).

C. *In Vitro* Systems

1. *In vitro replication systems.* There have been various attempts to isolate competent replication complexes from virus-infected plant material. The main problems are (1) the difficulty of separating complexes of proteins and nucleic acids from normal cell constituents; (2) membranes are an integral part of the replication complexes of (+)-strand RNA viruses, and the technology for isolating such components is not yet well developed; and (3) uninfected plant cells contain an endogenous RdRp, which is often enhanced on virus infection.

(continued)

---

2. *Primer extension.* Properties of replication complexes can be studied by adding nucleotide triphosphates under the appropriate conditions and assessing the resulting products from extension of primed strands on the existing template. The products can be analysed by incorporating a labelled nucleotide triphosphate (radioactive or fluorescent label) or by probing the product with a labelled probe. This approach can be used to study the optimum condition for the replicase enzymes.

3. *Enzyme activities.* The enzymes involved in replication have been purified by standard protein and enzyme purification techniques,

including size exclusion chromatography and ion exchange chromatography. The properties of these enzymes have been studied by standard enzymatological techniques and other techniques such as activity gels.

4. *Protein-protein and protein-nucleic acid interactions.* Virus replication often involves protein-protein and protein-nucleic acid interactions. These have been studied by techniques such as the yeast two-hybrid system, the yeast three-hybrid system, by cross linking using treatment with chemicals or with UV irradiation, by gel mobility shifts, and by sandwich blots.

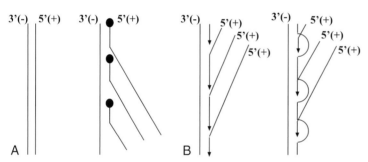

FIGURE 8.1   Replication of (+)-strand ssRNA. A. Forms of association between positive- and negative-sense viral RNA; left, replicative form (RF); right, replicative intermediate (RI). [Redrawn from Matthews (2002).] B. Possible structures of RI; left, semiconservative; right, conservative. [Redrawn from Buck (1999; *Phil. Trans. R. Soc. Lond. B.* **354**, 613–627).]

cells. The nature of the structures in the infected cell is unknown but is of great significance in view of the importance of dsRNA in the plant defense response (see Chapter 11). It is likely that these molecules are essentially single-stranded *in vivo*, the strands being kept apart either by compartmentalisation or by proteins bound to them. The (+) and (−) forms of the viral genome contain the signals that control both the specificity and timing of their replication.

It is likely that many of the control signals are at the 3′ ends of the template strands, though it should be recognised that there might be long-distance signals elsewhere in

the template, and/or the 3′ end interacts with other regions in the RNA. For (−)-strand synthesis, the 3′ terminus of the (+) strand has been found to be important. As described in Chapter 6, the three basic structures in plant viral RNA 3′ termini are tRNA-like structures, poly(A) tails, and non-tRNA heteropolymeric sequences. There is no real correlation between the 3′ terminal structure and the supergrouping of RdRps (described following), which might indicate similar roles for dissimilar 3′ termini.

The 5′ termini of the (+) strands (i.e., the 3′ termini of the (−)-strand templates for (+)-strand synthesis) are much more variable and offer no obvious directions as to the specific priming of (+)-strand synthesis. It is generally thought that the specificity is provided by the (−)-strand initiation and synthesis and that the (−)- and (+)-strand synthesis are coupled.

B. Replicase

Three or more virus-coded enzymatic activities can be involved in the replication of (+)-strand RNA viruses: the RNA-directed RNA polymerase (RdRp), a helicase, and a methyltransferase activity. These are collectively known as the viral replicase, but sometimes this term is used (incorrectly) for the RdRp.

1. RNA-Dependent RNA Polymerase

The RdRp catalyzes the synthesis of RNA using an RNA template. Two features have made this enzymic activity difficult to study. First, it is usually associated with membrane structures in the cell, and, on isolation, the enzyme(s) often become(s) unstable. Second, tissues of healthy plants may contain in the soluble fraction of the cell low amounts of a host enzyme with similar activities. The amounts of such enzyme activity may be stimulated by virus infection. Various properties of viral RdRps are outlined in Box 8.2.

The three-dimensional structure of the RdRp of poliovirus (supergroup 1) has been determined by X-ray crystallography (Figure 8.2). The overall structure of the enzyme is similar to those of other polymerases (DNA-dependent DNA polymerase, DNA-dependent RNA polymerase, and reverse transcriptase) that have been likened to a right hand. The palm domain contains the catalytic core and is similar to that of the other three polymerases. The thumb and finger domains differ from those of the other polymerases. Using the neural net PHD (for "Predict at Heidelberg") computer method the secondary structure of the RdRps of BMV, TBSV, and TMV has been predicted and compared with the poliovirus RdRp structure. This

analysis indicated that the RdRps of these supergroup 2 and 3 viruses have a similar structure to the supergroup 1 poliovirus enzyme, each containing a region unique to RdRps.

RdRp activity is also present in uninfected plants and at one time was thought to be involved in RNA virus replication. This activity has been isolated as a cDNA clone encoding a 128 kDa protein.

2. Helicases

Helicases are polynucleotide-dependent nucleoside triphosphate (NTP) phosphatases that possess ssDNA- and/or RNA-displacing activity. They play a pivotal role in genome replication and recombination by displacing complementary strands in duplex nucleic acids and possibly removing secondary structure from nucleic acid templates. Some helicases require a 3′ flanking single strand of nucleic acid and others a 5′ flanking single strand; these are known as 3′–5′ and 5′–3′ helicases, respectively. Some properties of viral helicases are outlines in Box 8.3.

Several plant virus genera appear to lack the characteristic NTP-binding motifs of helicases. Several possible reasons for this have been suggested:

1. The NTP-binding motifs may have diverged so much that they are not recognisable from the primary amino acid sequence.
2. The viral polymerase may have unwinding activity.
3. Unwinding may be effected by a helix-destabilising protein that uses the energy of stoichometric binding to ssRNA to melt the duplex in the absence of NTP hydrolysis.
4. The virus may coopt a host helicase.

Some plant viruses have additional helicase activities located elsewhere in their genomes. It is thought that the additional helicases in benyviruses, hordeiviruses, and potexviruses are involved in cell-to-cell movement.

BOX 8.2

SOME PROPERTIES OF VIRAL RdRps

RdRps have various characteristic amino acid sequence motifs. The most conserved of these is a Gly-Asp-Asp (GDD) motif, which is flanked by segments of mainly hydrophobic amino acids and is involved with Mg^{2+} binding. This suggests a structure of two antiparallel ß-strands connected by a short, exposed loop containing the GDD motif. Further alignments of RdRp sequences show eight motifs, which has led to the classification of RdRps into three "supergroups":

Supergroup 1 is sometime called the "picornavirus (-like)" supergroup.
Supergroup 2 is the "carmovirus (-like)" supergroup.
Supergroup 3 is the "alphavirus (-like)" supergroup.

The supergroups extend across viruses that infect vertebrates, plants, and bacteria, and there are representatives of plant viruses in each supergroup (Table). The members of each group have several properties in common.

Supergroups of RdRps

Supergroup 1 (Picornavirus supergroup)	Supergroup 2 (Carmovirus supergroup)	Supergroup 3 (Alphavirus supergroup)	
Comovirus	Aureusvirus	Alfamovirus	Potexvirus
Polerovirus	Carmovirus	Bromoviru	Tobamovirus
Potyvirus	Luteovirus	Closterovirus	Tobravirus
Sobemovirus	Tombusvirus	Hordeivirus	Tymovirus

1. Supergroup 1 members are characterised by usually having one genome segment that has a 5′ VPg and expresses its genetic information as a polyprotein.
2. Members of supergroups 2 and 3 have one to several genome segments, the RNA is often capped, and individual genes are translated.

---

3. Methyl Transferase Activity

The methyl transferase activity leading to 5′ capping of RNAs is described in Chapter 6.

4. Organisation of Functional Domains in Viral ORFs

Not all virus genera have the three functional domains. Some genera, such as *Tobamovirus* and *Cucumovirus*, do indeed have all three, but others, such as *Comovirus* and *Potyvirus*, lack the methyl transferase domain; the genomes of these viruses do not have a m^7G 5′ cap and therefore would not require this activity. Yet, others, like the *Tombusviridae* and the *Bromovirus* genus, lack the helicase domain (Box 8.3).

For all the viral genomes that express as polyproteins or fused protein (frameshift or read-through), the domains appear to be in the

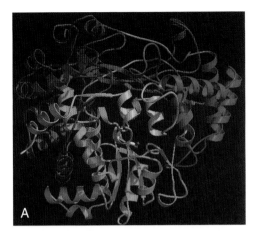
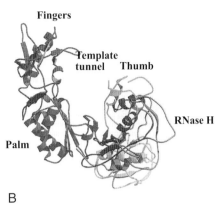
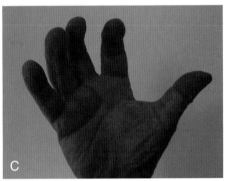

FIGURE 8.2 Structures of polymerases showing common features. A. Bacteriophage φ RdRp. Red indicates fingers, green the palm, blue the thumb, and yellow the priming domain. [From van Dijk *et al.* (2004). *J. Gen. Virol.* **85**, 1077–1093.] B. Moloney murine leukaemia virus reverse transcriptase. [This article was published in *Structure*, **12**, D. Das and M.M. Georgiadis, The crystal structure of the monomeric reverse transcriptase from *Moloney murine leukemia virus*, pp. 819–829, Copyright Elsevier Cell Press (2004).] C. A right hand for comparison.

order (N-terminal to C-terminal) methyl transferase, helicase, and the RdRp; in divided genomes, it is not possible to allocate the order. However, a feature of many of the virus genera is that the methyl transferase and helicase domains are physically separated from the RdRp domain. This can be by the methyl transferase and helicase domains being in one ORF and the RdRp being in a separately expressed ORF (e.g., *Bromovirus*), by them being in two adjacent ORFs separated by either a frameshift or read-through translational event (e.g., *Luteovirus*), or them being on a polyprotein and separated by protease activity (e.g., *Potyvirus*). However, in the Capilloviruses and

Marafiviruses, the protease domain lies between the methyl transferase and the other two domains. In the Trichoviruses, Vitiviruses, and Idaeoviruses, the three domains appear not to be separated.

For many viruses, it appears that the methyl transferase and helicase domains are on a single protein. However, although these two activities are expressed on the same ORF of *Beet yellows virus*, probing extracts from infected plants with monoclonal antibodies indicates that *in vivo* the 295 kDa protein is processed to a 63 kDa protein containing the methyl transferase domain and a 100 kDa protein containing the helicase domain.

BOX 8.3

SOME PROPERTIES OF VIRAL HELICASES

Based on conserved amino acid sequence motifs, helicases have been grouped into a number of superfamilies. Five superfamilies have been recognised, three of which have representatives in (+)-strand RNA viruses. Superfamilies I and II have seven conserved motifs, whereas superfamily III has three motifs. Two motifs common to all three superfamilies are variants of ATP-binding motifs and have the conserved sequences GXXXXGKT/S and MMMMD, where X is an unspecified amino acid and M is a hydrophobic residue. Most members of superfamilies I–III are 3′–5′ helicases. The superfamily designation for various plant virus genera is given in the Table.

Helicase superfamilies and plant virus genera

Supergroup I	Supergroup II	Supergroup III	No motif
Alfamovirus	*Bymovirus*	*Comovirus*	*Carmovirus*
Bromovirus	*Potyvirus*	*Sequivirus*	*Luteovirus*
Closterovirus			*Polerovirus*
Cucumovirus			*Sobemovirus*
Hordeivirus			*Tombusvirus*
Potexvirus			
Tobamovirus			
Tobravirus			
Tymovirus			

The crystal structure for the hepatitis C virus RNA helicase (superfamily II) has been determined and the mechanism for unwinding duplex RNA suggested. The structure comprises three domains forming a Y-shaped molecule. The RNA-binding domain is separated from the NTPase and other domain by a cleft into which ssRNA could be modeled. It is suggested that a dimer form of this protein unwinds dsRNA by passing one strand through the channel formed by the clefts of the two molecules and by passing the other strand outside the dimer. Because of the conserved motifs between the various superfamilies, it is likely that many of the features determined for the hepatitis C virus helicase are applicable to this enzyme from plant viruses.

C. Sites of Replication

Plant cells contain a range of membranes (see Box 9.1). It is generally assumed that the replication of (+)-strand RNA viruses involves association with cellular membranes. In many virus infections there is perturbation of membrane structures, frequently leading to the formation of vesicles, which in the proven cases have been shown to be associated with replication complexes. A variety of membranes are involved in vesicle formation (Table 8.1). There does

TABLE 8.1 Example of Membranes Possibly Associated with (+)-Strand RNA Virus Replication

Membrane	Virus[a]
Endoplasmic reticulum	BMV. TMV
Chloroplast outer membrane	TYMV, AMV
Vacuolar membrane	CMV
Peroxisome	TBSV
Mitochondria	CIRV, TRV

[a]Virus abbreviations: AMV, *Alfalfa mosaic virus*; BMV, *Brome mosaic virus*; CIRV, *Carnation Italian ring spot virus*; CMV, *Cucumber mosaic virus*; TBSV, *Tomato bushy stunt virus*; TMV, *Tobacco mosaic virus*; TRV, *Tobacco rattle virus*; TYMV, *Turnip yellow mosaic virus*.

not appear to be any correlation between the membrane involved and the supergroups of the replicase proteins or even with the virus family.

D. Mechanism of Replication

The RF and RI RNAs are considered to provide evidence on the mechanism of RNA replication. It is suggested that the RF could arise from the initial synthesis of a (−) strand on a (+)-strand template. RI RNAs usually contain more (+) strands than (−) strands, which is taken to indicate that each is a single (−) strand to which several (+) strands are attached (Figure 8.1B).

There are two hypotheses as to the mechanism of (+)-strand synthesis. The semiconservative mechanism involves total displacement of the newly synthesised strand by the oncoming strand (Figure 8.1B, left-hand structure), and in the conservative mechanism, it is suggested that the duplex RNA is only transiently unwound at the growing end of the nascent strands (Figure 8.1B, right-hand structure). The majority of evidence supports the semiconservative mechanism.

E. Discussion

Boxes 8.4. 8.5, and 8.6 detail the replication of three plant virus groups and show how (+)-strand viruses replicate. The evidence for the involvement of membranes is incontrovertible, but the reasons why different viruses use different membranes (Table 8.1) are not yet understood.

The replication complexes comprise several virus-coded proteins with different functions. These are assembled onto the relevant membrane by a membrane-binding protein or domain that then interacts with the other components. The coordination of assembly and the composition of the replication complexes are controlled not only by the interactions between the component proteins (and nucleic acids) but also by the way they are expressed. Thus, in some cases, some of the component proteins are expressed from frameshift or read-through, and in other cases, they are expressed from a polyprotein processed in a defined manner.

Various host proteins are involved in replication complexes. The involvement of translation initiation factors (Boxes 8.4–8.6) is of particular interest in indicating coordination between translation and replication. However, these two functions operate along the template RNA in different directions, translation being $5' > 3'$ and replication $3' > 5'$, and thus, there must be controls to prevent interference between the two processes. Other cell processes, such as phosphorylation, play a role in control of replication.

Most of the RNA elements involved in replication operate in *cis*, showing that the template RNA is an integral part of the replication complex. The elements at the $3'$ end of the template RNA that initiate (−)-strand synthesis appear to be well defined. This is in contrast to those at the $5'$ end of the genomic RNA, which initiate (+)-strand synthesis. It seems likely that (−)-strand and (+)-strand synthesis are highly coordinated and that once the (+)-strand

BOX 8.4

REPLICATION OF TOBACCO MOSAIC VIRUS

In protoplasts, the synthesis of (–) strands of TMV RNA ceases 6–8 hours after inoculation, whereas (+)-strand synthesis continues for a further 10 hours. The 5′ ORF of TMV encodes a 126 kDa protein, the stop codon of which is read through to give a 183 kDa protein (see Profile 14 for genome organisation); thus, both proteins have the same N-terminal sequence, the 183 kDa protein having a unique C-terminal sequence. The proteins are in a 1:1 ratio, and both have the methyl transferase and helicase motifs; the 183 kDa protein also contains the RdRp motif. Efficient replication requires the 183 kDa protein to form a heterodimer with the 126 kDa protein, which is probably bound to the template RNA. However, there may be still other functions for the 126 kDa protein. As noted in Chapter 7, it is produced in about ten times the amount of the read-through product, the 183 kDa protein, yet the two proteins form a 1:1 heterodimer.

There are various *cis*-acting factors involved in the replication of TMV RNA. The 3′ untranslated region (UTR) can be folded into three structural domains (see Figure 6.2), a 3′ domain mimicking a tRNA acceptor branch, an analogue of a tRNA anticodon branch, and an upstream domain comprising three pseudoknots, each containing two double-helical segments. It is thought that the secondary structure rather than the primary structure is important in binding the replicase.

Sequences at the 5′ end of TMV RNA are also important for replication. Large deletions in the

5′ region and deletion of nucleotides 2–8 abolished replication, but other small deletions in the 5′ UTR did not. This suggests that the 5′ replicase binding site may be complex.

Various host proteins have been found associated with TMV replication complexes. These include a 56 kDa protein that is immunologically related to the 55 kDa (GCD10) subunit of translation initiation factor eIF3 from yeast and wheat germ (56 kDa) and that interacts specifically with the methyl transferase domain shared by the 126 and 183 kDa TMV proteins. The association of a GCD10-like protein with the TMV replication complex suggests that, *in vivo*, replication and protein synthesis may be closely connected.

The cellular site for TMV replication is associated with cytoplasmic inclusions or viroplasms that enlarge to form what are termed "X bodies" composed of aggregate tubules embedded in a ribosome-rich matrix.

The use of the green fluorescent protein (GFP) as a reporter for TMV expression identified irregular shaped structures in cells that contained the viral replicase (and movement protein); these structures were derived from the endoplasmic reticulum (ER). Using *in situ* hybridisation and immunostaining, it was shown that TMV RNA, the viral replicase, and the viral movement protein colocalised at the ER, including the perinuclear endoplasmic reticulum; these molecules are associated with ER-related vesicles.

template has been "captured" by the *in vivo* replication complex, the full round of RNA replication will occur. The lack of (+)-strand synthesis in most *in vitro* replication systems indicates either that an important factor is lost during extraction or that there are conformational constraints imposed by the location of the complex *in vivo*.

BOX 8.5

REPLICATION OF BROMOVIRUSES

A. *Brome Mosaic Virus*

In barley protoplasts infected *in vitro* with BMV, RNAs 1 and 2 (see Profile 3 for genome organisation) are detected 6 hours after infection. All four RNAs are present at 10 hours, and maximum RNA synthesis is from 16 to 25 hours. Virus particle formation is greatest between 10 and 25 hours after inoculation.

BMV replication complex contains proteins 1a and 2a that have the motifs for methyl transferase (1a) and RdRp (2a). Protein 1a binds to protein 2a, the interaction being between a 115 amino acid region at the N-terminus of protein 2a and a 50 kDa region of protein 1a encompassing the helicase domain. The yeast two-hybrid system has shown further features of interactions between and within the 1a and 2a proteins. The direct interaction between these two proteins is stabilised by the presence of the centrally conserved RdRp domain of 2a. There are both intramolecular interactions between the capping and helicase domains of protein 1a and intermolecular 1a-1a interactions involving the N-terminal 515 residues of that protein. This has led to a model for the assembly of BMV replicase (Fig.); it is suggested that this model is applicable to other members of the *Bromoviridae*.

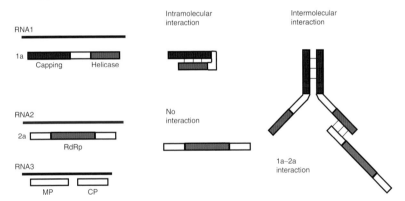

Fig. Model of the assembly of the BMV 1a-2a replication complex. On the left-hand side are shown the three genomic RNAs (see Profile 3); RNA 1 encodes protein 1a which has the capping (methyl transferase) and helicase domains; RNA2 encodes protein 2a which contains the RdRp domain; RNA 3 encodes the movement protein (MP) and coat protein (CP) which are not involved in the 1a-2a complex. In the middle is shown the intramolecular interaction between the capping and helicase domains of protein 1a which inhibits interaction with protein 2a. On the right-hand side is the shown the intermolecular interaction between the capping domains of two molecules of protein 1a which enables the N-terminal region of protein 2a to interact with the helicase domain of 1a, thus forming the 1a-2a complex.

The replication complex obviously also involves BMV RNA. *Cis*-acting signals are located in the 5′ and 3′ UTRs and in the intercistronic region (ICR) of RNA3 sites, where (–)-strand, (+)-strand, and sgRNA synthesis are initiated. The 3′ UTR of BMV is highly structured, mimicking tRNAs and forming a pseudoknot (see Chapter 6). The 3′ 134 bases of BMV RNAs containing the tRNA-like sequence (Figure 6.2B)

(continued)

BOX 8.5 *(continued)*

have been identified as the minimum sequence requirement for *in vitro* (–)-strand production. Initiation *in vitro* is primer independent and is at the penultimate C residue of the 3′-CCA. This indicates that the terminal A residue is not templated but is added to the (+) strands by tRNA nucleotidyl transferase after replication. The tRNA-like sequence contains the replicase-binding site, the specificity for the interaction with the replicase being determined by a stem-loop structure in the tRNA-like domain.

A transition between initiation to elongation of (–)-strand synthesis occurs when nascent RNAs of 10 nucleotides or longer remain associated with the replication complex; it can then be extended into full-length RNAs whereas shorter RNAs were released from the complex.

The 5′ UTRs of bromovirus RNAs share sequence similarity, the most notable region matching the box B recognition sequence of RNA polymerase III promoters and thus also the conserved TΨC loop of tRNAs; there is also a box B consensus in the intercistronic region of RNA3. It is suggested that this sequence might have a role in interacting with host factors.

BMV RNA3 encodes two proteins, the downstream one being expressed from an sgRNA. As well as the box B consensus sequence just described, BMV RNA3 has an oligo (A) tract in its intercistronic region.

Various host proteins have been found associated with BMV replication complexes, including a protein antigenically related to the wheat germ eukaryotic translation initiation factor 3 (eIF-3) that binds strongly and specifically to BMV protein 2a. The development of the yeast system for studying BMV replication has led to the identification of further host proteins that appear to be involved.

Ultrastructural changes induced by BMV involve a proliferation and modification of parts of the ER. Double-label immunofluorescence and confocal microscopy of BMV-infected barley protoplasts and in yeast cells show that 1a and 2a proteins colocalise throughout infection to defined cytoplasmic spots adjacent to, or surrounding, the nucleus and that BMV-specific RNA synthesis coincides with these spots. The BMV replication complexes were tightly associated with markers for endoplasmic reticulum and not with the medial Golgi or later compartments of the cellular secretory pathway. In yeast cells, protein 1a localises to the endoplasmic reticulum in the absence of protein 2a. As it interacts with the 2a protein, it recruits that protein to the endoplasmic reticulum, and thus it is considered to be a key organiser of the replication complex.

B. *Alfalfa Mosaic Virus*

The genome organisation of *Alfalfa mosaic virus* (AMV) is similar to that of BMV (see Profile 3), but there is a requirement of coat protein, either from the virus particle or expressed from sgRNA4 for replication. One to three coat protein dimers bind through their N-terminal amino acids to a specific site near the 3′ terminus of RNAs 1–3. There is a homologous sequence at the 3′ termini of the three genomic RNAs.

Most of the basic features of the replication of AMV are similar to those just described for BMV. The major differences are the structure of the 3′ terminus of the genomic RNAs and the necessity for binding of coat protein for genome activation. Binding of coat protein to the 3′ end of AMV RNA inhibits the production of (–)-strand RNA by interfering with the formation of the pseudoknot similar to that of BMV. It is proposed that the coat protein binding has two roles, first in very early stages of virus replication, most probably in translation, and later in infection in shutting off (–)-strand RNA synthesis.

BOX 8.6

REPLICATION OF POTYVIRUSES

Although most of the protein products of poty-viruses are involved in some way in viral replication, there is a set of core replication proteins: the CI helicase, the $6K_2$, the NIa VPg proteinase, and the NIb RdRp. As can be seen from the genome map (Profile 10), these form a block of proteins. These proteins are analogous in terms of gene order and sequence motifs to the *Poliovirus* 2C, 3A, 3B, 3C, and 3D proteins, respectively. Thus, it is likely that there are similarities in functions and in the replication strategy.

The CI proteins of potyviruses have amino acid motifs indicative of helicases and RNA binding; NTPase and helicase activities have been demonstrated. The protein is membrane associated and mutagenesis demonstrated that it is involved in virus replication. The NIa processing product comprises two regions: the N-terminal VPg and the C-terminal protease; the properties of the VPg are described in Chapter 6. The NIb protein has an RdRp sequence motif, and nucleic acid binding has been demonstrated for this protein.

Various studies have revealed a series of interactions between several of these potyviral replication-associated proteins. The NIb interacts with NIa, the interaction being between the protease region for some viruses and the VPg region for others. The NIa and VPg proteins stimulate the NIb-associated RNA polymerase activity, the stimulation being mainly attributable to the VPg.

Thus, a picture is being built up of the composition and assembly of potyviral replication complexes. This involves both interactions between the gene products and control of the processing of the polyprotein. Three polyproteins containing the 6K protein have been detected in *Tobacco etch virus* (TEV)-infected cells, CI/6K, 6K/VPg, and 6K/NIa (VPg plus proteinase).

It is thought that the VPg is involved in the initiation of RNA synthesis in a manner similar to that proposed for picornaviruses. It is likely that host proteins and other potyviral proteins are also recruited into the complex. The VPg interacts with the host translational eukaryotic initiation factor (iso) 4E and the Nib protein with a host poly(A) binding protein. The $6K_2$ protein associates with large vesicular structures derived from endoplasmic reticulum (ER). On infection with TEV, the ER network appears to collapse into discrete aggregated structures, and viral RNA in replication complexes is associated with ER-like membranes.

Building on the interactions between the proteins it is suggested that the replication complex assembles on ER-like membranes, initially by the binding of the 6K protein to those membranes. This also brings in the NIa protein, as the membrane-binding activity of the 6K protein overrides the nuclear localisation signal of NIa. NIa brings the viral RNA into the complex by its RNA-binding ability. Then the 6K-NIa complex recruits the NIb product through the NIa-NIb interaction, delivering the polymerase to the RNA. The controlled processing of the potyviral polyprotein enables the elements of the complex to be assembled and then released when they have completed their function. Thus, NIa, having recruited NIb, is released from the complex, and its nuclear-localisation takes over targeting it to the nucleus.

IV. REPLICATION OF NEGATIVE-SENSE SINGLE-STRANDED RNA VIRUSES

Plant rhabdoviruses resemble animal rhabdoviruses in having large membrane-bound particles that contain a single species of (–)-sense ssRNA (see Profile 13 for general description and genome organisation). Basically, the (–)-strand virion RNA is associated with the nucleocapsid protein (N) to form coiled nucleocapsid; a large protein (L), considered to be the replicase, is also associated with the nucleocapsid. The nucleocapsid is encased in the matrix (M) protein to form a core that, in turn, is enveloped in a membrane to form the bacilliform particle. As with animal rhabdoviruses, the (–)-strand genome of plant rhabdoviruses has two functions: as the template for transcription of mRNAs for individual genes (described in Chapter 7) and as the template for replication via a full-length (+) strand [(–)-sense ssRNA > (+)-sense ssRNA > (–)-sense ssRNA]. The polymerase complex undertakes both functions.

All vertebrate rhabdoviruses replicate and assemble in the cytoplasm as do some plant rhabdoviruses (the cytorhabdoviruses), but other plant rhabdoviruses (the nucleorhabdoviruses) replicate in the nucleus. The replication and transcription of both genera of plant rhabdoviruses that lead to the formation of nucleocapsid cores occur in viroplasms, which are discrete bodies formed from accumulations of proteins and nucleic acids. The new nucleocapsid cores mature by budding through cellular membranes to acquire the outer viral membrane.

When examining rhabdoviruses in the cell, it must be remembered that the outer nuclear membrane is contiguous with the endoplasmic reticulum (ER). Thus, nucleorhabdoviruses budding through the inner nuclear membrane into the perinuclear space (see Figure 2.5C) may further be included in vesicles derived from the outer membrane and be found in the cytoplasm. Similarly, cytorhabdoviruses that associate with the ER may affect the outer nuclear membrane, giving an appearance of a nuclear involvement. The replication cycle of plant nucleorhabdoviruses and cytorhabdoviruses is outlined in Figure 8.3.

It must be remembered that plant rhabdoviruses also replicate in their insect vectors. The membrane involved in the maturation of the virus particle depends on the insect cell type in which replication is taking place. The replication of tospoviruses resembles that of other bunyaviruses, with the particles maturing in association with the Golgi stack membranes.

V. REPLICATION OF DOUBLE-STRANDED RNA VIRUSES

Plant members of the *Reoviridae* family have either 10 or 12 dsRNA genome segments, depending on the genus. The genome organisation of the 10 segments of *Rice ragged stunt virus* is outlined in Profile 11.

The replication is dsRNA > (+)-sense ssRNA > dsRNA. Plant reoviruses replicate in the cytoplasm, as do those infecting mammals. Following infection, densely staining viroplasms appear in the cytoplasm of both infected plant cells and those of various tissues in the insect vector. Much of the viroplasm is made up of protein—most likely viral proteins. Viral RNA appears to be synthesised in the viroplasm, where the mature particles are assembled. The mature particles then migrate into the cytoplasm. Little is known about the detailed molecular aspects of plant reovirus replication, but it is likely to be similar to that of animal reoviruses. It must also be remembered that as with (–)-sense ssRNA viruses just discussed, the dsRNA viruses also replicate in their insect vector. The problem with packaging the dsRNAs of reoviruses is discussed in Chapter 5.

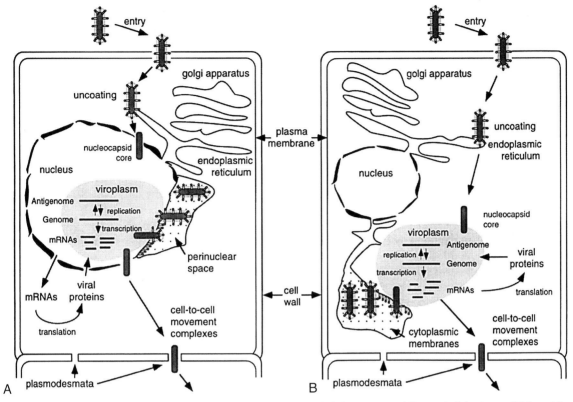

FIGURE 8.3 Models for the replication cycle of A, plant nucleorhabdoviruses, and B, cytorhabdoviruses. [This article was published in *Encyclopedia of virology*, A.O. Jackson, M. Goodin, I. Moreno, J. Jackson, and D.M. Lawerence (A. Granoff and R.G. Webster, Eds.), Plant rhabdoviruses, pp. 1531–1541, Copyright Elsevier Academic Press (1999).]

VI. REPLICATION OF REVERSE TRANSCRIBING VIRUSES

A. Introduction

The *Caulimoviridae is* the only family of plant viruses with dsDNA genomes, and it comprises six genera that differ in genome organisation but have essentially the same replication methods. Most experimental work has been carried out on *Cauliflower mosaic virus* (CaMV) and *Rice tungro bacilliform virus* (RTBV). The replication of members of the *Caulimoviridae* is DNA > RNA > DNA and thus resembles that of retroviruses; however, it does differ from that of retroviruses in several important points:

- The replication does not involve integration into the host genome for transcription of the RNA. This is done from an episomal minichromosome.
- The virus does not encode an integrase gene.
- The virion DNA is circular dsDNA and not the linear DNA with long terminal repeats characteristic of retroviruses.
- The DNA phase of the replication cycle is encapsidated rather than the RNA phase, which is encapsidated in retroviruses.

Thus, the *Caulimoviridae* are known as pararetroviruses.

As with retroviruses, the replication cycle of pararetroviruses has two phases: a nuclear phase where the viral DNA is transcribed by host DNA-dependent RNA polymerase and a cytoplasmic phase where the RNA product of transcription is reverse transcribed by virus-encoded RNA-dependent DNA polymerase or reverse transcriptase (RT) to give DNA. In retroviruses, the RT activity is part of the *pol* gene, which also includes the RNase H activity that removes the RNA moiety of the RNA:DNA intermediate of replication. The *pol* gene is part of the gag-pol polyprotein that is cleaved by an aspartate proteinase, the gag being analogous to coat protein. In pararetroviruses, the reverse transcriptase and RNaseH activities are closely associated. In badnaviruses, the coat protein and *pol* are expressed from the same ORF, but in caulimoviruses they are expressed from separate ORFs (see genome maps in Profile 4). All plant pararetroviruses encode an aspartate proteinase.

B. Reverse Transcriptase

The structure of retrovirus RT is shown in Figure 8.2B, and it is thought that pararetrovirus RT has a similar structure; the structure has been likened to a right hand as previously described for RdRp. RT has a characteristic motif of tyrosine-isoleucine-aspartic acid-aspartic acid (YIDD), and several amino acid motifs identify the RNase H domain; these are found in caulimoviruses.

C. Replication of "Caulimoviruses"

1. Replication Pathway

The replication pathway is outlined in Figure 8.4. In the first phase of replication, the dsDNA of the infecting particle moves to the cell nucleus, where the overlapping nucleotides at the gaps are removed, and the gaps are covalently closed to form a fully dsDNA.

The covalently closed DNA associates with host histones to form minichromosomes that are the template used by the host enzyme, DNA-dependent RNA polymerase II, to transcribe two RNAs of 19S and 35S, as described in Chapter 7.

The two polyadenylated RNA species migrate to the cytoplasm for the second phase of the replication cycle that takes place in inclusion bodies (see Figure 2.6). The 19S RNA is the mRNA for gene VI that is translated in large amounts to produce the inclusion body protein.

To commence viral DNA synthesis on the 35S RNA template, a plant methionyl tRNA molecule forms base-pairs over 14 nucleotides at its 3′ end with a site on the 35S RNA corresponding to a position immediately downstream from the D1 discontinuity in the α-strand DNA (see following). The viral reverse transcriptase commences synthesis of a DNA (−) strand and continues until it reaches the 5′ end of the 35S RNA, with the RNase H activity removing the RNA moiety of the RNA:DNA duplex, giving what is termed "strong stop" DNA. At this point, a switch of the enzyme to the 3′ end of the 35S RNA is needed to complete the copying. The switch is made possible by the 180-nucleotide direct repeat sequence at each end of the 35S RNA, which enables the 3′ end of the strong stop DNA to hybridize with the 3′ end of the 35S RNA. When the template switch is completed, reverse transcription of the 35S RNA continues up to the site of the tRNA primer, which is displaced and degraded to give the D1 discontinuity in the newly synthesised DNA (Figure in Profile 4).

The rest of the used 35S template is removed by an RNase H activity. In this process, two polypurine tracts (PPT) of the RNA are left near the position of discontinuities D2 and D3 (Figure in Profile 4) in the second DNA strand (the (+)-strand). Synthesis of the second (+) strand of the DNA then occurs, initiating at these two RNA primers. The growing (+) strand

FIGURE 8.4 Diagram of the replication cycle of CaMV. [From Hull (2002).]

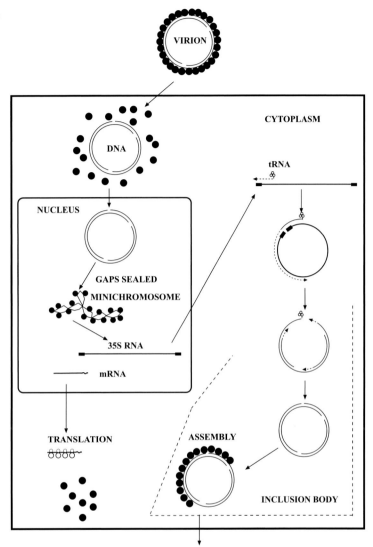

has to pass the D1 gap in the (–) strand, which again involves a template switch.

The presence of RT activity in inclusion bodies and virus particles and of replication intermediates in virus particles indicates that, as with retroviruses, the reverse transcription of CaMV occurs in particle-like proviral structures.

2. Inclusion Bodies

CaMV (and other caulimoviruses) induce characteristic inclusion bodies in the cytoplasm of their host cells. There are two forms of inclusion bodies: electron-dense ones that are made up of ORF VI product (see Figure 2.6A) and electron-lucent ones (see Figure 2.6B) that are made up of ORF II product; virus particles are found in both types of inclusion bodies. The electron-dense inclusion bodies are the site for progeny viral DNA synthesis and for the assembly of virus particles. As we will see in Chapter 12, the electron-lucent inclusion bodies are involved in aphid transmission of the virus. Most virus particles are retained within the inclusion bodies.

VII. REPLICATION OF SINGLE-STRANDED DNA VIRUSES

The replication of members of the *Geminiviridae* is ssDNA > dsDNA > ssDNA. Most of the geminivirus genera have monopartite genomes, but many of the begomoviruses have bipartite genomes; however, DNA A of bipartite begomoviruses contains all the information necessary for virus replication, except for the genes on DNA B encoding proteins involved in movement of particles to the nucleus.

Geminivirus particles usually accumulate in the nucleus, and with some, such as *Maize streak virus* (MSV), large amounts of virus accumulate there. In some infections, fibrillar rings, which must be part of a spherical structure, appear in the nucleus.

A. Geminivirus Replication

Geminiviruses use the rolling-circle mechanism of replication similar to that described for viroids in Chapter 3. This is a two-step process. In the first phase, the ss (+) strand is the template for the synthesis of (−) strand to generate a ds, replicative form (RF). This RF acts as the template for both transcription, as described in Chapter 7, and (+)-strand synthesis generating free ssDNA. The priming of (−)-strand synthesis is usually by an RNA molecule that is generated through RNA polymerase or DNA primase activity. (+)-strand synthesis is primed by a site-specific nick in the (+) strand of the RF.

The elucidation of the replication cycle has revealed several aspects of the normal cell cycle, since geminivirus replication depends on many host functions. Of especial interest is that geminiviruses replicate in differentiated cells that are in the G phase and have shut down most of their DNA replication activities. Thus, geminiviruses reactivate the replication activities that they require and convert the cell back to S phase (Box 8.7). The geminivirus replication cycle is outlined in Figure 8.5.

Minus-strand synthesis of MSV occurs in the nucleus and is primed by a small RNA oligonucleotide complementary to the 3′ intergenic region, which is extended by host DNA-dependent DNA polymerase.

The priming of geminivirus (+)-strand synthesis is through a DNA cleavage at a specific site *in vivo* in the intergenic common region. The geminivirus Rep protein is a site-specific endonuclease that nicks and ligates (+)-strand viral DNA at the same position *in vitro*.

B. Geminivirus Rep Proteins

The Rep protein is the only geminiviral protein that is essential for replication. In begomoviruses, Rep is encoded by ORF C1, and in mastreviruses, it is expressed from ORFs C1:C2 through a spliced mRNA; unspliced RNA gives RepA from ORF C1 (for geminivirus genome organisation, see Profile 7).

Rep and RepA are multifunctional proteins in that they do the following:

1. Localise within the nucleus.
2. Have specific DNA recognition sites.
3. Have site-specific endonuclease and ligation activity for (+)-strand viral DNA (see preceding).
4. Have ATP/GTPase activity.

BOX 8.7

GEMINIVIRUS CONTROL OF CELL CYCLE

Geminiviruses replicate in differentiated plant cells in which host DNA replication has ceased. The viral replication is dependent on host DNA replication factors, and thus the cell cycle has to be modified. The viral Rep protein is involved in this modification. Rep proteins (or RepA of mastreviruses) bind to retinoblastoma (Rb) proteins from a variety of sources including plants.

Animal Rb proteins regulate cell growth most probably through control of the transition of the G0/G1 into S phase of the cell cycle. It is thought that the plant analogues of Rb proteins have a similar function. Various animal DNA viruses control their host cell cycle through the binding of a virus-encoded protein with the host Rb protein through a LXCXE motif. Most geminivirus Rep (and RepA) proteins have this LXCXE motif and bind Rb proteins from various sources. Thus, the suggestion is that the binding inhibits the Rb protein control that maintains the host cell in the G phase of the cell cycle, enabling it to return to S phase and produce the factors required for viral replication. However, for this to occur, Rep must be expressed from the incoming virus. Therefore, there must be enough capability in the newly infected cell to initiate (–)-strand synthesis to give the dsDNA for transcription of the mRNA for Rep. The C4 protein of *Beet curly top virus* also controls the G2/M checkpoint.

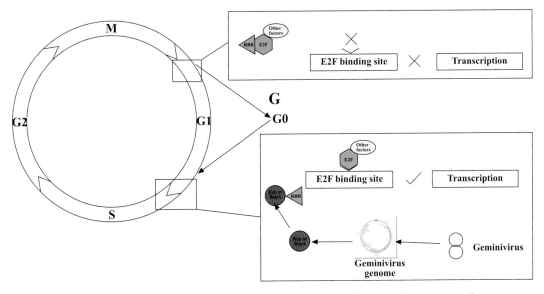

Fig. The left part of the diagram shows the cell cycle: M, mitosis phase; G1, (growth) interphase; G0, quiescent phase; S, DNA synthesis phase; G2, interphase. The top box to the right shows how the retinoblastoma (RBR) protein interacts with the E2F protein, preventing it from binding to its binding site that is necessary for transcription leading to the G1 phase. The bottom right-hand box shows how the viral Rep (or RepA) protein prevents the RBR protein from interacting with E2F, thus enabling E2F to bind to the DNA, overcoming the G1/S phase checkpoint, and initiating transcription.

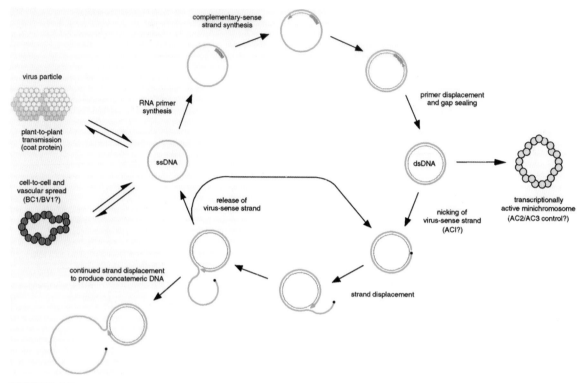

FIGURE 8.5   Diagram of the replication of a geminivirus. [Kindly provided by J. Stanley. From Hull (2002).]

5. Activate the promoter for the coat protein gene mRNA.
6. Interact with retinoblastoma proteins (see Box 8.7).
7. For the begomoviruses, Rep can repress its own promoter and can stimulate the expression of the proliferating cell nuclear antigen.

VIII. FAULTS IN REPLICATION

The two main ways by which faults arise in replication are by mutation and recombination.

A. Mutation

As noted in Chapter 4, replication mutations can be base substitutions, base additions, or base deletions. In discussing mutations, one has to distinguish between *mutation frequency* and *mutation rate*. Mutation frequency is the proportion of mutants (averaged for an entire sequence or specific for a defined site) in a genome population. Mutation rate is the frequency of occurrence of a mutation event during genome replication.

The rate of mutational errors depends on the mode of replication, the nucleotide sequence context, and environmental factors. As shown in Figure 4.5, nucleic acids that replicate DNA→DNA have much lower mutation rates than those that replicate by other pathways. This is because DNA-dependent DNA polymerase has a proofreading ability that checks that the correct nucleotide has been added, whereas the other polymerases (DNA-dependent RNA polymerase, RdRp, and RT) do not. The crystal structures of RNA replicases and RT do not reveal the 5′ to 3′ exonucleolytic proofreading domain present in DNA-dependent DNA polymerases.

B. Recombination

Recombination is the formation of chimeric nucleic acid molecules from segments that were previously separated on the same molecule or are present in different parental molecules. It usually, but not always, takes place during replication and can be a repair mechanism for aberrations resulting from mutation.

In many of the experiments on recombination, the design is to restore an important function by recombination between two nucleic acids with lost or depleted function. In this approach there is strong selection for the recombination event that may distort measurement of recombination frequency. A more realistic picture of the "natural" situation is given if one performs the experiments under reduced or nonselective conditions. Thus, although it is recognised that the rates of recombination, especially that of RNA, are high there are few estimates as to the actual values under "natural" conditions. The mechanisms of recombination between DNA and RNA viruses have both differences and similarities.

1. DNA Virus Recombination

The two basic forms of recombination in DNA viruses are homologous recombination, which occurs between two DNA sequences that are the same or very similar at the crossover point, and nonhomologous or illegitimate recombination, which occurs at sites where there is either microhomology or no obvious homology; the latter usually happens during double-strand break repair. In animal and bacterial viruses, nonhomologous recombination is a rare event and is usually mediated by a virus- or host-encoded protein. Homologous recombination can require specific host or viral proteins but can also be due to template switching during replication.

Recombination is common among geminiviruses and is a major driving force in the evolution of this virus family (e.g., see Box 4.5). It has been found both within and between geminivirus species, and there is strong evidence for both homologous and nonhomologous recombination. Recombination is also common in CaMV and probably in all the *Caulimoviridae*. Both DNA and RNA recombination have been implicated in CaMV.

2. RNA Virus Recombination

Three classes of RNA recombination have been recognised (Box 8.8). Various mechanisms have been proposed for RNA recombination. The most widely accepted is the replicase-driven template switching model, which involves four elements: three RNAs [the primary RNA template (donor strand), the strand synthesised from the primary strand (nascent strand), and the acceptor strand] and the replicase complex (Figure 8.6).

Synthesis of the nascent strand on the donor strand is halted or slowed temporarily, which enables either the RdRp or nascent strand to interact with the acceptor strand, leading to template switching. Thus, there are two types of signal on the donor or nascent strand: one (pausing or arrest signal) that halts the RdRp but from which it can escape and one (terminator signal) that releases the RdRp from the RNAs. It is thought that these signals may be similar to those involved in template switching

RNA RECOMBINATION

Class 1: Similarity-Essential Recombination

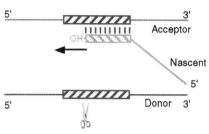

RdRp Internal pausing/termination
or RNA breakage

Class 2: Similarity-Nonessential Recombination

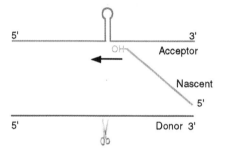

Class 3: Similarity-Assisted Recombination

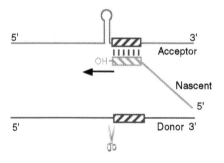

Fig. Three classes of RNA recombination. Replicase-mediated RNA synthesis after the template-switch event is shown by an arrow. The hairpin structure shown on the acceptor RNA symbolically represents various RNA structures that are required for class 2 and 3 recombination. [This article was published in *Virology*, **235**, P.D. Nagy and A.E. Simon, New insights into the mechanisms of RNA recombination, pp. 1–9, Copyright Elsevier (1997).]

There are three forms of RNA recombination.

Class 1, termed *similarity-essential recombination*, has substantial sequence similarity between parental RNAs. There can be two types of products: precise and imprecise recombinants. For BMV, five features that influence this form of recombination have been identified:

- The length of sequence identity of the common region between the donor and acceptor RNA needs to be 15 nucleotides or more to effect efficient homologous crossovers.
- The extent of sequence identity is important.
- The AU content of the common region with $\geq$ 61–65 percent AU supporting frequent homologous recombinations.
- The relative position of the AU-rich region.
- The inhibitory effect of GC-rich sequences on upstream hotspot regions while present on the acceptor RNA. Thus, a recombination hotspot has a GC-rich common region followed by an AU-rich region. It is thought that the AU-rich region causes the replicase to pause and promotes RdRp slippage, leading to the incorporation of nontemplated nucleotides; nontemplated nucleotides are found in BMV recombinants.

Class 2 recombination, *similarity-nonessential recombination*, occurs when there are no similar regions between the parents. It is thought that features such as transesterification, RdRp binding sites, and secondary structure play a role in the recombination event. In BMV this occurs at about 10 pecent of the rate of class 1 recombination.

Class 3 recombination, *similarity-assisted recombination*, combines features from both classes 1 and 2 recombinations. In this class, there are sequence similarities between the parental RNAs, but additional RNA determinants on only one of the parental RNAs are required for efficient recombination.

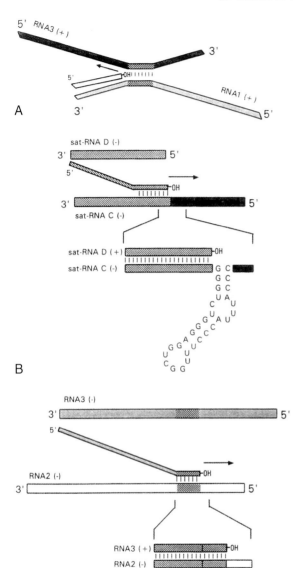

A

B

C

FIGURE 8.6 Models for replicase-mediated template switching during recombination of BMV and of *Turnip crinkle virus* (TCV). A. Heteroduplex-mediated recombination between (+) strands of BMV RNAs 1 and 3. B. Recombination between satellite RNAs associated with TCV. The sequence of the required hairpin motif is shown. C. Recombination events between the identical regions of BMV RNAs 2 and 3. Recombination is favoured when GC-rich and AU-rich are located as shown. [This article was published in *Virology*, **235**, P.D. Nagy and A.E. Simon, New insights into the mechanisms of RNA recombination, pp. 1–9, Copyright Elsevier (1997).]

by DNA-dependent DNA polymerases and RT and to be the sequence and/or secondary structure (e.g., stable hairpin) of the donor or nascent RNA.

Recombination is thought to occur in most, if not all, RNA viruses. The evidence for recombination is chimeric molecules, defective (D), and defective interfering (DI) molecules, which are discussed in Chapter 3. However, some viruses, such as the bromoviruses and tombusviruses, appear to have higher rates of recombination than others do, and these have been the subjects for detailed studies.

Mutagenesis of the replicase also gives some information on the mechanism. Mutations within the helicase domain of BMV 1a protein increased the frequency of recombination and shifted the recombination sites to energetically less stable parts of the heteroduplex. This suggests that the mutations reduced the processivity of the replication complex facilitating template switching at higher frequencies. On the other hand, mutations in the RdRp (2a protein) decreased the frequency of recombination.

3. Recombination and Integrated Viral Sequences

Some *Caulimoviridae* and *Geminiviridae* viral sequences have been found to be naturally integrated into plant genomes. It is not altogether surprising that nuclear-located viral DNA sequences are inserted into the host chromosomes by illegitimate recombination. However, cases exist in which episomal infection can arise from such integrated sequences, as exemplified in Box 8.9.

IX. VIRUSES OF OTHER KINGDOMS

The replication of plant viruses is very similar to that of viruses in other kingdoms. There are a few differences, which were discussed in this chapter.

BOX 8.9

EPISOMAL INFECTION FROM INTEGRATED BANANA STREAK VIRUS (BSV) SEQUENCES

In certain banana (*Musa*) cultivars, there has been apparently spontaneous outbreaks of BSV, especially during tissue culture and breeding programmes. The best studied situation is with variety Obino l'Ewai (AAB genome) and the tetraploids following crossing with variety Calcutta 4 (AA genome). The evidence for episomal infections that arise from integrated sequences is as follows:

- Many of the tetraploid lines had up to 100 percent infection after crossing symptomless parent plants. Tissue culture plantlets of Obino l'Ewai from symptomless mother plants had lower rates of infection, and those of Calcutta 4 no infection.
- PCR of total DNA from Obino l'Ewai using BSV primers and Southern blotting of that DNA probed with BSV sequences gave positive results even though no virus could be detected by immunoelectron microscopy or by immune-capture PCR.
- Sequencing of genomic clones from Obino l'Ewai revealed a complex insert of BSV (see following), the sequence of which was >99 percent homologous to that of the episomal virus.
- The cloned products prepared by the sequence-specific amplification polymorphism (S-SAP) approach fell into three classes, one of which comprised *Musa* sequence interfacing BSV sequence.
- Fluorescent in situ hybridisation revealed a major and a minor locus.
- Fibre-stretch hybridisation, in which chromosomes are denatured, spread, and then hybridized with fluorescent probes showed that the integrants were complex (see following).

The picture that has been derived for BSV in Obino l'Ewai is that there are two integration sites. In each, there are tandem replicates of BSV sequence interspersed with *Musa* sequence. The major locus comprises six repeats over 150 kb, and the minor locus three repeats over 50 kb. The interface of the BSV and *Musa* sequence is at the same site as the 5' end of the 35S RNA transcript. From there, a segment of BSV sequence is uninterrupted for about 5 kbp. There then follows a region of short BSV-derived sequences in reverse and forward orientation, followed by a segment of BSV in reverse orientation. Downstream of this is a further region of BSV in forward orientation covering the part of the genome missing from the first segment. The model is that the 7.4 kbp episomal viral genome is derived from the integrant by two recombination events (Fig.), which are triggered by the stresses induced by crossing and/or tissue culture.

(continued)

BOX 8.9 *(continued)*

Fig. Model for recombination from integrated BSV sequence to give episomal virus. A. Structure of an integrated BSV sequence in *Musa* Obino l'Ewai nuclear genomic DNA. Filled arrows and numbers represent the BSV sequence and their directions relative to the episomal virus. X is a rearranged assortment of short BSV sequences; open arrows represent direct repeats. B. Hypothetical intermediate after a first recombination event between 280 bp direct repeats at BSV 5530–5810. C. Episomal BSV sequence produced after a second recombination between 98 bp direct repeats at BSV 7265–7363. Predicted ORFs are shown. [This article was published in *Virology*, **255**, T. Ndowora, G. Dahal, D. LaFleur, G. Harper, R. Hull, N.E. Olszewski, and B. Lockhart, Evidence that badnavirus infection in *Musa* can originate from integrated pararetroviral sequences, pp. 214–220, Copyright Elsevier (1999).]

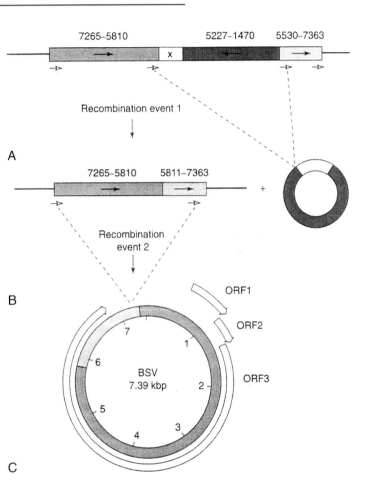

X. SUMMARY

- A range of host functions including nucleotides and amino acids and structural components of the cell are used by viruses.
- Plant virus replication is studied using both *in vivo* and *in vitro* systems.

- (+)- and (−)-sense single-stranded RNA viruses and double-stranded RNA viruses are replicated by a virus encoded polymerase. There are various *cis*-acting control signals on the viral nucleic acid.
- Plant viruses that replicate using a virus-coded reverse transcriptase are termed

pararetroviruses. Their replication is basically similar to that of animal retroviruses but differs in several important points.

- Plant single-stranded DNA viruses use the host DNA replicase. They encode a protein that alters the cell cycle from the quiescent to the DNA synthesis stage.

- Faults in viral nucleic acid replication are caused by mutation and recombination.

Further Reading

Bujarski, J.J. (2008). Recombination. *Encyclopedia of Virology*, Vol. 4, 374–381.

Cann, A.J. (2008). Replication of viruses. *Encyclopedia of Virology*, Vol. 4, 406–411.

Hull, R. (2002). *Matthew's plant virology*. Academic Press, San Diego.

Nagy, P.D. (2008). Yeast as a model host to explore plant-host interactions. *Annu. Rev. Phytopathol.* **46**, 217–242.

For RNA virus replication

Beauchemin, C., Boutet, N., and Laliberté, J.-F. (2007). Visualisation of the interaction between the precursors of VPg, the viral protein linked to the genome of *Turnip mosaic virus*, and the translation eukaryotic initiation factor iso 4E in planta. *J. Virol.* **81**, 775–782.

Jakubiec, A. and Jupin, I. (2007). Regulation of positive-strand RNA virus replication: The emerging role of phosphorylation. *Virus Res.* **129**, 73–79.

Nagy, P.D. and Pogany, J. (2006). Yeast as a model host to dissect functions of viral and host factors in tombus-virus replication. *Virology* **344**, 211–220.

Noueiry, A.O. and Ahlquist, P. (2003). Brome mosaic virus RNA replication: Revealing the role of the host in RNA virus replication. *Annu. Rev. Phytopathol.* **41**, 77–98.

Reichert, V.L., Choi, M., Petrillo, J.E., and Gehrke, L. (2007). Alfalfa mosaic virus coat protein bridges RNA and RNA-dependent RNA polymerase *in vitro*. *Virology* **364**, 214–226.

Sanfaçon, H. (2005). Replication of positive-strand RNA viruses in plants: contact points between plant and virus components. *Can. J. Bot.* **83**, 1529–1549.

For rhabdoviruses

Jackson, A.O., Dietzgen, R., Goodin, M.M., Bragg, J.N., and Deng, M. (2005). Biology of plant rhabdoviruses. *Annu Rev. Phytopathol.* **43**, 623–660.

Jackson, A.O., Dietzgen, R.G., Fang, R.-X-, Goodin, M.M., Hogenhout, S.A., Deng, M., and Bragg, J.N. (2008). Plant rhabdoviruses. *Enclopedia of Virology*, Vol. 4, 187–196.

For geminivirus replication

Gutierrez, C. (2002). Strategies for geminivirus DNA replication and cell cycle interference. *Physiol. Molec. Plant Pathol.* **60**, 219–230.

Hanley-Bowdoin, L., Settlage, S.B., and Robertson, D. (2004). Reprogramming plant gene expression: A prerequisite to geminivirus DNA replication. *Molec. Plant Pathol.* **5**, 149–156.

For pararetrovirus replication

Hohn, T. and Richert-Pöeggeler, K.R. (2006). Replication of plant pararetroviruses. In *Recent advances in DNA virus replication* (K.L. Hefferon, Ed.), pp. 289–319. Research Signpost 37/661, Kerala, India.

HOW DO PLANT VIRUSES WORK?

Virus-Host Interactions — Plant Level

To induce a disease, the virus must spread throughout and replicate in much of the plant. At this stage, the viral genome and the host genome confront one another, with the virus attempting to establish infection and the host attempting to resist it.

I. MOVEMENT AND FINAL DISTRIBUTION

As outlined in Table 9.1, a plant responds in many different ways to the introduction of a virus into the initial cell. (We will examine total resistance in Chapter 10 and recovery in Chapter 11.) For local and systemic infection, virus movement is closely coupled with virus replication and is a dynamic regulated cascade of events.

The full systemic infection of a plant is shown in Figure 9.1. From the initially infected cell, the virus moves locally to adjacent cells and then to the vascular system, enabling full systemic spread to distal parts of the plant (for relevant details of plant anatomy, see Box 2.2). An exception to this is that initial cell-to-cell movement may be bypassed in phloem-limited viruses, which are injected directly into the phloem by their vector (see Chapter 12).

TABLE 9.1 Types of Response by Plants to Inoculation with a Virus

---

IMMUNE (nonhost). Virus does not replicate in protoplasts or in the initially inoculated cells of the intact plant. Inoculum virus may be uncoated, but no progeny viral genomes are produced.

INFECTIBLE (host). Virus can infect and replicate in protoplasts.

   Resistant (extreme hypersensitivity). Virus multiplication is limited to initially infected cells because of an ineffectual virus-coded movement protein, giving rise to *subliminal infection.* Plants are *field resistant.*

   Resistant (hypersensitivity). Infection limited by a host response to a zone of cells around the initially infected cell, usually with the formation of visible necrotic local lesions. Plants are *field resistant.*

   Susceptible (systemic movement and replication)

   Sensitive. Plants react with more or less severe disease.

   Tolerant. There is little or no apparent effect on the plant, giving rise to *latent* infection.

A. Intracellular Movement

In Chapter 8, we saw the association of replication of viruses with cell membrane systems. Plant viruses use the cytoskeleton and membrane systems (Box 9.1) to move from the sites of replication to the periphery of the cell to enable infectious units to pass to adjacent cells.

Details of the process of intracellular movement of viruses are difficult to separate from those of intercellular movement, but the general picture is that various virus proteins such as movement proteins (see below) and coat proteins are involved. The actual proteins used in the process differ between different virus groups, but the basic process appears to be that a complex of viral nucleic acid and protein binds to microtubules and/or microfilaments and is translocated to the plasmodesmata; the complex, including cytoskeleton components, may also bind to the endoplasmic reticulum.

B. Intercellular Movement

The cell-to-cell (or short-distance) movement is from the initially infected cell(s), which are usually epidermal or mesophyll cells, to the vascular bundle. In the majority of cases there are two major barriers to movement: movement from the first infected cell and movement out of parenchyma cells into vascular tissues.

1. Plasmodesmata

Since the virus cannot cross the cell wall directly, it must use plasmodesmata, which are cytoplasmic connections between adjacent cells (Box 9.2). However, plant virus particles, or even free, folded viral nucleic acids, are too large to pass through unmodified plasmodesmata (Figure 9.2). Thus, the plasmodesmatal size exclusion limit (SEL) has to be increased, and viral movement proteins (MPs) facilitate this.

Of special note to virus movement are plasmodesmata between the bundle sheath and phloem parenchyma, between the phloem parenchyma and companion cells, and between the companion or intermediary cells and the sieve elements, giving different tissue boundaries, at least for some viruses. Plasmodesmata between the bundle sheath and phloem parenchyma differ from mesophyll plasmodesmata in that they require additional modification before some viruses can enter the phloem of minor veins. The systemic infection of several viruses, such as *Brome mosaic virus* (BMV), appears to be controlled by the bundle sheath–vascular bundle interface. This tissue boundary also appears to be demonstrated by phloem-limited viruses (e.g., luteoviruses) that are unable to spread across the bundle sheath to mesophyll cells (see following). Plasmodesmata between companion cells and sieve elements have a special structure comprising a single pore on the sieve element wall and a branched arrangement in the adjoining companion cell wall.

FIGURE 9.1 Diagram showing the spread of TMV through a medium young tomato plant. The inoculated leaf on the left is marked by hatching, and systemically infected leaves are shown in black. [From *Ann. Appl. Biol.* **21**, G. Samuel, The movement of *tobacco mosaic virus within the plant*, 90–111, Copyright Wiley-Blackwell (1934).]

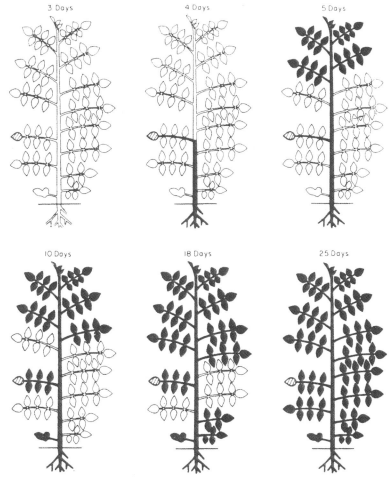

There is considerable variation in plasmodesmatal SELs, depending on factors such as type of plasmodesma and the sink-source transition of photoassimilate production, and measurements have ranged from less than 10 kDa to more than 50 kDa. Plasmodesmata occur in groups on cell walls, and it must not be assumed that those within one group are all the same. Similarly, they will vary with time and possibly with condition, such as virus infection. Thus, the system of interconnections between plant cells is dynamic and changing with the different functional demands that are placed on this symplastic system.

2. *Movement Proteins (MPs)*

Most plant viruses encode MPs, which have three functional characteristics:

- They are associated with and/or have the ability to increase the SEL of plasmodesmata.
- They have the ability to bind to either ssRNA or ssDNA.
- They have the ability to transport themselves and/or viral nucleic acid from cell to cell.

The wide variety of virus-encoded movement proteins fall into four main types. Type 1

BOX 9.1

PLANT ENDOMEMBRANE AND CYTOSKELETAL SYSTEMS

The cell contains a membrane and cytoskeleton system extending through the cytoplasm and linking various organelles and other features.

a. The Endoplasmic Reticulum

The endoplasmic reticulum (ER) system is a pleomorphic and multifunctional organelle found in all eukaryotic cells offering a large membrane surface with different functional domains. At least 16 types of ER domain have been recognised (Fig.). Of these, seven are currently recognised as having, or potentially having, an involvement with plant viruses.

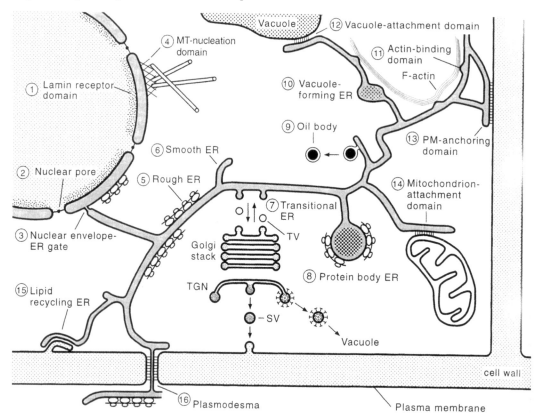

Fig. Schematic diagram of a plant cell depicting the various forms of the ER domain. MT, microtubule; PM, plasma membrane; TV, transport vesicle; TGM, trans-Golgi network. [From *Plant J.* **11**, A. Staehelin, The plant ER: A dynamic organelle composed of a large number of discrete functional domains, pp. 1151–1165, Copyright Wiley-Blackwell (1997).]

(continued)

BOX 9.1   *(continued)*

The outer membrane of the nuclear envelope is continuous with the ER and has membrane-bound ribosomes. The outer nuclear membrane joins the inner membrane at nuclear pore complexes (2 in Fig.) that mediate the directed transport of proteins and nucleic acids between the nucleus and cytoplasm. Plant rhabdoviruses mature by budding through either the inner nuclear membrane or the ER (see Chapter 8). The outer membrane of the nuclear envelope is also the major microtubule organising centre (MTOC) in plant cells (4 in Fig.). The two classical types of ER are the rough (sheet) and smooth ER, which are distinguished by the presence (rough) or absence (smooth) of attached ribosomes (5 and 6 in Fig.).

As we saw in Chapter 8, the membrane of tospoviruses is derived by budding into the Golgi cisternae though the replication and expression of the viral genome takes place in the cytoplasm. The transport of proteins from the ER to the Golgi bodies is by transport vesicles from the transitional ER (7 in Fig.). The ER has a structural association with filaments in plant cells suggesting actin-binding domains (11 in Fig.). The seventh domain with viral associations is the part of the ER that passes through plasmodesmata (16 in Fig.).

b. The Cytoskeleton

The two main elements of the plant cytoskeleton are the microtubules and the microfilaments. Microtubules are made up of tubulin. In plant cells, the α and ß tubulins that form the α/ß dimmers, which are the basic units for constructing microtubules and γ tubulin, are found in association with all microtubule arrays. Associated with microtubules are specific proteins, microtubule-associated proteins (MAPs), some of which are mechano-chemical motors that move various components along microtubules. Other MAPs modulate the assembly of microtubules at the MTOCs (4 in Fig.) and the disassembly at their distal end; this forms a "treadmill" that moves any protein associated with the microtubule from the MTOC end to the distal end. These functions are probably involved in cytoplasmic streaming and intracellular transport of molecules and macromolecules.

Microfilaments are composed primarily of actin, a helical assembly of globular units (G-actin) that form a ropelike filament (F-actin). The endomembrane system is surrounded by a sheath of cytoskeleton, a primary constituent of which is F-actin. This endomembrane sheath confers both basic structure and the structural dynamism to the ER and is suggested to be the framework on, and within which, many of the metabolic reactions within the cell are organised and operate. The sheath links the ER to various organelles, such as the plasma membrane, via various molecules. Actin is associated with plasmodesmata and is suggested to be involved in regulation of the size exclusion limit.

is exemplified by TMV and has been studied in most detail. The MP of 30 kDa (P30; see Profile 14 for TMV genome organisation) has a predicted secondary structure suggesting a series of ß-elements flanked by an α-helix at each end. The MP binds to the viral RNA, melting the tertiary and secondary structure to form a ribonucleoprotein (RNP) complex of 1.5–3 nm diameter. The MP also modifies the structure of the plasmodesmata to give a diameter of 3–4 nm.

Transport of the RNP complex to and through the plasmodesmata is considered to be an active process involving interaction with actin microfilaments and possibly the ER. Various functional domains have been identified on TMV P30, including targeting the MP to the plasmodesmata, increasing the SEL and binding ssRNA.

BOX 9.2

PLASMODESMATA

Plasmodesmata are cytoplasmic connections through the wall of adjacent plant cells and form an important route for communication between the cells. They regulate cell-to-cell communication, thus enabling the differentiation of plant organs and tissues. Developmental changes in their structure, frequency, and size exclusion limit (SEL) can lead to establishment of symplastic domains, within which the metabolism and functions of cells is probably synchronised.

The basic structure of plasmodesmata consists of two concentric membrane cylinders—the plasma membrane and the endoplasmic reticulum (appressed ER or desmotubule)—that traverse the cellulose walls between adjacent plant cells (Fig. panel A). The annulus between the two membrane cylinders gives continuity of the cytosol between cells. High-resolution electron microscopy reveals proteinaceous particles about 3 nm in diameter embedded in both the plasmamembrane and the appressed ER and connected by spokelike extensions. The spaces between these protein particles are thought to form tortuous microchannels about 2.5 nm in diameter. Injection of dyes of various molecular radii indicates that these microchannels have a basal SEL allowing passive diffusion of molecules of about 1 kDa.

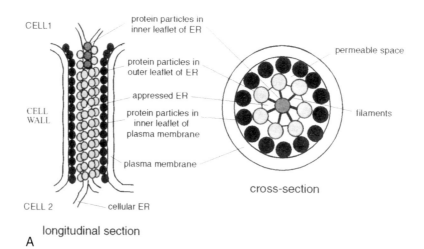

Fig. The structure of plasmodesmata. A. Diagram of the structure of a simple plasmodesma. [From Ghoshroy *et al.* (1997; *Annu. Rev. Plant Physiol. Plant Mol. Biol.* **48**, 27–50). Reprinted, with permission, from the *Annual Review of Plant Physiology and Plant Molecular Biology*, Volume 48 © 1997 by Annual Reviews www.annualreviews.org.]

(continued)

BOX 9.2   *(continued)*

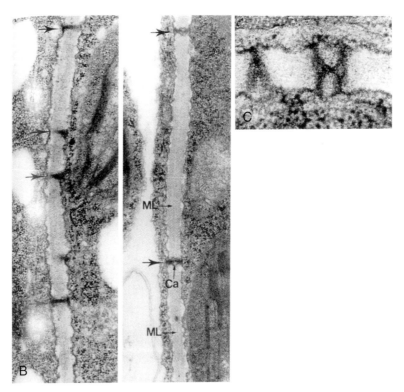

[Electron micrographs of sections showing B. primary plasmodesmata and C. secondary plasmodesmata [(c)]. ML, middle lamella of cell wall; Ca, central cavity in a plasmodesma. [From Ding *et al.* (1992; *Plant Cell* **4**, 915–928).]

Plasmodesmata come in many varieties (Fig. panel B). Primary plasmodesmata formed in a new cell wall are simple but then undergo modification to give complex structures (secondary plasmodesmata) with branched channels and a conspicuous central cavity that are formed in a basipetal pattern as leaves undergo expansion growth. Of especial note to virus movement are plasmodesmata between the bundle sheath and phloem parenchyma, between the phloem parenchyma and companion cells, and between the companion or intermediary cells and the sieve elements, giving different tissue boundaries, at least for some viruses.

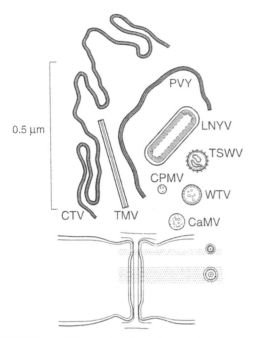

FIGURE 9.2   The relative sizes of some plant virus particles (above) compared with the size of a plasmodesma (below): CaMV, *Cauliflower mosaic virus*; CPMV, *Cowpea mosaic virus*; CTV, *Citrus tristeza virus*; LNYV, *Lettuce necrotic yellows virus*; PVY, *Potato virus Y*; TMV, *Tobacco mosaic virus*; TSWV, *Tomato spotted wilt virus*; WTV, *Wound tumor virus*. [From Gibbs (1976; in *Intercellular communication in plants: Studies on plasmodesmata*, B.E.S. Gunning and R.W. Robards, Eds., pp. 149–164, Springer-Verlag, Berlin).]

In several groups of viruses (type 2)—for instance, the hordeiviruses and the potexviruses—the MP activity is found in three overlapping or closely located ORFs known as the triple-gene block (TGB) encoding TGBp1, TGBp2, and TGBp3 (see Profiles 1 and 9, respectively, in Appendix for genome organisations).

In type 3, the potyviruses appear not to have a dedicated MP or MPs but involve several virus-coded proteins that also have other functions to that of cell-to-cell movement (see Profile 10 for potyvirus genome organisation). The viral RNA forms a complex with the coat protein and the HC-Pro; both these proteins

increase the SEL of plasmodesmata. The CI protein is a helicase and is found near plasmodesmata early in infection. It is considered to be an ancillary protein to the movement process probably facilitating delivery to and alignment of the HC-Pro/Coat protein/viral RNA complex at the plasmodesmatal pore.

In type 4 proteins, some viruses, such as comoviruses, caulimoviruses, and tospoviruses, produce tubules that extend through plasmodesmata or cell walls (Figure 9.3). Electron microscopy shows that those of comoviruses and caulimoviruses contain virus particles, and those of tospoviruses contain presumed non-enveloped nucleocapsids. The 58K/48K MP of *Cowpea mosaic virus* (CPMV) is encoded by the RNA 2 (see Profile 6 for CPMV genome organisation), the 48K moiety forming virion-containing tubules (35 nm diameter) replacing the desmotubules. CaMV MP is the product of ORFI (P1), a 38 kDa protein forming the tubule and having nucleic acid binding activity *in vitro*.

The model for this form of virus movement (Figure 9.4) is that MPs localise to the plasmodesmata, where they induce the removal of the desmotubule and assemble into tubules extending unidirectionally into the adjacent plant cell. Virions assembled in the cytoplasm are escorted to the tubular structures through interactions with their MP and are then transported to the adjacent cell.

Thus, there are two basic strategies for cell-to-cell movement (Figure 9.4). In strategy 1 (types 1, 2, and 3), the plasmodesmata are gated open at the infection front and close after, thus maintaining the integrity of normal intercellular communication. In most, if not all, cases the infection unit that passes through is a nucleoprotein complex and not the discrete virus particle. In strategy 2 (type 4), the plasmodesmata used for viral transport are permanently modified, usually by the movement protein forming a tubule through which virus particles move (except in the case of tospoviruses).

FIGURE 9.3 Virus particles in tubules. A. Plasmodesma containing a tubule with virus particles in a section of a zinnia leaf infected with the cauli-movirus, *Dahlia mosaic virus*; v = virus particles. [This article was published in *Virology*, **37**, E.W. Kitajima and J.A. Lauritis, Plant virions in plasmodes-mata, pp. 681–685, Copyright Elsevier (1969).] B. Tubules extending from a pro-toplast infected with CPMV and probed with an antibody to the 48/58 kDa viral gene product linked to a fluorescence probe; bar = 5 µm. C. Electron micro-graph of CPMV tubule in a partially pur-ified fraction; bar = 100 nm. [From Kasteel (1999; PhD Thesis, University of Wageningen, The Netherlands).]

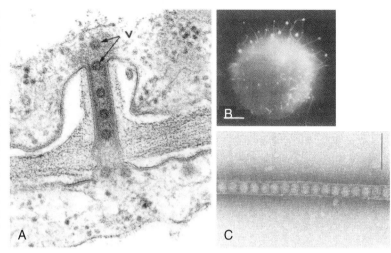

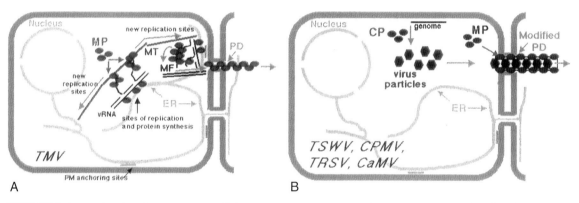

FIGURE 9.4 Models for plant virus intracellular movement. A. TMV. Viral genomic RNA complexed with MP moves along microtubules from ER-associated sites of replication and protein synthesis (viral factories) to establish additional viral factories at other ER sites. From these sites the RNA-MP complex associates with actin filaments and is delivered to the plasmodesmata. B. Viruses that form MP tubules. The viral genome encapsidated in virus particles moves through MP-containing tubules that pass through highly modified plasmodesmata. CP, coat protein; ER, endoplasmic reticulum; MF, microfilament; MP, movement protein; MT, microtubule; PD, plasmodesmata; vRNA, viral RNA genome. [Modified from Lazarowitz and Beachy (1999; *Plant Cell*, **11**, 535–548).]

3. *What Actually Moves*

For many viruses, movement proteins are not the only viral gene product involved in cell-to-cell movement (Table 9.2). For strategy 1, shown in Figure 9.4, five different combinations of proteins have been recognised. Group 1 viruses require only the movement protein, whereas group 2 viruses also require the viral coat protein. The ssDNA begomoviruses (group 3) require a nuclear shuttle protein as well as the movement protein. Groups 4 and 5 viruses require multiple genes for movement.

4. *Cell-to-Cell Movement of Viroids*

The rapid movement of *Potato spindle tubor veroid* from cell to cell is mediated by a sequence-specific or structural motif (see Chapter 3).

TABLE 9.2   Viral Factors Involved in the Local and Systemic Spread of Plant Viruses

Virus	Viral Factor	Known or Possible Site of Effect
TMV	MP	Cell-to-cell movement
	Coat protein	Possible for entrance to CC/SE complex
	126/183 kDa proteins	Entrance into CC and possible entry into and exit from SE
TEV	Coat protein, CI	Cell-to-cell movement
	HC-Pro	Possible entry into or exit from SE
	VPg	Possible entry into or exit from SE

TEV, *Tobacco etch virus*; TMV, *Tobacco mosaic virus*; CI, *cytoplasmic inclusion*; VPg, genome-linked protein; CC, companion cell, SE, sieve element.

5. Complementation

The presence of a virus with a compatible MP can complement the cell-to-cell spread of another virus in an apparent nonpermissive host. For example, *Southern cowpea mosaic virus* (a sobemovirus with isometric particles) does not accumulate in leaves of bean (*Phaseolus vulgaris*) when inoculated by itself, but it does when coinoculated with *Sunn-hemp mosaic virus* (a tobamovirus with rod-shaped particles). Complementation can also lead to the invasion of cells outside the bundle sheath by phloem-limited viruses. For instance, joint infection with a mesophyll cell-infecting virus can lead to luteoviruses also infecting mesophyll cells.

6. Rate of Cell-to-Cell Movement

To measure the movement out of the initially infected cell, the epidermis was stripped from *N. sylvestris* at various times after inoculation with TMV; this showed that virus moved into the mesophyll in 4 hours at 24–30°C. *Tobacco rattle virus* microinjected into trichome cells of *N. clevelandii* took about 4 hours to move out of inoculated cells. The rates of subsequent cell-to-cell spread vary with leaf age, with different cell types, and in different directions within the leaf. The measured rates range from 4 to 13 µm/hour (roughly one cell every 1 to 3 hours). There are generally fewer plasmodesmata per unit area on the vertical walls of mesophyll cells than on the walls that are more or less parallel to the leaf surface. Furthermore, there tend to be lines of mesophyll cells linked efficiently together and ending in contact with a minor vein. Viruses may spread more rapidly along such routes than in other directions within the mesophyll.

C. Systemic Movement

1. Steps in Systemic Movement

The long-distance transport is through the plant vascular tissue, usually the phloem sieve-tubes (see Box 2.2 for plant structure; details of vein structure are given in Box 9.3), following the flow of metabolites from the source leaves to the sink leaves or tissues. Then, further cell-to-cell movement establishes systemic infection of the young leaves.

There are five distinct and consecutive steps from localised cell-to-cell movement of a virus to full systemic invasion of the plant (Figure 9.5).

1. The first event of the movement of a virus into the vascular system is to pass the bundle sheath cells. It is likely that viruses enter the bundle sheath cells from the mesophyll by the cell-to-cell movement processes just described. However, there is strong evidence that a different process is

BOX 9.3

VEIN AND PHLOEM STRUCTURE

Veins are defined as minor or major, based on their structure, location, branching pattern, and function. For dicotyledons, major veins are enclosed in parenchyma tissue, forming a rib rising above the leaf surface and branch, usually no more than twice. They function in long-distance transport of water, inorganic nutrients, photoassimilates, and other organic matter and are thought to be involved with photoassimilate unloading. Minor veins that are are embedded in mesophyll cells are the result of three or more branchings from the first-order veins and function in loading of photoassimilates in mature leaves. Veins have also been divided into five classes, classes I to III being major veins and IV and V being minor veins (Fig.). The difference between major and minor veins is less apparent in monocotyledons. The structure of veins and vein-associated cells changes as the leaf develops from a sink to a source. This can have a significant effect on virus movement.

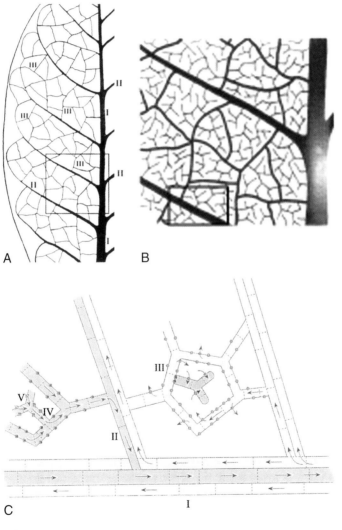

Fig. See legend on next page

BOX 9.3   *(continued)*

Fig. Veins in plant leaves. A. Major veins in a leaf. B. Enlargement of boxed area in A showing minor veins. C. Schematic model of sink source transition in a developing leaf (enlargement of boxed area in B). The blue region indicates the exporting (source) part of the leaf and the yellow the importing (sink) part. The numbers show the vein class and the arrows the direction of movement of photoassimilates. [From Roberts *et al.* (1997; *Plant Cell* **9**, 1381–1396).]

The vascular tissue is surrounded by the bundle sheath and comprises parenchyma (in major veins), sieve tubes with companion cells, and xylem elements (see Box 2.2).

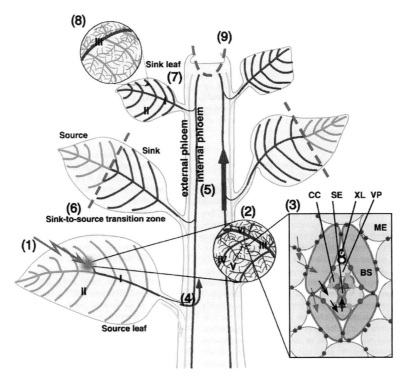

FIGURE 9.5   Cellular route for systemic movement of plant viruses. Leaf veins used by viruses for systemic transport shown in red and yellow; those not used shown in green. Blue dotted lines indicate cellular boundaries that viruses are unable to cross—for example, sink/source transition zones and apical meristem. 1 and 2. Viral infection initiates with mechanical damage (red jagged arrow) of mesophyll cells of a lower source leaf. The virus spreads from cell to cell and reaches the vascular system, where it is trafficked toward systemic organs but not locally. 3. To enter the phloem, the virus must cross from mesophyll cells (ME) through bundle sheath cells (BS) and phloem parenchyma (VP) into companion cells (CC) and then into the sieve elements (SE) of the phloem. This movement involves viral MPs (blue arrows) and is through plasmodesmata (pink circles). Additional factors (black arrows) are required for movement BS > VP > CC > SE. 4 and 5. Once within the SEs, virus passes out of the inoculated leaf and spreads rapidly upward using the internal, adaxial (red) phloem, and slowly downward in the external, abaxial (yellow) phloem. 6, 7, and 8. Virus enters the sink, but not the source, zones, unloading only from major veins (classes 1–3, indicated in red). 9. Virus does not enter the apical meristem. [From Waigmann, E., Ueki, S., Trutnyeva, K., *et al.* (2004). The ins and outs of nondestructive cell-to-cell and systemic movement of plant viruses. *Crit. Rev. Plant. Sci.* **23**, 195–250, reprinted by permission of the publisher (Taylor & Francis Ltd., http://www.tandf.co.uk/journals).]

III. HOW DO PLANT VIRUSES WORK?

required for the infection unit to pass out of the bundle sheath cell into the vascular parenchyma; little is known about this process. Luteoviruses are introduced into the vascular system by their insect vector and do not move into mesophyll cells. However, as just mentioned, this lack of movement can be complemented by a coinfection with a virus capable of crossing the bundle sheath. Thus, the bundle sheath forms a major boundary for systemic virus movement.

2. Having crossed the vascular parenchyma, the virus reaches the vascular parenchyma/ companion cell boundary. The companion cells supply the enucleate sieve elements with most cellular maintenance functions through specialised plasmodesmata. There is evidence that some, but not all, viral movement proteins facilitate crossing this boundary.

3. When the infection unit penetrates the sieve element tubes of inoculated leaves, it can move throughout the plant. In many plant species there are two structural types of phloem: external (abaxial, on the underside of the leaf), which transport metabolites downward to the roots, and internal (adaxial, on the upper side of the leaf), which transport to the upper leaves and apex of the plant. Many viruses are transported in both directions, slowly downward in the external phloem and rapidly upward in the internal phloem.

4. Virus exit from the phloem is not simply a reverse process of the entry into the phloem but is probably by a different mechanism. For example, TMV loads into both major and minor veins in the source leaves but unloads only from the major veins (class III and larger; see Box 9.3) in the sink leaves of *N. benthamiana*.

5. It is likely that the passage of virus across the bundle sheath cell of the importing leaf involves the same mechanisms as those in the exporting leaf. Thus, luteoviruses are retained within the vascular tissue.

2. *Form in Which Virus Is Transported*

Most viruses require coat protein for long-distance transport, but some, like *Barley stripe mosaic virus* (BSMV), are capable of long-distance movement without a coat protein; BSMV forms a nucleoprotein complexed with TGB1 nonstructural protein. It is not clear if all the viruses that require coat protein for long-distance transport move in the sieve elements as virus particles. There is good evidence for virus particle involvement for some such as TMV, but for others it is difficult to distinguish between a requirement of coat protein for cell-to-cell movement necessary for phloem loading and after unloading, from that for actual sieve element transport.

Other viral proteins are important for vascular-dependent accumulation, including the MP of TMV, the CMV gene 2b, and the HC-Pro of potyviruses. However, care must be taken in attributing protein function directly to long-distance spread, as they may be involved in, for example, suppression of host defense mechanisms.

3. *Rate of Systemic Movement*

The time at which infectious material moves out of the inoculated leaf into the rest of the plant varies widely, depending on such factors as host species and virus, age of host, method of inoculation, and temperature. After transmission by aphids, *Barley yellow dwarf virus* may move out of the inoculated leaf within 12 hours. TMV moves out of tobacco leaves 32–48 hours after mechanical inoculation, and *Cucumber mosaic virus* can spread systemically 24–30 hours after inoculation. In the classic experiments of Samuel (Figure 9.1), one terminal leaflet of a tomato plant was inoculated with TMV and the spread of virus with time then followed by cutting up sets of plants into many pieces at various times. He incubated the pieces to allow any small amount of virus present to increase and then tested for the presence of virus by infectivity.

Virus moved first to the roots and then to the young leaves. It was some time before the middle-aged and older leaves became infected. In very young plants, older leaves did not become infected, even after several months.

Once virus enters the phloem, movement can be very rapid. Values of about 1.5–8 cm/hour have been recorded for TMV. However, when examined in detail, systemic invasion by a virus is affected by a complex of factors and especially the source-sink status of individual leaves and parts of leaves.

The time at which unrelated viruses move from the inoculated leaf of the same individual host plant may be different. For example, in tobacco plants inoculated with a mixture *of Potato virus X* (PVX) and *Potato virus Y* (PVY), PVY moved ahead of PVX and could be isolated alone from the tip of the plant.

4. Movement in the Xylem

Few viruses move long distance through the xylem. For example, *Southern bean mosaic virus* (SBMV) and other sobemoviruses move as virus particles through dead tissue, which implicates xylem vessels. For the sobemoviruses, movement in the xylem has been correlated with transmission by beetles.

D. Final Distribution in the Plant

Whether or not a virus will move systemically at all depends on events in or near the local infection. If a virus forms a local lesion, it may be retained to only a small group of cells. However, some viruses "escape" from the lesions and spread systemically. It is often assumed that viruses that do move systemically become fairly evenly distributed throughout the plant, but, in fact, this seldom happens. Several factors can result in a very unequal distribution, including vascular connections between the initially infected leaf and the rest of the plant, host genes, viral genes, the host defense system and environmental conditions;

these and other factors affect the rate and extent of virus movement through the plant. A common situation is shown in Figure 9.1, where the first systemically infected leaf is above the inoculated leaf (the same phyllotaxis), and the young sink leaves become systemically infected before older leaves to which the virus spreads by cell-to-cell movement. As leaves undergo sink-source transition in photoassimilate import, there is a progressive basipetal decline in the amount of photoassimilate and virus entering the lamina so that in more mature sink leaves only the base of the leaf becomes infected.

Many of the viruses giving a general systemic infection apparently recover from infection in newly produced young leaves or even go through cycles of recovery and reinfection (see Chapter 11). In a plant that is infected with a systemic virus for a length of time, the concentration of virus may not be uniform in different organs. With most mosaic-type viruses that have been investigated, virus reaches a much higher concentration in the leaf lamina than in other parts of the plant.

The distribution of some viruses is limited to certain tissues. Luteoviruses and other phloem-limited viruses are found usually only in phloem parenchyma, companion cells, and sieve elements. Dark green areas in the mosaic patterns of diseased leaves usually contain very little virus compared with yellow or yellow-green areas. This has been found for viruses that differ widely in structure and that infect both mono- and dicotyledonous hosts. The phenomenon may therefore be a fairly general one for diseases of the mosaic type. Possible reasons for the low concentration of virus in dark green areas are discussed in Chapter 11.

Most viruses do not invade apical meristematic tissues. With many virus-host combinations, there appears to be a zone of variable length (usually about 100 μm but up to 1,000 μm) near the shoot or root tip that is free of virus or that contains very little virus. It seems likely that the plant defense system of gene silencing (see Chapter 11) is enhanced in meristematic cells, and this inhibits the presence and replication of the virus.

E. Outstanding Questions on Plant Virus Movement

Here are some of the questions that remain to be answered about cell-to-cell movement of viruses:

How do MPs gate plasmodesmata open at the infection front?

What are the forms of infection unit that pass through plasmodesmata, and what are the forces that move them from the infected to the uninfected cell?

How do plasmodesmata close or return to normal signaling control after the passage of the infection unit?

Do viruses pass through existing plasmodesmata, or are new ones formed? If they pass through existing plasmodesmata, are certain ones amenable to gating?

What are the features of plasmodesmata that determine tissue specificity and symplastic domains?

There are also many questions to be answered about the details of long-distance transport:

What are the exact routes of movement from the phloem parenchyma to the sieve elements? Are there different routes specific for individual viruses or groups of viruses? If so, what are the factors that determine individual routes?

What is the sieve element route from the source leaf to the sink leaf? As noted earlier in this chapter, there is some evidence for switching from the external phloem from the source leaf to the internal phloem leading to the sink leaf.

Does this occur for all systemically moving viruses in all plant species?

What are the forms in which the infection units are transported? Many viruses require coat protein for long-distance transport, but do they form particles?

What are the factors that lead to the infection units being unloaded from the sieve elements?

These last two questions are difficult to answer, as it may be only a minority of the infection unit that is unloaded and establishes systemic infection. Furthermore, for individual viruses there may be differences in the preferred unloading (and loading) form under different conditions.

II. EFFECTS ON PLANT METABOLISM

All of the various macroscopic and microscopic symptoms of disease that are discussed in Chapter 2 must originate from biochemical aberrations induced by the virus.

A. Nucleic Acids and Proteins

It is widely assumed that the small RNA viruses have little effect on host-cell DNA synthesis, but there are very few, if any, definitive experiments addressing this question. However, they do have an effect on ribosomal RNA synthesis and the concentration of ribosomes that differs with the virus, strain of virus, time after infection, and the host and tissue concerned. In addition, 70S and 80S ribosomes may be affected differently. For example, in Chinese cabbage leaves that are chronically infected with *Turnip yellow mosaic virus* (TYMV), the concentration of 70S (chloroplast) ribosomes in the yellow-green islands in the mosaic is greatly reduced compared to that in dark green islands in the same leaf; there is little effect on the concentration of cytoplasmic ribosomes in such yellow-green islands of tissue. The extent of this reduction depends on the strain of TYMV, and it also becomes more severe with time after infection. Loss of 70S ribosomes more or less parallels the loss of chlorophyll, with "white" strains causing the most severe loss.

A somewhat different result is obtained if the effect of TYMV infection with time in a young systemically infected leaf is followed. Chloroplast ribosome concentration falls markedly as the virus concentration reaches a maximum. About the same time, there is a significant increase in cytoplasmic ribosome concentration, which is mainly due to the stunting effect of infection. On the other hand, if the effects of virus infection on these components for the plant as a whole are considered, a different picture emerges. Infection reduces both cytoplasmic and chloroplast ribosomes.

The coat protein of a virus such as TMV can come to represent about half the total protein in the diseased leaf. This can occur without marked effects on the overall content of host proteins. Most other viruses multiply to a much more limited extent. Effects on host protein synthesis are not necessarily correlated with amounts of virus produced. A reduction in the amount of the most abundant host protein— ribulose bisphosphate carboxylase-oxygenase (rbcs or rubisco)—is one of the most common effects of viruses that cause mosaic and yellowing diseases.

TMV infection has been estimated to reduce host protein synthesis by up to 75 percent during the period of virus replication. Infection does not alter the concentration of host polyadenylated RNA or its size distribution, which suggests that infection may alter host protein synthesis at the translation stage rather than interfering with transcription.

Virus infection can result in some host genes being shut-off, some induced, and others not affected (Box 9.4). In this, plant viruses resemble in some respects various animal viruses but differ in that all three phenomena are associated with the same virus. These observations raise many questions. Does the virus shut-off the expression of genes that could be considered "competitive" but enhance others that may be helpful? Is this phenomenon restricted to cotyledons, or is it a general mechanism at the infection front throughout the plant? If it is general, could it be associated with symptom expression?

B. Lipids

The sites of virus synthesis within the cell almost always contain membrane structures (see Box 9.1). TYMV infection alters the ultrastructure of chloroplast membranes, and rhabdovirus and tospovirus particles obtain their outer membrane by budding through some host-cell membrane.

C. Carbohydrates

Some viruses appear to have little effect on carbohydrates in the leaves, whereas others may alter both their rate of synthesis and rate of translocation. These changes may be illustrated in a simple manner.

Leaves of *Cucurbita pepo* that have been inoculated several days previously with *Cucumber mosaic virus* that does not cause necrotic local lesions are harvested in the morning or after some hours in darkness, decolourised, and treated with iodine, which stains starch. Cryptic local lesions may show up as dark-staining areas against a pale background, indicating a block in carbohydrate translocation. On the other hand, if the inoculated leaves are harvested in the afternoon after a period of photosynthesis, decolourised, and stained with iodine, local lesions, termed starch lesions, may show up as pale spots against the dark-staining background of uninfected tissue. Thus, virus infection can decrease the rate of accumulation of starch when leaves are

EFFECT ON CELLULAR GENES OF INFECTION WITH *PEA SEED-BORNE MOSAIC VIRUS*

In a study on the mRNAs of various host genes at the infection front of *Pea seed-borne mosaic virus* in pea cotyledons (Fig.), three situations were identified:

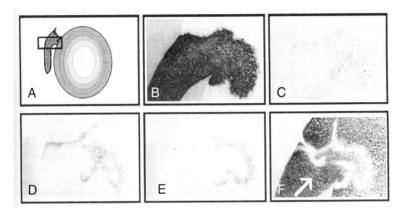

Fig. Induction and shut-off of cellular genes associated with *Pea seed-borne mosaic virus* (PSbMV) infection identified by in situ hybridization on near-consecutive section of PSbMV-infected pea tissue. Positive detection of specific RNAs is shown by dark staining. A. Diagrammatic representation of area of pea seed with box indicating region shown in other panels. B. Detection of genomic viral RNA. C. Detection of (–)-strand viral RNA showing regions where virus is replicating. D–F. Detection of HSP70, polyubiquitin, and lipoxigenase RNAs, respectively. Arrow in F shows recovery from shut-off (restoration of lipoxigenase expression) within the affected area. [This article was published in *Virology*, **243**, M. Aranda and A. Maule, Virus-induced host gene shutoff in animals and plants, pp. 261–267, Copyright Elsevier (1998).]

1. *Inhibition of host gene expression.* The expression of at least 11 host genes was suppressed—for example, lipoxigenase 1 (Fig., panel F).
2. *Induction of host gene expression.* The expression of the heat shock protein HSP70 and of polyubiquitin was induced in association with viral replication (Fig. panels D and E, respectively).
3. *No effect on host gene expression*—for example, actin and ß-tubulin.

Probing the infection front with antibodies showed that host protein accumulation was also affected in a similar manner.

exposed to light. A detailed analysis of infected *C. pepo* cotyledons showed that activity of some enzymes (e.g., sucrose synthase and ATP-dependent phosphofructokinase) rose, that of other enzymes (e.g., ADP glucose pyrophorylase and rubisco) dropped, and that of others (e.g., chloroplastic fructose bisphosphatase and hydroxypyruvate kinase) were not affected at all.

Infection with some viruses, such as *Beet yellows virus*, can induce damage to the phloem that restricts translocation of photoassimilates. These accumulate in the leaf lamina and possibly give rise to some of the symptoms like yellowing, thickened leaves, and leaf rolling (see Profile 5).

From the few diseases that have been examined in any detail, it is not possible to make very firmly based generalisations about other carbohydrate changes, but the following may be fairly common effects:

- A rise in glucose, fructose, and sucrose in virus-infected leaves
- A greater rise in these sugars caused by mild strains of a given virus compared with severe strains
- Effects of infection on mesophyll cells, not yet understood, may reduce translocation of carbohydrates out of the leaves.

D. Photosynthesis

Virus infection usually affects the process of photosynthesis, with reduction in carbon fixation being the most common effect in leaves showing mosaic or yellows diseases. This reduction usually becomes detectable some days after infection.

Photosynthetic activity can be reduced by changes in chloroplast structure, reduced content of photosynthetic pigments or rubisco, or reduction in specific proteins associated with the particles of photosystem II. However, such changes appear to be secondary, occurring some time after infection when much virus synthesis has already taken place.

Overall, during the period of rapid virus replication, infection may cause a diversion of the early products of carbon fixation away from sugars and into pathways that lead more directly to the production of intermediates for the synthesis of nucleic acids and proteins. The most general result of virus infection is a reduction in photosynthetic activity, but there may be no overall effect in relatively early stages of infection. Any reduction in photosynthesis is likely to arise from a variety of biochemical and physical changes. The relative importance of different factors varies with the disease.

E. Respiration

For many host-virus combinations where necrosis does not occur, there is a rise in respiration rate, which may begin before symptoms appear and continue for a time as disease develops. In chronically infected plants, respiration is often lower than normal. In host-virus combinations where necrotic local lesions develop, there is an increase in respiration as necrosis develops. This increase is accounted for, at least in part, by activation of the hexose monophosphate shunt pathway.

F. Transpiration

In chronically virus-infected leaves, transpiration rate and water content are generally lower than in corresponding healthy tissues. The reported effects over the first one to two weeks after inoculation vary.

G. Low-Molecular-Weight Compounds

There are numerous reports on the effects of virus infection on concentration of low-molecular-weight compounds in various parts of virus-infected plants. The analyses give rise to large amounts of data, which vary with different hosts and viruses and are impossible to interpret in relation to virus replication. The following are some of the effects:

- A consistent increase in one or both of the amides, glutamine, and asparagine.
  A general deficiency in soluble nitrogen compounds compared with healthy leaves may occur during periods of rapid virus synthesis.
- Phosphorus is a vital component of all viruses and as such may come to represent about one-fifth of the total phosphorus in the leaf. In spite of this, there is little or no effect of infection on host phosphorus metabolism.
- Virus infection frequently involves yellow mosaic mottling or a generalised yellowing of the leaves. Such changes are obviously due to a reduction in leaf pigments, such as carotene and xanthophyll. The reduction in amount of leaf pigments can be due either to an inhibition of chloroplast development or to the destruction of pigments in mature chloroplasts. The first effect probably predominates in young leaves that are developing as virus infection proceeds.
- Virus infection usually appears to affect only the vacuolar anthocyanin pigments in flowers. The pigments residing in chromoplasts may not be affected.

In summary, the physiological and biochemical changes most commonly found in virus-infected plants are a decrease in rate of photosynthesis, often associated with a decrease in photosynthetic pigments, chloroplast ribosomes, and rubisco; an increase in respiratory rate; an increase in the activity of certain enzymes, particularly polyphenoloxidases; and decreased or increased activity of plant growth regulators.

Many of the changes in host plant metabolism are probably secondary consequences of virus infection and not essential for virus replication. A single gene change in the host, or a single mutation in the virus, may change an almost symptomless infection into a severe disease. Furthermore, metabolic changes induced by virus infection are often nonspecific. Similar changes may occur in disease caused by cellular pathogens or following mechanical or chemical injury. In many virus diseases, the general pattern of metabolic change appears to resemble an accelerated aging process. Because of these similarities, rapid diagnostic procedures based on altered chemical composition of the virus-infected plant must be used with considerable caution.

III. PROCESSES INVOLVED IN SYMPTOM INDUCTION

There are various processes involved in the induction of disease. At the present time we are unable to implicate specific virus- or host-coded proteins with the initiation of any of these processes.

A. Sequestration of Raw Materials

The diversion of supplies of raw materials into virus production, thus making host cells deficient in some respect, is an obvious mechanism by which a virus could induce disease symptoms. Sometimes an increase in severity of symptoms is associated with increased virus production. However, in well-nourished plants, there is no general correlation between amount of virus produced and severity of disease. Except under conditions of specific preexisting nutritional stress, it is unlikely that the actual sequestration of amino acids and other materials into virus particles has any direct connection with the induction of symptoms.

The rate of virus replication in individual cells could be an important factor in influencing the course of events. Very high demand for key amino acids or other materials over a very short period, perhaps of a few hours, could lead to irreversible changes with major long-term effects on the cell and subsequently on tissues and organs. Another consideration is the possibility that the "switching off" of host genes at the infection front (see Box 9.3) is a general phenomenon and that this could lead to depletion of key proteins at an important time.

B. Effects on Growth

Virus infection has various effects on the growth of plants. The stunting of growth could be due to a change in the activity of growth hormones, a reduction in the availability of the products of carbon fixation by direct effects on chloroplast structure or translocation of fixed carbon, and a reduction in uptake of nutrients.

Most plant viruses belonging to the *Reoviridae* induce galls or tumours in their plant hosts but not in the insect vectors, in which they also multiply. There is a clear organ or tissue specificity for the different viruses. For example, tumours caused by *Wound tumor virus* (WTV) predominate on roots and, to a lesser extent, stems. We can be reasonably certain that some function of the viral genome induces tumour formation, but we are quite ignorant as to how this is brought about. Wounding plays an important role as an inducer or promoter of tumours caused by WTV. Hormones released on wounding may play some part in this process.

In leaves showing mosaic disease, the dark green islands of tissue frequently show blistering or distortion. This is due to the reduced cell size in the surrounding tissues and the reduced size of the leaf as a whole. The cells in the dark green island are much less affected and may not have room to expand in the plane of the lamina. The lamina then becomes convex or concave to accommodate this expansion.

C. Effects on Chloroplasts

TYMV infection in Chinese cabbage causes clumping of chloroplasts and some other changes, such as the formation of large vesicles (sickling) and fragmentation. In the sickling process a large, clear vesicle appears in chloroplasts, the chlorophyll-bearing structures being confined to a crescent-shaped fraction of the chloroplast volume. The vesicle is bounded by a membrane that appears to arise from stroma lamellae. Red and blue light are equally effective inducers of sickling, which does not occur in the dark.

D. Mosaic Symptoms

Mosaic symptoms are very common in plant virus infections. They only develop in sink leaves to which the virus has spread by long-distance movement. There are various factors involved in mosaic symptoms.

Virus strain. In some infections, such as TMV in tobacco, the disease in individual plants appears to be produced largely by a single strain of the virus. However, occasional bright yellow islands of tissue in the mosaic contain different strains of the virus, which probably arise by mutation and, during leaf development, come to exclude the original mild strain from a block of tissue.

Leaf age at time of infection. By inoculating plants at various stages of growth, it has been demonstrated for TYMV and TMV that the type of mosaic pattern that develops in a leaf at a particular position on the plant depends not on its position but on its stage of development when infected by the virus. There is a critical leaf size at the time of infection above which mosaic disease does not develop. This critical size is about 1.5 cm (length) for tobacco leaves infected with TMV.

Gradients in size of colour islands. Although the size of macroscopic islands in a mosaic is very variable, there tends to be a definite gradient up the plant. Leaves that were younger when systemically infected tend to have a mosaic made up of larger islands of tissue. Even within one leaf there may be a relationship between both the number and size of islands and the age of different parts of the lamina as determined by frequency of cell division.

Patterns in the macroscopic mosaic. Patterns in the macroscopic mosaic are often so jumbled that an ontogenetic origin for them cannot be deduced. Occasionally, patterns that are clearly derived from the apical initials have been observed. For example, Chinese cabbage leaves (or several successive leaves) have been observed to be divided about the midrib into two islands of tissue—one dark green and the other containing a uniform virus infection.

The microscopic mosaic. The overall mosaic pattern may be made up of microscopic mosaics. In Chinese cabbage leaves infected with TYMV, areas in the mosaic pattern that macroscopically appear to be a uniform colour may be found on microscopic examination to consist of mixed tissue in which different horizontal layers of the mesophyll have different chloroplast types. A wide variety of mixed tissues can be found. For example, in some areas, both palisade and the lower layers of the spongy tissue may consist of dark green tissue, whereas the central zone of cells in the lamina is white or yellow green. This situation may be reversed, with the central layer being dark green and the upper and lower layers consisting of chlorotic cells. These areas of horizontal layering may extend for several millimeters, or they may be quite small, grading down to islands of a few cells or even one cell of a different type. The junction between islands of dark green cells and abnormal cells in the microscopic mosaic is often very sharp.

Genetic control of mosaics. In spite of the fact that the most striking effects of TYMV infection in *Brassica* spp. are on the chloroplasts, the response of these organelles to infection is under some degree of nuclear control. Certain varieties of *Brassica rapa* respond to all the strains isolated from *B. chinensis* with a mild diffuse mottle. In reciprocal crosses between *B. rapa* and *B. chinensis,* all the progeny gave a *B. chinensis* type of response to the strains.

The nature of dark green tissue. Dark green islands in the mosaic pattern are cytologically and biochemically normal as far as has been tested. They contain low or zero amounts of infectious virus and no detectable viral protein or viral dsRNA. Dark green islands are resistant to superinfection with the same virus or closely related viruses.

Various factors can influence the proportion of leaf tissue that develops as green islands in a mosaic. These include leaf age, strain of virus, season of the year, and removal of the lower leaves on the plant. The dark green islands of tissue may not persist in an essentially virus-free state for the life of the leaf. "Breakdown" leading to virus replication usually takes place after a period of weeks or after a sudden elevation in temperature. Dark green islands are further discussed in Chapter 11.

E. Role of Membranes

Typical leaf cells are highly compartmentalised, and there is an increasing awareness of the vital roles that membranes play in the functioning of normal cells. Virus replication involves the induction of new or modified membrane systems within infected cells. Both animal and bacterial viruses alter the structure and permeability properties of host cell membranes. Infection of insect vector cells by plant viruses may alter the physical properties of the plasma membrane.

- Various virus-induced responses, some of which were discussed in earlier sections, indicate ways in which changes in membrane function may have far-reaching consequences for the plant: Virus infection may affect stomatal opening (stomata are the breathing pores in plant leaves). Net photosynthesis may be limited by stomatal opening, which in turn depends on guard cell membrane function.

- A virus-induced increase in stomata resistance leads to increased internal ethylene concentration followed by an epinastic response. The change in stomatal opening is probably induced by changes in properties of the guard cell plasma membranes.

- Sucrose accumulation in infected leaves appears to depend on membrane permeability changes rather than altered phloem function.

- Virus-induced starch accumulation in chloroplasts may be due in part at least to alterations in permeability of chloroplast membranes.

- There is little doubt that ethylene is involved in the induction of some symptoms of virus disease.

- Living cells maintain an electrochemical potential difference across their plasma membrane, which is internally negative. There is an interdependence between this potential difference and ion transport across the membrane. Infection of *Vigna sinensis* cells by *Tobacco ringspot virus* alters their transmembrane electropotentials.

IV. OTHER KINGDOMS

In the previous chapters we noted the similarities between plant viruses and those of other kingdoms in the expression and replication of their genomes. However, there are major differences between how plant viruses and those of other kingdoms move around their host, which reflect differences in the cellular structure of the organism. As shown in Box 9.2, plant cells are connected by cytoplasmic links (plasmodesmata), whereas those of animals and bacteria are separate being surrounded by a liquid medium (see also

Box 2.2). Thus, as far as a virus is concerned, a plant can be considered as a single cell, whereas an animal or bacterium is a collection of cells in which the virus has to pass across the outer cell membrane into the liquid medium and enter through the cell membrane of the next cell.

There are similarities and differences in the effects that plant viruses have on the host metabolism when compared with animal and bacterial viruses. These reflect similarities in the basic host metabolism, such as synthesis of nucleotides and amino acids, and in the translation machinery. Most of the differences focus on carbohydrate production, with plants fixing carbon through photosynthesis and animals and bacteria acquiring carbohydrates from external sources.

There are basic differences in the processes involved in symptom production, again reflecting differences in the hosts. In plants many of the symptoms are associated with metabolic perturbations or damage to cell organelles, especially chloroplasts. In animals and bacteria there is some metabolic perturbation, but many of the symptoms are associated with damage to the cells themselves.

V. SUMMARY

- The three phases in the establishment of full systemic infection of plants are intracellular movement, intercellular movement, and systemic movement of the infection unit.

- Intracellular movement of the infection unit involves the cell membrane and cytoskeleton system and is integrated to virus replication.

- Intercellular movement is via cytoplasmic connections through the cell wall (plasmodesmata) and involves virus-encoded movement proteins (MPs). The MPs either temporally increase the size exclusion limit of the plasmodesmata or form tubules through the plasmodesmata.

- The five steps in systemic movement are movement across the bundle sheath into the vascular system, movement into the companion cells, movement into the phloem sieve tubes (or xylem vessels for a few viruses), exiting from the phloem, and movement in the newly infected leaf from the vascular system across the bundle sheath into the mesophyll.
- The movement across the bundle sheath and into and out of the companion cells involves modifying various types of plasmodesmata. The movement in the phloem follows the source-sink flow of normal plant metabolites (e.g., carbohydrates).
- The final distribution of a virus in a plant usually does not involve all tissues. Frequently, viruses are excluded from meristematic tissues.
- Virus infections have considerable effects on cell metabolism such as photosynthesis, respiration, and transpiration. The effects include increasing the activity of some enzymes, decreasing the activity of others, and not affecting the activity of yet others.
- Symptom induction is primarily by the perturbation of the cell metabolism and damage to cell organelles such as chloroplasts.

Further Reading

Harries, P.A. and Nelson, R.S. (2008). Movement of viruses in plants. *Encyclopedia of Virology*, Vol. 3, 348–355.

Hull, R. (2002). *Matthew's plant virology*. Academic Press, San Diego.

Lucas, W.J. (2006). Plant viral movement proteins: Agents for cell-to-cell trafficking of viral genomes. *Virology* **344**, 169–184.

Verchot-Lubicz, J., Ye, C-M., and Bamunusinghe, D. (2007). Molecular biology of potexviruses: Recent advances. *J. Gen. Virol.* **88**, 1643–1655.

Waigmann, E., Ueki, S., Trutnyeva, K., and Citovsky, V. (2004). The ins and outs of nondestructive cell-to-cell and systemic movement of plant viruses. *Crit. Rev. Plant Sci.* **23**, 195–250.

Virus-Plant Interactions:
1. Molecular Level

The ultimate interactions between a virus and its host occur at the molecular level where the virus genome meets the host genome. It is these interactions that dictate the success of the virus infection and the symptoms that result.

I. INTRODUCTION

The application of methods based on DNA technology for studying the role of viral gene products in disease induction is providing increasing amounts of information on the interactions involved in the full virus infection of a plant. Important procedures include the construction of infectious clones of viruses, site-directed mutagenesis of the viral genome, switching genes between viruses and virus strains, the construction of transgenic plants that express only one or a few viral genes, and the isolation and sequencing of genes from host plants, especially *Arabidopsis thaliana*. Even more important are the recent conceptual advances in the interplay between the viral and host genomes. Essentially, the host is attempting to restrict the virus infection and the virus is attempting to overcome these restrictions.

II. HOST RESPONSES TO INOCULATION

The term *host* is defined in Chapter 2, but the terms for describing the various kinds of responses to inoculation with a virus that plants display have been used in ambiguous and sometimes inconsistent ways. The current definitions used by plant virologists are given in Table 9.1; some of these differ from those in other branches of virology. For example, *latent* used in reference to bacterial or vertebrate viruses usually indicates that the viral genome is integrated into the host genome. There are very few plant viruses that integrate into the host genome, and the integrant is activated to give an episomal infection (Box 8.9); unlike bacterial and vertebrate integrating viruses, integration is not an essential part of the plant viral replication cycle. Thus, the three basic responses of a plant to inoculation with a virus are total immunity; nonpermissive infection, in which the plant reacts to contain the virus infection; and permissive infection, in which the plant does not contain the infection.

It had been assumed for many years that virus-host cell interactions must involve specific recognition or lack of recognition between host and viral macromolecules. Interactions might involve activities of viral nucleic acids, or specific virus-coded proteins, or host proteins that are induced or repressed by viral infection. The effects of virus infection on host proteins are discussed in Chapter 9. An increasing understanding of interactions involved in the induction of disease is accruing from the study of various forms of virus resistance and the realisation of how similar some of these are to responses to attack by other microbes and by pests.

Natural resistance to viruses in plants is conferred by two types of genes: dominant genes and recessive genes (Table 10.1), as well as the inherent defence against foreign nucleic acids discussed in Chapter 11. The majority of dominant resistance genes are monogenic and resemble genes that confer resistance to other pathogens in the "gene-for-gene" process in which the host *R* gene interacts with the pathogen's *Avr* gene (Box 10.1). Most recessive resistance genes interfere with the virus replication cycle, preventing the expression or replication of the viral genome.

A. Immunity

Most plant species are resistant to most plant viruses; this is termed nonhost resistance.

TABLE 10.1   Plant Resistance Gene to Viruses

	Host			Virus		
Gene	Species	R type[a]	Genotype	Virus[b]	Avr	Type of resistance
Rx1	Potato	CC-NBS-LRR	Dominant	PVX	Coat protein	Immunity
Tm-1	Tomato	-	Dominant	ToMV	Replicase	Immunity
N	Tobacco	TIR-NBS-LRR	Dominant	TMV	Replicase/helicase	HR
Tm-2	Tomato	CC-NBS-LRR	Dominant	ToMV	Movement protein	HR
N'	Tobacco	-	Dominant	TMV	Coat protein	HR
Pat-1	*Lycopersicum* spp.	Translation factor eIF4E	Recessive	PVY	VPg	Immunity

[a] See Box 10.2 for types of *R* gene.
[b] Virus abbreviations: PVX, *Potato virus X*; PVY, *Potato virus Y*; TMV, *Tobacco mosaic virus*; ToMV, *Tomato mosaic virus*.

BOX 10.1

FLOR'S GENE-FOR-GENE MODEL FOR RESISTANCE OF A PLANT TO A PATHOGEN OR PEST

In his studies on the inheritance of the resistance of flax (*Linum usitatissimum*) to the flax rust pathogen (*Melampsora lini*), Flor (1971) revealed the classic "gene-for-gene" model in which the host resistance (*R*) gene interacts with the pathogen's avirulence (*Avr*) gene.

The model proposes that for resistance to occur, complementary pairs of dominant genes, one in the host and the other in the pathogen, are required. A loss or alteration in the host *R* gene or in the pathogen *Avr* gene leads to disease or compatibility (Fig). Basically, the interaction between the *R* and *Avr* genes leads to both a local and systemic signal cascade. The local signalling cascade triggers a host response that contains the pathogen infection to the primary site; the systemic cascade primes defence systems in other parts of the plant.

The gene-for-gene hypothesis of Flor (1971). The reaction of the host with either dominant, R, or recessive, r, gene to the pathogen with either dominant avirulence, AA, or recessive avirulence, Aa, gene is indicated as: +, susceptible; −, resistant.

Host phenotype	Pathogen phenotype		
	Avirulent AA	Avirulent Aa	Virulent aa
Susceptible rr	+	+	+
Resistant Rr	−	−	−
Resistant RR	−	−	+

There are likely to be many mechanisms but these are known in a few cases where there are both resistant and susceptible varieties of a plant species whose genetics can be studied.

Extreme resistance to *Potato virus X* (PVX) in potato is provided by the dominant *Rx1* and *Rx2* genes, which are located on chromosomes XII and V, respectively. *Rx* resistance is elicited by the viral coat protein in a strain-specific manner, and, by comparison of PVX sequences and mutagenesis, amino acid residue 121 in the coat protein gene was recognised as the major determinant of resistance-breaking activity. This extreme resistance is not associated with a hypersensitive response (HR).

The primary structure of the *Rx1* and *Rx2* gene products is similar to the LZ-NBS-LRR class of *R* genes (Box 10.2). Thus, *Rx* is similar to *R* genes that give resistance to bacteria and fungi by conferring an HR; in contrast, phenotypic analysis showed that *Rx*-mediated resistance is independent of an HR. However, transgenic expression of the *Rx* gene showed that there is the potential for an *Rx*-mediated HR but that this potential is not realised when the coat protein is expressed from the PVX genome during viral infection.

The multiplication of ToMV in tomato plants is inhibited by the presence of the *Tm-1* gene, the inhibition being more effective in the homozygote (*Tm-1/Tm-1*) than in the heterozygote (*Tm-1/+*). *Tm-1* resistance is expressed in protoplasts even in the presence of actinomycin D and thus can be classed as being extreme.

BOX 10.2

Numerous plant *R* genes have been identified, cloned, and sequenced. Many of them share striking structural similarities, and they are now placed into five groups based on their structures (Fig.); a sixth group is of *R* genes that do not fit the other five groups. Many have a leucine-rich repeat region (LRR) and a nucleotide-binding site (NBS). Other common features include a serine-threonine kinase domain, a leucine zipper or structures found in insects (the developmental gene *Toll* of *Drosophila*), and the immune response gene *TIR* from mammals.

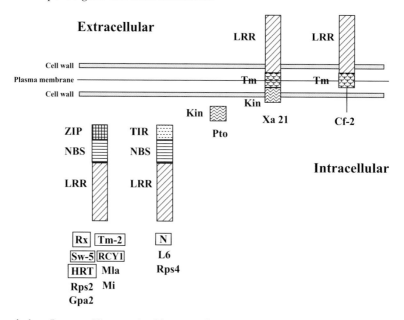

Fig. Structure of plant *R* genes: Kin = serine/threonine kinase; LRR = leucine-rich repeats; NBS = nucleotide binding site; TIR = Toll and interleukin receptor domain. Below each structure are given some examples of *R* genes, those controlling viruses being boxed; *N* against TMV; *Rx* against PVX; *Sw5* against *Tomato spotted wilt virus*; *HRT* against *Turnip crinkle virus*; *RCY1* against *Cucumber mosaic virus*; *Tm-2* against *Tomato mosaic virus*. Others are examples of *R* genes controlling fungi (*Mla* against powdery mildew, *Erysiphe graminis* f. sp. *hordei*; *L6* against flax rust, *Melampsora lini*; *Cf2* against *Cladosporium fulvum*), bacteria (*Rps2*, *Rps4*, and *Pto* against *Pseudomonas syringae* pv. tomato; *Xa21* against *Xanthomonas oryzae* pv *oryzae*), against nematodes (*Gpa2* against cyst nematode; *Mi* against root-knot nematodes and the aphid *Macrosiphum euphorbiae*).

Plant *R* genes seem to encode receptors that interact directly or indirectly with elicitors produced by the pathogen *Avr* genes. It is likely that the LRR is the pathogen recognition domain, and recognition of the *Avr* gene product prompts a signal transduction cascade, the precise mechanisms of which are poorly understood but that possibly involve salicylic acid, jasmonic acid, and ethylene.

It is interesting to note that plant *R* genes against viruses have the same structure as those against fungi, bacteria, nematodes, and aphids. Genome mapping shows that *R* genes against all these pathogens cluster on the chromosome.

Comparison of the sequences of a virulent with an avirulent strain of TMV revealed two base substitutions resulting in amino acid changes in the replicase 130 and 180 kDa proteins (Gln to Glu at position 979 and His to Tyr at 984). Mutagenesis of infectious transcripts suggested that the two concomitant base substitutions, and possibly also the resulting amino acid changes, were involved in the recognition of this *Avr* gene by the tomato *Tm-1* gene.

Immunity to many potyviruses is given by recessive genes (Table 10.1). The interaction between the host eukaryotic translation initiation factor 4E (eIF4E) and its isoform [eIF(iso)4E] and the viral VPg (see Chapter 6 for VPg) is central to this form of resistance (Box 10.3). This interaction is important in the translation and replication of potyviral genomes (see Box 8.6). The resistance gene in plants corresponds to *eIF4E* or *eIF(iso)4E,* mutations of which prevent the interaction with the VPg. Thus, examples exist of immunity being controlled both by dominant and recessive genes.

B. Subliminal Infection

As noted in Box 9.1, some viruses can replicate in the initially infected cell but cannot spread to adjacent cells because they cannot pass through the plasmodesmata; this is termed subliminal infection and can be shown by the ability of the virus to replicate in protoplasts of a plant species but not in the whole plant.

C. Nonpermissive Infection

The host may respond to virus infection either by containing the virus to the inoculated leaf (local infection) or by reacting to systemic infection by the virus.

1. Local Infection

A number of virus-host combinations induce an HR around the site of infection, which shows as local lesions (figure in Profile 14). The localisation of virus replication in tissue near the site of infection is important in

---

BOX 10.3

INTERACTION BETWEEN POTYVIRAL VPG AND EUKARYOTIC TRANSLATION INITIATION FACTOR

The 5′-linked protein, VPg, (see Chapter 6) of several potyviruses binds to the eukaryotic translation initiation factor 4E (eIF4E) or to its isoform [eIF(iso)4E] in yeast two-hybrid and *in vitro* binding systems. In the eukaryotic translation initiation complex, eIF4E binds the 5′ cap of mRNAs, bringing it into close proximity to 3′ poly(A) sequence, which is bound to the complex by the poly(A)-binding protein. It is thought that the binding of the VPg by eIF4E is involved in the translation and/or replication of the potyviral genome. Potyviruses differ in their ability to use eIF4E isoforms from a given host plant with some requiring one specific isoform for their replication cycle and others using several isoforms. Mutations in the translation initiation factor confer recessive resistance. These mutations are nonconservative substitutions of a few amino acids clustering in two neighbouring regions of the eIF4E 3D structure located near the cap-binding pocket and at the surface of the protein.

agriculture and horticulture as the basis for field resistance to virus infection.

Genes that induce an HR in intact plants or excised leaf pieces fail to do so when isolate protoplasts are infected. In the whole plant, virus particles are found in, but restricted to, the region immediately surrounding the necrotic lesion. The *N* gene of tobacco (e.g., cv Samsun NN) that confers resistance to TMV (Box 10.4)

belongs to the TIR-NB-LRR class of *R* genes (Box 10.2). The response is elicited by the helicase domain of the TMV replicase.

The *N'* gene, originating from *Nicotiana sylvestris*, controls the HR directed against most tobamoviruses, except U1 (vulgare) and OM strains of TMV, which move systemically and produce mosaic symptoms in *N'*-containing plants. The TMV coat protein gene is involved

---

BOX 10.4

THE N GENE OF TOBACCO

The *N* gene was isolated by transposon tagging using the maize activator transposon and characterised. The gene encodes a protein of 131.4 kDa with an amino-terminal domain similar to that of the cytoplasmic domain of the *Drosophila* Toll protein and the interleukin-1 receptor in mammals, a nucleotide binding site, and 14 imperfect leucine-rich regions (Fig.). The *N* gene is expressed from two transcripts, N_S and N_L, via alternative splicing pathways. The N_S transcript codes for the full-length N protein and is more prevalent before and for 3 hours after TMV infection. The N_L transcript codes for a truncated N protein (N^{tr}), lacking 13 of the 14 leucine-rich repeats and is more prevalent 4 to 8 hours after infection. A TMV-sensitive tobacco variety transformed to express the N protein but not the N^{tr} protein is susceptible to TMV, whereas transgenic plants expressing both N_S and N_L transcripts are completely resistant. However, the ratio of N_S to N_L mRNAs before and after TMV infection is critical, as the expression of either one mRNA alone or the two at a 1:1 ratio gives incomplete resistance. It is suggested that the relative ratio of the two N messages is regulated by TMV signals.

The HR response of the *N* gene to TMV is temperature sensitive, being inactivated above 28°C. It is suggested that at higher temperatures the

interaction between the viral elicitor and the host surveillance mechanism that leads to HR is weakened.

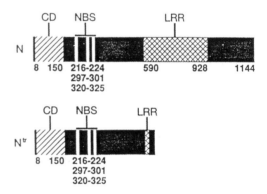

Fig. Schematic diagram of the N and N^{tr} proteins showing the various domains. CD, putative cytoplasmic domain of N with sequence similarity to Toll, interleukin-1R, and MyD88; NBS, putative nucleotide-binding site; LRR, leucine-rich repeat region consisting of 14 imperfect tandem leucine-rich repeats. The N^{tr} protein arises from alternative splicing of the chromosomal N gene. [From *Proc. Natl. Acad. Sci. USA.*, 92, S.P. Dinesh-Kumar, S. Whitham, D. Choi, R. Hehl, C. Corr, and B. Baker, Transposon tagging of *tobacco mosaic virus* resistance gene N: Its possible role in the TMV-N-mediated signal transduction pathway, pp. 4175–4180, Copyright (1995) National Academy of Sciences, U.S.A.]

---

in the induction of the *N'* gene HR. The elicitor active site covers approximately 600Å^2 of the right face of the coat protein a-helical bundle and comprises 30 percent polar, 50 percent nonpolar, and 20 percent charged residues. The *N'* gene specificity is also dependent on the three-dimensional fold of the coat protein, as well as on specific surface features within the elicitor active site.

Two allelic genes in tomato, *Tm-2* and *Tm-2²*, give an HR to certain strains of ToMV (see Table 10.2). As well as being determined by virus strain, the HR is also dependent on genotype of tomato and on environmental conditions, especially temperature. The response can vary from a very mild necrotic lesion giving apparent subliminal infection, through the normal necrotic local lesion to systemic necrosis. The 30 kDa movement protein (MP) of TMV has been identified as being the inducer of the HR in both *Tm-2* and *Tm-2²* plants.

HR in potato to PVX is controlled by the *Nb* gene, which has been mapped to a resistance gene cluster in the upper arm of chromosome V. The *Nb* avirulence determinant was mapped to the PVX 25 kDa gene encoding the MP.

a. Host Protein Changes in the Hypersensitive Response.

One of the earliest detectable events in the interaction between a plant host and a pathogen that induces necrosis is a rapid increase in the production of ethylene, which is a gaseous plant stress hormone. In the hypersensitive response to viruses, there is an increased release of ethylene from leaves. The fact that ethepon (a substance that releases ethylene) introduced into leaves with a needle can mimic the changes associated with the response of Samsun NN (containing the *N* gene) to TMV is good evidence that ethylene is involved in the initiation of this HR. An early burst of ethylene production is associated with the virus-localising reaction, but the increase in ethylene production is not determined by the onset of necrosis but by a much earlier event.

The HR involves a series of complex biochemical changes at and near the infection site

TABLE 10.2   Genetic Interactions Between ToMV-Resistant Tomato Plants and Strains of the Virus[a]

Host genotype[b]	Virus genotype					
	0	1	2	2²	1.2	1.2²
Wild type	M[c]	M	M	M	M	M
Tm-1	R	M	R	R	M	M
*Tm-2**	R	R	M	R	M	R
*Tm-2²**	R	R	R	M	R	M
Tm-1/Tm-2	R	R	R	R	M	R
Tm-1/Tm-2²	R	R	R	R	R	M
Tm-2/Tm-2²	R	R	R	R	R	R
Tm-1/Tm-2/Tm-2²	R	R	R	R	R	R

[a]Modified from Fraser [1985; in *Mechanisms of resistance to plant disease* (R.S.S. Fraser, Ed.), pp. 62–79, Martinus Nijhoff/Junk, Dordrecht].
[b]Plants with genotype marked
*May show local and variable systemic necrosis rather than mosaic when inoculated with virulent strains.
[c]M = systemic mosaic; R = resistance.

III. HOW DO PLANT VIRUSES WORK?

that include the accumulation of cytotoxic phytoalexins, the deposition of callose and lignin in the cell walls, and the rapid death of plant cells forming the necrotic lesion. The regulation of HR is equally complex, involving interplay of many potential signal transducing molecules including reactive oxygen species, ion fluxes, G proteins, jasmonic and salicylic acids, protein phosphorylation cascades, activation of transcription factors, and protein recycling by the polyubiquitin system.

b. Local Acquired Resistance.
A high degree of resistance to TMV develops in a 1 to 2 mm zone surrounding TMV local lesions in Samsun NN tobacco. The zone increases in size and resistance for about 6 days after inoculation. Greatest resistance develops in plants grown at 20–24°C; resistance is not found in plants grown at 30°C.

2. Systemic Infection

On occasions, the necrosis induced by virus infection is not limited to local lesions but spreads. This is usually from expanding local lesions that reach veins and result in systemic cell death. The systemic necrosis can range from necrosis in a few areas of upper leaves or sporadic necrotic spots mixed with mosaic symptoms to widespread necrosis leading to death of the plant. The systemic necrotic symptoms are dependent on host genotype, virus strain, and environmental conditions. Ring spot symptoms, in which necrotic rings spread in an apparently diurnal manner are described in Chapter 2.

3. Systemic Acquired Resistance

Systemic acquired resistance (SAR) is a whole-plant resistance that occurs in response to an earlier localised exposure to a pathogen, especially (but not only) one that causes tissue necrosis. It is the activation of defences in uninfected parts of the plant and gives a long-lasting and broad-based resistance. Plant viruses as well as bacteria and fungi induce SAR, usually following the development of necrotic local lesions.

One of the responses of plants to hypersensitivity is the production of a family of soluble proteins, termed pathogenesis-related (PR) proteins. PR proteins have been identified in a range of plant species and shown to be induced by a variety of microbial infections (viruses, viroids, bacteria, and fungi) and by treatment with certain chemical elicitors, notably salicylic acid (SA) and acetylsalicylic acid (aspirin). Their production inhibits the systemic infection by fungi and bacteria, but no specific PR protein appears to directly affect subsequent infections by viruses. The induction of PR proteins by HR due to fungi and bacteria requires the accumulation of endogenous SA. The application of SA affects the replication of some viruses but not others.

Thus, a conundrum arose in that the induction of PR proteins by the HR response to a virus was not associated with the SAR to that virus. However, the HR induced by TMV infection of N-gene tobacco has many features in common with HRs caused by fungi and bacteria. The reaction is mediated by a sustained burst of reactive oxygen species (ROS) followed by first local and then systemic accumulation of SA. Study of this conundrum revealed an alternative pathway that confers SAR to viruses but not to bacteria or fungi (Box 10.5). Therefore, the SAR pathway in plants differs between virus and fungi/bacteria.

The action at a distance involved in SAR presumably requires the translocation of some substance or substances. Substantial evidence exists that transport of a resistance-inducing material is involved. For example, when the midrib of an upper tobacco leaf is cut, resistance does not develop in the portion of the lamina distal to the cut. The nature of the material that migrates is unknown, as is the actual mechanism of resistance in the resistant

BOX 10.5

PATHWAY OF SYSTEMIC ACQUIRED RESISTANCE TO PLANT VIRUSES

The HR-induced systemic acquired resistance by TMV infection of *N*-gene tobacco has many features in common with that caused by fungi and bacteria through pathogenesis-related (PR) proteins. The reaction is mediated by a sustained burst of reactive oxygen (ROS) followed by first local and then systemic accumulation of salicylic acid (SA). However, the SA-induced resistance to TMV replication in tobacco is inhibited by salicylhydroxamic acid (SHAM), which does not inhibit the SA-induced synthesis of PR proteins; this suggests that there are two branches in the pathway to SA-induced resistance (Fig.) One branch leads to the production of PR proteins that confer resistance to fungi and bacteria, and the other induces resistance to viral replication and movement.

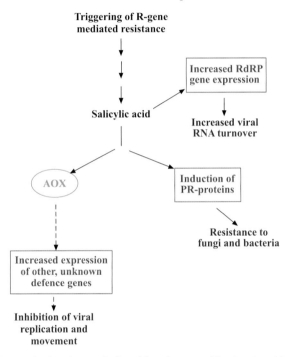

Fig. Pathways for systemic acquired resistance induced by viruses and by fungi and bacteria.

SHAM is a relatively selective competitive inhibitor of the alternative oxidase (AOX) in the mitochondrial electron flow pathway in plants. The SHAM-sensitive pathway, induced by SA and potentially by AOX, is critical in the early stages of *N*-gene mediated resistance to TMV in tobacco.

Salicylic acid also increases the turnover of the host RNA-dependent RNA polymerase (RdRp), which could be involved in RNA silencing.

Thus, the current model for the HR-induced SAR of TMV in *N*-gene tobacco is that the interaction between the viral avirulence gene (the helicase domain of the RdRp) and the *N* gene triggers the production of salicylic acid, which opens up two pathways. One increases the activity of the AOX in the mitochondria, which leads to the increased expression of other (unknown) defence genes in the nucleus. The other increases the RNA silencing defence system.

uninfected leaves. The migrating material might involve SA, ethylene, jasmonic acid, nitrous oxide, or even small peptides such as systemin.

4. *Programmed Cell Death*

Multicellular organisms have mechanisms for eliminating developmentally misplaced or unwanted cells or sacrificing cells to prevent the spread of pathogens. This is termed programmed cell death (PCD), or apoptosis; apoptosis is a specific case of PCD with a distinct set of physiological and morphological features. Although much of the work on PCD has been done in animal systems, there is increasing interest in this process in plants. The HR response to plant pathogens has various features in common with PCD. Certain animal viruses can inhibit PCD. It will be interesting to see if plant viruses have similar properties.

D. Permissive Infection

1. *Systemic Host Response*

As described in Chapter 2, there is a wide range of systemic symptoms induced by viruses, the most common and characteristic of which is the mosaic symptom. The mosaic symptom comprises areas of the leaf showing various degrees of chlorosis, together with areas that remain green and are termed "dark green islands." The dark green, light green, and even chlorotic patches that make up mosaics range from relatively large (e.g., TMV, TYMV, and AbMV) to relatively small, giving a fine mosaic (e.g., CPMV in cowpea, AMV in tobacco). These areas are often delimited by the vein structure of the leaf, giving streak or stripe symptoms in monocotyledons. The development of mosaic symptoms is described in more detail in Chapter 9.

Very little is known about the detailed molecular biology of the plant in the development of mosaic symptoms. However, it is clear

that the RNA silencing host defence system plays a significant role (see Chapter 11).

2. *Virus Genes Involved*

Because virus genomes are relatively easy to manipulate, much more is known about virus determinants of symptoms. This can be shown by two examples. By making *in vitro* recombinants between various strains of *Cauliflower mosaic virus* (CaMV), the input of various parts of the genome to symptom production were determined (Figure 10.1). The effects of some of the genes, such as the movement protein (ORF1 product), on virus spread were obvious, but that of others (e.g., ORFI/II, movement protein + aphid transmission factor and ORFV, replicase on stunting) are not clear.

Specific amino acid substitutions and deletions in the coat protein of TMV affect the production of chlorotic symptoms. Mutants that retained the C-terminus of the coat protein induce the strongest chlorotic symptoms in tobacco in both expanded and developing leaves. The chlorotic symptom formation is related to the concentration of TMV capsid proteins, which form aggregates in infected cells but do not accumulate in chloroplasts. In contrast to this, in infections with YSI/1, a naturally occurring chlorotic mutant of the U1 strain of TMV, coat protein, is found in the chloroplasts associated with the thylacoid membrane fraction in both the stroma and membrane fractions of the chloroplasts of infected cells. The coat protein of YSI/1 differs from that of U1 in two nucleotides, one of which gives an Asp to Val change at amino acid 19 is responsible for the chlorotic phenotype. The coat protein of another natural chlorotic TMV mutant, flavum, also has a substitution at amino acid 19, but this time Asp to Ala. However, these severe chlorotic symptoms are unusual in TMV infections and only associated with mutants. In natural infections, the chlorotic element of the mosaic is usually light green and is not accompanied by the accumulation of coat protein bodies or with coat protein in chloroplasts.

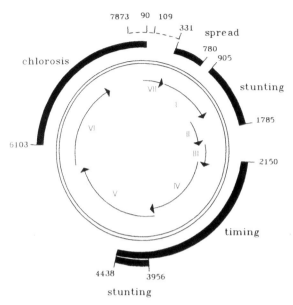

FIGURE 10.1 Location of CaMV genome domains containing strain-specific symptom determinants. Arrows with Roman numerals represent viral genes. The outer black boxes show the regions determining symptom differences between two CaMV strains; the dotted line is a regions whose specific role was not determined. [This article was published in *Virology*, **172**, R. Stratford and S.N. Covey, Segregation of *cauliflower mosaic virus* symptom genetic determinants, pp. 451–459, Copyright Elsevier (1989).]

Even though these examples identify symptom determinants in viral genomes, they do not reveal the interactions between the viral and host genome that lead to the expression of specific symptoms. The in planta role of some of the protein products of viral genes involved in disease induction may be very difficult to study for several reasons:

- The proteins may be present in very low concentration, as a very few molecules per cell of a virus-specific protein could block or depress some host-cell functions.
- Such proteins may only be present in the infected cells for a short period relative to that required for the completion of virus synthesis.
- The virus-specified polypeptide may form only part of the active molecule in the cell.
- The virus-specified polypeptide may be biologically active only in situ, such as in the membrane of some particular organelle.

- Many macroscopic disease symptoms may be due to quite unexpected side effects of virus replication. Here is a hypothetical example. Consider the 49 kDa proteinase (22 kDa VPg + 27 kDa nuclear inclusion a; see Profile 10 for genome map) coded for by *Tobacco etch virus*. The amino acid sequences that function as substrate recognition signals have been identified. In the usual host under normal conditions, this proteinase can accumulate to high levels within infected cells without causing cell death. This must mean that the proteinase does not significantly deplete the amount of any vital host protein. Suppose that we change to another host species or to different environmental conditions. In the new situation, some host-coded protein that is essential for cell function might be sensitive to cleavage, with a new pattern of disease developing as a consequence.

III. HOW DO PLANT VIRUSES WORK?

However, as we will see in Chapter 11, one viral gene product, the suppressor of RNA silencing, obviously plays a major part in symptom production. The final expression of many symptoms reflects the balance between RNA silencing and the suppression of this defence mechanism

III. INTERACTIONS BETWEEN VIRUSES

The joint infection of a plant with two (or more) viruses can lead to responses that differ from those of infection by the individual viruses. The response can depend on whether the viruses are related or unrelated and whether the infection is simultaneous or sequential.

A. Interactions Between Related Viruses

Sequential infection by strains of viruses, or even related viruses, can lead to the second virus being suppressed. This has been termed **cross protection,** which is defined as the protection conferred on a host by infection with one strain of a virus that prevents infection by a closely related strain of that virus. For instance, tobacco plants inoculated with a mild strain of PVX are immune from subsequent inoculation with severe strains of the virus, even if inoculated after only 5 days later. They are not immune to infection with the unrelated viruses, TMV and PVY. This phenomenon, which has also been called antagonism, or interference, occurs very commonly among related virus strains. It is most readily demonstrated when the first strain inoculated causes a fairly mild systemic disease and the second strain causes necrotic local lesions or a severe disease. Interference between related strains can also be demonstrated by mixing the two viruses in the same inoculum and inoculating

to a host that gives distinctive lesions for one or both of the two viruses or strains.

Strains of *Alfalfa mosaic virus* (AMV) differ in the aggregation bodies that their particles form in infected cells. Electron microscopy was used to investigate the cross-protection interactions between two strains of AMV. When the challenging strain was inoculated at the same time as the protecting strain or a short time after (about 4 hours), the two aggregation types were found side-by-side merging into each other in the same cell. When there was a longer interval between inoculation of the protecting and challenging strains (about 7 hours), the two strains were found in separate parts of the cytoplasm of the same cell and after an interval of about 10 hours in separate cells. Only when cross-protection was complete as assessed by back inoculation to an indicator host could the aggregation bodies of the challenging strain not be found.

Several mechanisms have been proposed for cross-protection. However, the most rational explanation of this phenomenon is the RNA silencing mechanism induced by plant virus infection (see Chapter 11). Cross-protection with mild virus strains is used as a control measure (see Box 14.4).

Coinoculation can lead to concurrent protection in which there is a reduction in challenging virus infection rate and/or titre induced by a protecting virus that does not accumulate or induce symptoms in that host plant. The cowpea line Arlington shows no symptoms after inoculation with *Cowpea mosaic virus* (CPMV), and no infectivity or accumulation of capsid antigen can be detected. Coinoculation of Arlington with CPMV and *Cowpea severe mosaic virus* (CPSMV) reduces the numbers of CPSMV-induced lesions. Inoculation of an isogenic line derived from Arlington with CPMV also protects it against infection with *Cherry leaf roll virus* and *Southern Cowpea mosaic virus*. This protection is elicited by CPMV RNA 1.

B. Interactions Between Unrelated Viruses

1. Complete Dependence for Disease

For some virus combinations, there is complete dependence of one virus on the other for its replication. The dependent virus is termed a satellite virus and is fully described in Chapter 3.

2. Incomplete Dependence for Disease

This situation exists between two viruses where both are normally associated with a recognised disease in the field. For example, the important tungro disease of rice is normally caused by a mixture of RTBV, a reverse transcribing DNA virus, and RTSV an RNA virus. As described in Chapter 2, RTSV on its own is transmitted by the rice green leafhopper but causes few or no symptoms. RTBV causes severe symptoms in rice, but no vectors are known for this virus on its own; it requires the presence of RTSV for transmission. Thus, in the disease complex, RTSV gives the transmission and RTBV most of the symptoms.

Most, if not all, umbraviruses are associated with luteoviruses that provide their insect transmission. An even more complex system is that of groundnut rosette disease which involves three agents, an umbravirus (GRV), a luteovirus (GRAV), and a satellite RNA (GRV sat-RNA) as described in Box 2.1.

3. Synergistic Effects on Virus Replication

Joint infection of tobacco plants with PVX and PVY is characterised by severe veinal necrosis in the first systemically infected leaves (see Box 11.3). Leaves showing this synergistic reaction contain up to 10 times as much PVX as with single infections but only the same amount of PVY. Ultrastructural studies and fluorescent antibody staining show that both viruses are replicating in the same cells and that the increased production of PVX is due to an increase in virus production per cell rather

than an increase in the number of cells supporting PVX replication. The level of PVX (–)-strand RNA increases disproportionately to that of (+)-strand RNA in doubly-infected tissues suggesting that the synergism involves an alteration in the normal regulation of the relative levels of the two RNA strand polarities during viral replication. This is due to the activity of the HC-Pro product of PVY, which is a suppressor of RNA silencing (see Chapter 11).

4. Effects on Virus Movement

As discussed in Chapter 9, infection and systemic movement by one virus in a particular host may complement the cell-to-cell and systemic movement of an unrelated virus that normally would not move from the initially infected cells in that host. Similarly, a fully systemically infecting virus can complement the movement of a tissue-restricted virus (e.g., phloem-limited virus) out of that tissue.

C. Interactions Between Viruses and Other Plant Pathogens

Virus infection can affect resistance to fungal and bacterial infection of plants. For example, infection with *Phytophthora infestans* develops less rapidly in potato plants infected with a number of viruses and infection of a hypersensitive tobacco cultivar with TMV induced systemic and long-lived resistance against *Phytophthora parasitica*, *Peronospora parasitica*, and *Pseudomonas tabaci*. The development of resistance of this sort probably involves the PR proteins discussed earlier in this chapter. Indeed, fungicidal compounds have been isolated from plants reacting with necrosis to virus infection.

On the other hand, virus infection may increase the susceptibility of a plant to fungal infection. For example, sugar beet plants in the field infected with *Beet mild yellows virus* have greatly increased susceptibility to *Alternaria* infection.

Alternatively, fungal infections can affect the susceptibility to viral infections. For example, pinto bean leaves were heavily inoculated with the uredinial stage of the rust fungus, *Uromyces phaseoli*, on one half-leaf and then later with TMV over the whole leaf. Subsequent estimations of the amounts of TMV showed the presence of up to 1,000 times as much virus infectivity in the rusted as in the nonrusted half-leaves.

Other fungi may induce resistance or apparent resistance to viral infection. Xanthi tobacco plants that had been injected in the stem with a suspension of spores of *Peronospora tabacina* produced fewer and smaller necrotic local lesions when inoculated with TMV three weeks later.

IV. VIRUSES OF OTHER KINGDOMS

There are both similarities and differences in the interactions of viruses with plant and animal hosts reflecting the fact that, although the cells of both are eukaryotic, their distribution within the organism and their interconnectivity differ. Thus, nonhost resistance to viruses in animals can be attributed to the possibility that the virus can not enter the initial cell as it can not interact with the cell surface receptor(s). As noted in Chapter 12, plant viruses do not interact with cell surface receptors but have to be introduced directly into the cell (usually by a vector); thus, nonhost resistance reflects the fact that the virus can not replicate in that cell.

The ways that the outcome of successful infection is viewed in animals (see Dimmock *et al.* 2007, Chapter 14) differ from those of plant virus infections. This makes it difficult to compare these outcomes in detail, but as noted above, reflect differences in organisational differences in the two organisms. However, at the single cell level there are likely to be many similarities in the molecular interactions involved in the expression and replication of the viral genomes.

The differences in interactions of viruses with plants and bacteria reflect the fact that the latter are usually single celled organisms and that their cell organisation is prokaryotic.

V. SUMMARY

- Three basic responses to inoculation of a plant with a virus: total immunity, nonpermissive infection, permissive infection.
- Extreme immunity is usually either nonhost resistance or conferred by one or more genes which may be either dominant or recessive.
- Nonpermissive infection is the containing of the virus to the inoculated leaf or preventing full systemic infection of the plant.
- Local infection is usually contained by a hypersensitive response (HR) which is usually controlled by a single gene.
- Systemic infection may be controlled by a signal moving from a local response (systemic acquired resistance: SAR). The SAR pathway in response to virus infection differs from that to fungal or bacterial infection.
- Both viral and host genes are involved in the establishment of permissive infection and symptom production. Among the most important viral genes is/are that (those) that suppress RNA silencing (see Chapter 11).
- Viruses can interact to give synergistic effects where the overall symptoms are more severe than those of the individual viruses combined.

References

Dimmock, N.J., Easton, A.J., and Leppard, K.N (2007). *Introduction to modern virology*. Blackwell Publishing, Oxford, UK.

Flor, H.H. (1971). Current status of the gene-for-gene concept. *Annu. Rev. Phytopathol.* **9**, 275–296.

Further Reading

Caranta, C. and Dogimont, C. (2008). Natural resistance associated with recessive genes. *Encyclopedia of Virology*, Vol. 4, 177–186.

Foster, G.D., Johansen, I.E., Hong, Y., and Nagy, P.D. (2008). *Plant virology protocols: From viral sequence to protein function*, 2nd ed., Humana Press, Forlag.

Kang, B.-C., Yeam, I., and Jahn, M.M. (2005). Genetics of plant virus resistance. *Annu. Rev. Phytopathol.* **43**, 581–621.

Maule, A.J., Caranta, C., and Boulton, M.I. (2007). Sources of natural resistance to plant viruses: Status and prospects. *Molec. Plant Pathol.* **8**, 223–231.

Moffett, P. and Klessig, D.F. (2008). Natural resistance associated with dominant genes. *Encyclopedia of Virology*, Vol. 4, 170–176.

Robaglia, C. and Caranta, C. (2006). Translation initiation factors: A weak link in plant RNA virus infection. *Trends Plant Sci.* **11**, 40–45.

Ryals, J.A., Neuenschwander, U.H., Willits, M.G., Molina, A., Steiner, H.Y., and Hunt, M.D. (1996). Systemic acquired resistance. *Plant Cell*, **8**, 1809–1819.

Singh, D.P., Moore, C.A., Gilliland, A., and Carr, J.P. (2004). Activation of multiple antiviral defence mechanisms by salicylic acid. *Molec. Plant Pathol.* **5**, 57–63.

Soosaar, J.L.M., Burch-Smith, T.M., and Dinesh-Kumar, S.P. (2005). Mechanisms of plant resistance to viruses. *Nature Reviews: Microbiology* **3**, 789–798.

11

Virus-Plant Interactions:
2. RNA Silencing

RNA silencing is a major defence system in plants that successful viruses must overcome.

I. INTRODUCTION

As we will see in Chapter 15, attempts were made to confer protection against viral infection by transforming plants to express viral gene products. However, the results showed various inconsistencies, especially with protection being given by constructs that would express a transcript but not the viral protein. These and other observations were in accord with an increasing number of cases in which transformation with homologues of endogenous plant genes led to both the transgene and endogenous gene expression being cosuppressed. The cosuppression is due to either transcriptional gene silencing (TGS) or posttranscriptional gene silencing (PTGS) or, possibly, a combination of the two. This led to the realisation that plants have a defence system against "foreign" nucleic acids and subsequently this, or a similar defence system, has been recognised in vertebrates, invertebrates, and fungi. As well as being termed PTGS in plants, it been called RNA interference (RNAi) in animals and gene quelling in fungi. Since the mechanism is similar in all organisms, this book uses the term *RNA silencing*.

The RNA silencing system not only provides defence against viruses but also against activation of transposons and transgenes. Furthermore, it is involved extensively in developmental control through the microRNA (miRNA) pathway (see following). Thus, it is a generic endogenous system, one function of which is to respond to virus infections.

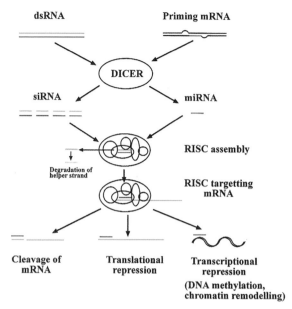

FIGURE 11.1   RNA silencing pathways. The red line indicates the guide strand and the green line the antiguide (helper) strand and the target mRNA.

II. MECHANISM OF SILENCING

A. The Basic Pathway

A common feature of RNA silencing is that it involves highly structured or double-stranded (ds) RNA, which is an unusual molecule in eukaryotic cells. The basic pathway is shown in Figure 11.1. The dsRNA is cleaved into small interfering (si) RNA fragments of 20–25 base pairs by a ribonuclease called DICER. The strands of the siRNA fragments are separated to give the *guide strand*, which is complementary to the target mRNA (the RNA that gave the original dsRNA) and the *antiguide strand* (helper strand), which is further degraded. The guide strand is then incorporated into the RNA-induced silencing complex (RISC), which targets it to the cognate mRNA, forming a duplex with that RNA. The cognate mRNA strand complementary to the bound siRNA is cleaved by the slicer activity of RISC to give further siRNA fragments. This leads to three control pathways: further cleavage of mRNAs, translational repression, and transcriptional

repression. All three pathways are involved in the plant response to virus infections.

B. Components of the System

1. dsRNA

As just noted, the silencing system targets several pathways that have dsRNA or highly structured RNA as a common starting element (Figure 11.2).

RNA Viruses. ssRNA viruses replicate via a complementary RNA (see Chapter 8) and thus go through a dsRNA stage (the replicative form or intermediate). Furthermore, the secondary structure of their single-stranded genomes can contain significant regions of base-pairing that may or may not be accessible to DICER; it is likely that both the replicative form and the secondary structure are targets for RNA silencing. The genomes of dsRNA viruses are, by their nature, double-stranded. The extensive secondary structure of viroids (see Chapter 3) is a dsRNA target for the silencing system.

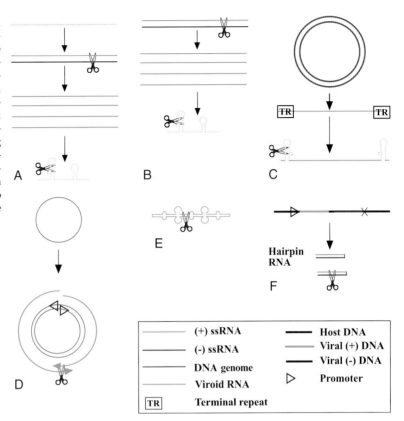

FIGURE 11.2 Sources of dsRNA for the silencing pathway. A. Replicating ssRNA virus; note both the replicative intermediate and secondary structure in the ss form can be targets. B. Replicating dsRNA virus. C. Retro- and pararetro-virus. The secondary structure in the terminal repeats (TR) can be a target. D. Replicating ssDNA virus. The overlapping 3′ ends of transcripts from complementary strands can be targets. E. Viroids. F. Structure of a hairpin construct for transformation into plant to give protection against the virus (see Chapter 15).

DNA Viruses. The genomes of DNA viruses (e.g., geminiviruses) are transcribed to give mRNAs (see Chapter 8), which are likely to have significant secondary structure. The reverse-transcribing DNA viruses—for example, caulimoviruses—replicate via RNA, which also has secondary structure (see Chapter 8).

Aberrant RNAs are transcripts that have faults due to, say, truncation of transcription or false transcription, giving ds regions and RNAs made double-stranded by host RNA-dependent RNA polymerase.

Transposons and Transgenes. Retrotransposons replicate by reverse transcription and thus have an RNA stage with secondary structure—for example, stem loops—in the terminal repeats. Transgenes are expressed through mRNA or aberrant RNAs and also may have inverted repeats.

Priming miRNA (pri-miRNA) is transcribed from intergenic and intronic regions of the host genome as stem-loop structure with imperfect base-pairing (Figure 11.3A) and with complementarity to one or more host mRNA. It is processed to give miRNA.

2. Dicer

Dicer and its homologues are RNaseIII enzymes that cleave dsRNA (see Box 11.1).

3. Products of Dicer

siRNA. Twenty to 25 nucleotide dsRNA fully base-paired with a 2 nucleotide 3′ overhang (Figure 11.3B). It is complementary only to the dsRNA from which it arose.

miRNA. Twenty-one to 23 nucleotide dsRNA produced from priming miRNA.

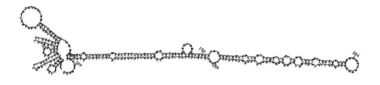

A

FIGURE 11.3   A. Structure of pri-miRNA (from www.en.wikipedia.org/ Image: MicroRNA). B. Diagram of the structure of an siRNA showing a ~19–21 base-pair RNA duplex with 2 nucleotide 3′ overhangs: OH, 3′ hydroxyl; P, 5′ phosphate (from www.en.wikipedia. org/Image: SiRNA).

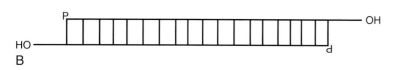

B

BOX 11.1

DICERS AND RISCS

Dicer

Dicers are a ~200 kDa family of proteins that usually contain an ATPase/RNA helicase domain, a PAZ domain, two RNaseIII domains, and a C-terminal dsRNA binding domain. The distance between the two RNase III domains determines the size of the siRNA that it produces. Dicer progressively cleaves dsRNA at 21–25 bp intervals to generate siRNA with 2-nt 3′ overhangs and phosphorylated 5′ termini (Figure 11-3B). It also cleaves pri-miRNA to give 20–25 bp fragments with blunt ends. There are various dicers that have distinct roles; those from plants are called dicer-like (DCL).

Dicer	Product	Function of Product
DCL1	miRNA	Developmental control
DCL2	22nt siRNA	Stress-related natural defence transcription
DCL3	24nt siRNA	Methylation of cognate DNA
DCL4	21nt siRNA	PTGS; systemic spread of signal

Humans and *Caenorhabditis* each have one dicer that processes both dsRNA (to give siRNAs) and pri-miRNA (to give miRNAs). *Drosophila* has two dicers: Dcr-1, which mainly produces miRNA, and Dcr-2, which mainly produces siRNA.

RISC

Besides the si- or mi-RNA, the RNA-induced silencing complex (RISC) includes members of the Argonaute (AGO) protein family and various accessory factors. The AGO proteins are ~100 kDa and have a PAZ domain and a PIWI domain, which is related to RNaseH endonuclease and functions in slicer activity. Various AGO proteins are probably associated with the variations of the basic silencing pathway.

Arabidopsis encodes 10 putative AGOs, with AGO1 being involved in the RISC. Vertebrate RISC has 4 AGOs (1–4), and *Drosophila* RISC has AGO-1 involved in the miRNA pathway and AGO-2 involved in the siRNA pathway. AGO-2 also processes the antiguide strand during RISC assembly.

4. *RISC*

RISC is a multiprotein complex containing either si- or mi-RNA that cleaves target mRNAs (see Box 11.1).

C. Results of the System

1. **Target RNA degradation**: Processing of the cognate viral RNA or mRNA targeted by si- or mi-RNA in the RISC (Figure 11.1)
2. **Translational repression**: Prevention of translation of mRNA due to binding by siRNA (or miRNA)
3. **DNA methylation**: Epigenetic effect caused by methylation of host (or viral) DNA because of the presence of siRNA (see also Section VI).

III. SYSTEMIC SILENCING

The silencing of a virus can spread systemically throughout the plant after induction in the initially infected cell. It is thought that the viral primary 21 nt siRNA produced by the dicer/RISC system in the initially infected cell moves through plasmodesmata to about 10–15 surrounding cells either with or ahead of the infection front (Figure 11.4). The host RNA-dependent RNA polymerase amplifies the siRNAs to give ds forms, which are then processed into secondary siRNAs. These then move to surrounding cells and are amplified. This reiterative process moves the siRNA signal to the vascular tissue from where it, possibly associated with small-RNA binding proteins, is distributed to other parts of the plant. From there, the signal spreads through the systemic leaves, once again in a reiterative manner. This systemic movement of the silencing signal ahead of viral movement primes the defence response prior to the arrival of the virus.

IV. OVERCOMING SILENCING

The RNA silencing defence system is a major problem facing a virus, especially one with an RNA genome. Thus, viruses have evolved a range of mechanisms to overcome this defence system, including suppression of the RNA-silencing pathway, avoidance of the defence system, and possibly outwitting the defence system.

A. Suppression of Silencing

Viruses and viroids have developed two basic strategies to suppress RNA silencing: encoding a protein that interferes with the silencing pathway or through a nucleic acid-mediated mechanism.

1. *Protein Suppressors of Silencing*

An increasing number of plant (and animal) viral gene products have been identified as suppressors of RNA silencing (Table 11.1). These proteins often also have other functions and frequently are identified as being "pathogenicity determinants." However, these other functions, such as replication enhancement, determining virus movement, and symptom production, can be attributed to the suppression of silencing. Some viruses, such as *Citrus tristeza virus* (see Profile 5 for genome organisation), have more than one gene product involved in suppression of silencing; these proteins are thought to have different modes of action in silencing suppression.

A common feature of many of the suppressors that have been characterised is that they bind dsRNA, some binding both long and short dsRNAs and others just binding ds siRNA (Table 11.1). The p19 of tombusviruses (see Box 11.2), the p20 of closteroviruses, the p15 of *Peanut clump virus*, the TGB1 of *Barley stripe mosaic virus*, and the HC-Pro of potyviruses each just bind ds siRNA and inhibit an intermediate step of RNA silencing. It is thought that

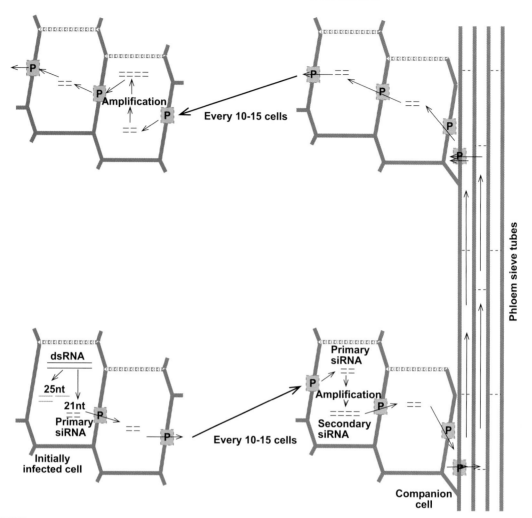

FIGURE 11.4 Systemic spread of silencing signal. The signal is generated in the initially infected cell (bottom, left hand) and spreads to about 10–15 adjacent cells, where it is amplified. It moves out of the initially infected leaf via the phloem sieve tubes and then spreads throughout systemic leaves being amplified at various times. P, plasmodesma.

these silencing suppressors inhibit the assembly of RISC by competing more efficiently for the ds siRNA than the RISC assembly complex. HC-Pro, p19, and p21 suppressors bind 21-nt siRNA duplexes more efficiently than 24-nt siRNA duplexes; however, p21 and HC-Pro require a 2-nt 3′ end overhang, whereas p19 does not. The aureusvirus p14 and *Turnip crinkle virus* coat protein bind both long and short

dsRNA, which could suggest inhibiting at an earlier stage in the silencing pathway.

Some evidence exists that not all suppressors function by binding ds RNA and thereby inhibiting the assembly of RISC. For instance, *Cucumber mosaic virus* (CMV)-encoded 2b suppressor protein interacts with Argonaute 1 protein in RISC and inhibits its cleavage activity, and the Polerovirus P0 triggers ubiquitin-mediated proteolysis

TABLE 11.1   Some Viral RNA Silencing Suppressors

Virus Family	Virus	Suppressor(s)	Other Functions	Mechanism
Plant (+)-Strand RNA Viruses				
Carmovirus	Turnip crinkle virus	P38	Coat protein	Binds both long and si dsRNA
Closterovirus	Citrus tristeza virus	P20, P23, CP	Replication enhancer, Nucleic acid binding, Coat protein respectively	Suppress intra- and intercellular silencing, Suppress intracellular silencing, Suppress intercellular silencing respectively
Cucumovirus	Cucumber mosaic virus	2b	Host-specific movement, symptoms	Interacts with ARGONAUTE 1
Hordeivirus	Barley stripe mosaic virus	γb	Replication enhancer; movement; seed transmission; pathogenicity determinant	Binds si dsRNA
Pecluvirus	Peanut clump virus	P15	Movement	Binds both long and si dsRNA
Polerovirus	Beet western yellows virus	P0	Pathogenicity determinant	Targets ARGONAUTE 1
Potyvirus	Potato virus Y; Tobacco etch virus	HCPro	Movement; polyprotein processing; aphid transmission; pathogenicity determinant	Binds si dsRNA
Tombusvirus	Carnation Italian ringspot virus	P19	Movement; pathogenicity determinant	Binds si dsRNA
Plant (–)-Strand RNA Virus				
Tospovirus	Tomato spotted wilt virus	NSs	Pathogenicity determinant	Not determined
Plant DNA Virus				
Geminivirus (Begomovirus)	Tomato yellow leafcurl virus	AC2	Transcription activator protein	Activates host suppressor
Plant Reverse-Transcribing Viruses				
Caulimovirus	Cauliflower mosaic virus	P6	Transactivator of genome expression; symptom severity determinant	Not determined
Animal (+)-Strand RNA Virus				
Nodavirus	Flock House virus; Nodamura virus	B2	Plaque formation	dsRNA binding

Continued

TABLE 11.1   (*Continued*)

Virus Family	Virus	Suppressor(s)	Other Functions	Mechanism
		Animal (–)-Strand RNA Virus		
Orthobunya-virus	*La Crosse virus*	NSs		Not determined
Orthomyxo-virus	*Influenza virus A*	NS1	Poly(A)-binding; inhibitor of mRNA transport; PKR inhibitor	dsRNA binding
		Animal DNA Virus		
Adenovirus	*Adenovirus*	VA1 RNA	PKR inhibitor	Dicer binding
Poxvirus	*Vaccinia virus*	E3L	PKR inhibitor	dsRNA binding
		Animal Reverse-Transcribing Virus		
Lentivirus	*Human immunodeficiency virus-1*	Tat	Transcriptional activator	Interacts with dicer
Spumavirus	*Primate foamy virus - 1*	Tas	Transcriptional activator	Not determined

of Argonaute 1. Suppression by the begomovirus (geminivirus) AC2 gene product requires transactivation of host suppressor(s). Thus, there may be various mechanisms for the viral suppression of silencing.

2. *Nucleic Acid Suppressors of Silencing*

As we saw in Chapter 3, viroids do not encode any proteins. Yet, the finding of viroid-specific siRNAs shows that their highly structured RNAs are processed by the silencing pathway. That viroids successfully infect plants indicates that they must be able to suppress the silencing. It is suggested that the secondary structure may also have the property of suppressing silencing of the replication of, at least, some viroids.

B. Avoidance of Silencing

For some viruses no suppressor of silencing has yet been identified. The susceptible stage in the viral replication cycle to the silencing defence system is when dsRNA is exposed at stages such as RNA replication or translation. Thus, it is thought that some viruses avoid exposure to the defence system by replicating in inaccessible sites, such as vesicles. Furthermore, if a virus replicates and expresses rapidly and then safely encapsidated its genome, it may outcompete the defence system.

V. SILENCING AND SYMPTOMS

Silencing and suppression of silencing has a major influence on the symptoms that viruses produce in plants. It must be remembered that a productive infection is a balance between silencing and suppression of silencing. If there was no silencing or full suppression of silencing, it is most likely that the plant would die soon after infection. As pointed out in Chapter 4, rapid plant death is a selective disadvantage to the virus. On the other hand, if there was no suppression of silencing, the virus would not be able to establish an infection.

BOX 11.2

TOMBUSVIRUS SILENCING SUPPRESSOR

The silencing suppressor of several tombusviruses is a protein of 19 kDa (p19). It is responsible for *Tomato bushy stunt virus* pathogenesis. The crystal structure of p19 of a related virus, *Carnation Italian ringspot virus*, has been determined and shows that the protein homodimer acts as a molecular calliper binding a specific size (21nt) of siRNA duplex (Fig.).

Fig. The structure of the p19 silencing suppressor bound to siRNA. The p19 dimer binds one face of an siRNA duplex (brown). Contact between the "core" of the p19 dimer and the RNA phosphate groups contributes to the protein's high affinity for dsRNA, while a pair of tryptophan residues (red; Trp42 and Trp39) in the "reading head" measure siRNA length. Because each p19 monomer (blue and green) contribute a reading head, the protein has been described as a "molecular calliper" that sizes up the dsRNA so as to bind best to canonical siRNAs. [This article was published in *Curr. Biol.* **14**, P.D. Zamore, Plant RNAo: how a viral silencing suppressor inactivates siRNA, pp. R198–200, Copyright Elsevier (2004).]

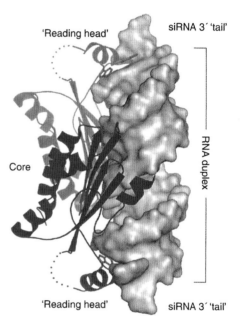

Although some symptoms are attributable to silencing and suppression not all are. As outlined in Chapter 9, the replication, movement, and accumulation of viruses in plants can cause upsets to the physiology and metabolism of a plant. A good example is the yellowing symptoms caused primarily by the presence of virus in the phloem upsetting the starch-sugar balance.

A. Recovery

As noted in Chapter 2, a virus-infected plant, especially with Nepoviruses, may show severe viral symptoms on the inoculated and first systemically infected leaves; however, new growth appears in which symptoms are milder or absent. For other viruses, such as *Alfalfa mosaic virus*, the virus content in the plant can increase

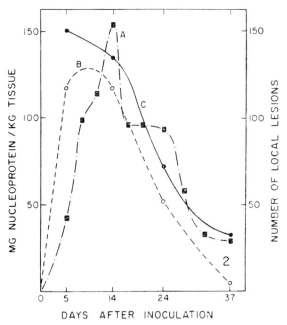

FIGURE 11.5 Concentration and specific infectivity of *Alfalfa mosaic virus* harvested from whole tobacco plants at different times after inoculation. Zero lesions are assumed for time 0. Curve A, amount of purified virus nucleoprotein (mg/kg total leaf wet weight); curve B, number of local lesions induced by sap inoculation; curve C, number of local lesions when purified sap samples were equalised spectrophotometrically. [This article was published in *Virology*, **15**, C.W. Kuhn and J.B. Bancroft, Concentration and specific infectivity changes of *alfalfa mosaic virus* during systemic infection, pp. 281–288, Copyright Elsevier (1961).]

and decrease (Figure 11.5) with concomitant changes in symptoms sometimes in a cyclical manner. These recovery phenomena likely reflect changes in the balance between silencing and suppression.

B. Dark Green Islands and Mosaics

One of the most common symptoms of virus infection is a mosaic of light and dark green areas on the leaves (Chapters 2 and 9). The mosaic pattern develops in leaves that are undergoing cell division and leaf expansion and can include well-defined dark green areas in which there is no detectable virus; these are

termed "dark green islands," and they are resistant to subsequent virus infection. It would appear that they result from one or a small group of cells in which there is silencing but no suppression and that divide to form a discrete tissue zone.

C. miRNA

As well as suppressing siRNA silencing, protein suppressors can affect the miRNA pathway. Not all suppressors impact upon this pathway, and not all miRNAs are affected. However, it is likely that the perturbation of the miRNA pathway leads to some symptom production. There is evidence for this in that some transgenic plants expressing suppressors of silencing have the phenotype of virus-infected plants even though there is no virus present.

Symptom differences between strains have been attributed to differences in the interactions between silencing protein and the miRNA pathway. The 2b protein of mild strains of CMV is a strong suppressor of RNA silencing but interacts weakly with miRNAs and thus is a weak inducer of symptoms.

D. siRNA Effects

Transgenic tomato plants expressing portions of *Potato spindle tuber viroid* contained viroid-specific siRNAs and had a phenotype similar to the symptoms of viroid infection. This suggests that the siRNAs might be responsible for the symptom production possibly by targeting plant genes for RNA silencing.

E. Synergistic Effects

The silencing suppressor of one virus can affect the replication of another virus in a joint infection leading to synergistic effects (Box 11.3).

BOX 11.3

SYNERGISM

Synergism is when the effects of the combined infection by two viruses are more severe than that of either of the constituent viruses alone. It can be caused by the infection with two viruses that belong to the same genus or by two viruses from different genera.

The Uganda variant of *African cassava mosaic virus* (UgACMV) caused a severe pandemic in cassava in Uganda in the 1990s (see Profile 7). It devastated the country's cassava production, leading to annual losses estimated of more than US$60 million between 1992 and 1997. The disease has subsequently spread to other surrounding countries. UgACMV is caused by synergism between two begomoviruses: *African cassava mosaic virus* (ACMV) and *East African cassava mosaic virus* (EACMV). The synergistic effect is mediated by the differential and complementary suppression of silencing by the *AC4* gene of ACMV and the *AC2* gene of EACMV.

The joint infection of tobacco plants with PVX and PVY is characterised by severe veinal necrosis in the first systemically infected leaves (Fig). Leaves showing this synergistic reaction contain up to 10 times as much PVX as with single infections but only the same amount of PVY. Ultrastructural studies and fluorescent antibody staining show that both viruses replicate in the same cells and that the increased production of PVX is due to an increase in virus production

per cell rather than an increase in the number of cells supporting PVX replication. The level of PVX (–)-strand RNA increased disproportionately to that of (+)-strand RNA in doubly infected tissues, suggesting that the synergism involves an alteration in the normal regulation of the relative levels of the two RNA strand polarities during viral replication. The HC-Pro silencing suppressor from PVY is responsible for the increase in the amount of PVX.

F. Other Activities of Silencing Suppressors

In Chapter 10, we saw some forms of resistance that plants can show to viruses. There are various complex interactions between RNA silencing and its suppression and some of these other forms of virus resistance. For instance, potyviral HC-Pro suppressor enhances other forms of host resistance to some viruses and reduces it to other viruses. In some cases, these changes in resistance involved a salicylic acid (SA)-dependent pathway, whereas in others an SA-independent pathway, and it could be reversed by the addition of the CMV 2b suppressor.

VI. TRANSCRIPTIONAL AND TRANSLATIONAL REPRESSION

Figure 11.1 shows that there are two other pathways besides the siRNA/RISC-directed cleavage of mRNA—namely, translational repression and transcriptional repression (see also Section II, B, 3). In translational repression the (−) strand of the siRNA binds to the mRNA preventing translation. In the context of a virus infection, it is difficult to differentiate this from mRNA cleavage, as in the absence of active suppressors, there will only be viral genome in the initially infected cell(s), and any suppressor will operate earlier at the RISC or pre-RISC stage. Transcriptional repression could result from the siRNA inducing DNA methylation or remodelling of the chromatin. There is circumstantial evidence for methylation of both geminiviral DNA and of modification of chromatin structure.

VII. EVOLUTIONARY ASPECTS

It is suggested that RNA silencing has a long evolutionary history. The diversity of viral suppressor protein sequences indicates that this counterdefence strategy has arisen several times. Some could have been derived and modified from endogenous host suppressors and others from nucleic acid–binding proteins. It is obvious that the selection pressure for suppressor proteins has been paramount in the development of most successful viruses.

As suggested earlier, evolution of the silencing/suppressor system has been to achieve a balance between host defence and viral counterdefence. As the prime target for silencing is dsRNA, there must be selection against excessive secondary structure in viral genomes. Thus, there must be a balance between the needs for secondary structure, such as replication and translational control, packaging in isometric particles, and the potential exposure to the silencing pathway.

VIII. RNA SILENCING IN ANIMAL AND OTHER VIRUSES

The silencing pathway appears to be present in most, if not all, eukaryotes, and it seems as if it controls viruses in both invertebrate and vertebrate hosts. In vertebrates silencing may not be as important as the immune response to virus infection; on the other hand, its importance may not be fully recognised. Several viruses of both invertebrates and vertebrates encode suppressors of silencing (Table 11.1), which indicates that this defence system operates in these hosts. Since the silencing pathway is not found in prokaryotes, there appears to be no such defence system against bacterial viruses.

IX. SUMMARY

- The RNA-silencing defence system is found in eukaryotes.
- The defence system is activated by dsRNA that is cleaved to give small RNA molecules that target the cognate RNA.
- There are two main sources of dsRNA: perfectly base-paired from viruses, transposons, transgenes, and aberrations in transcription, which are processed to give small interfering (si)RNA; and imperfectly base-paired priming-micro (pri-mi)RNA, which is processed to give micro (mi)RNAs that control host development.

- The dsRNAs are processed through three pathways that have a common basic section that cleaves the dsRNA and targets the si- or miRNA to its cognate mRNA.
- The three pathways lead to degradation of mRNA, translational repression of mRNA, and transcriptional repression. Defence against RNA viruses is usually by the first pathway and against DNA viruses by the first and third pathways. The second pathway is frequently used by miRNA.
- In plants the silencing response spreads systemically and primes the defence ahead of the virus infection front.
- Viruses encode suppressors of silencing that act as a counterdefence against the silencing system.
- Silencing and suppression are also found to be associated with viruses of invertebrates and vertebrates.

Further Reading

Bisaro, D.M. (2006). Silencing suppression by geminivirus proteins. *Virology* 344, 158–168.

Bortolamiol, D., Pazhouhandeh, M., Marrocco, K., Genschil, P., and Ziegler-Graff, V. (2007). The polerovirus F box protein P0 targets ARGONAUTE1 to suppress RNA silencing. *Current Biology* 17, 1615–1621.

Díaz-Pendón, J.A. and Ding, S.W. (2008). Direct and indirect roles of viral suppressors of RNA silencing in pathogenesis. *Annu. Dev. Phytopathol.* 46, 303–326.

Itaya, A., Zhong, X., Bundschub, R., Qi, Y., Wang, Y., Takeda, R., Harris, A.R., Molina, C., Nelson, R.S., and Ding, B. (2007). A structured viroid RNA serves as a substrate for dicer-like cleavage to produce biologically active small RNAs but is resistant to RNA-induced silencing complex-mediated degradation. *J. Virol.* 81, 2980–2994.

Lakatos, L., Csorba, T., Pantaleo, V., Chapman, E.J., Carrington, J.C., Liu, Y.-P., Dolja, V.V., Calvino, L.F., López-Moya, J., and Burgyán, J. (2006). Small RNA binding is a common strategy to suppress RNA silencing by several viral suppressors. *EMBO J.* 25, 2768–2780.

Lewsey, M., Robertson, F.C., Canto, T., Palukaitis, P., and Carr, J.P. (2007). Selective targeting of miRNA-regulated plant development by a viral counter-silencing protein. *Plant J.* 50, 240–252.

Li, F. and Ding, S.-W. (2006). Virus counterdefense: Diverse strategies for evading the RNA-silencing immunity. *Annu. Rev. Microbiol.* 60, 503–531.

MacDiarmid, R. (2005). RNA silencing in productive virus infections. *Annu. Rev. Phytopathol.* 43, 523–544.

Olson, K.E., Keene, K.M., and Blair, C.D. (2008). An innate immune response to viruses. *Encyclopedia of Virology,* Vol. 3, pp. 148–154.

Padmanabhan, M.S. and Dinesh-Kumar, S.P. (2008). Virus induced gene silencing (VIGS). *Encyclopedia of Virology,* Vol. 5, 375–379.

Shams-Bakhsh, M., Canto, T., and Palukaitis, P. (2007). Enhanced resistance and neutralisation of defense responses by suppressors of RNA silencing. *Virus Research* 130, 103–109.

Szittya, G., Dalmay, T., and Burgyan, J. (2008). Gene-silencing pathway. *Encyclopedia of Virology,* Vol. 4, 141–148.

Trinks, D., Rajeswaran, R., Shivaprasad, P.V., Akbergenov, R., Oakeley, E.J., Veluthambi, K., Hohn, T., and Pooggin, M.M. (2005). Suppression of RNA silencing by a geminivirus nuclear protein, AC2, correlates with transactivation of host genes. *J. Virol.* 79, 2517–2527.

Voinnet, O. (2005). Induction and suppression of RNA silencing: Insights from viral infections. *Nature Reviews Genetics* 6, 206–220.

Xie, Q. and Guo, H.-S. (2006). Systemic antiviral silencing in plants. *Virus Res.* 118, 1–6.

Zhang, Z.-H., Xie, Q., and Guo, H.-S. (2007). Antiviral defense in plants. *Plant Viruses* 1, 21–26.

Zhang, X., Yuan, Y.-R., Pei, Y., Lin, S.-S., Tuschl, T., Patel, D.J., and Chua, N.H. (2006). Cucumber mosaic virus-encoded 2b suppressor inhibits Arabidopsis Argonaute1 cleavage activity to counter plant defense. *Genes and Development* 20, 3255–3268.

PLANT VIRUSES IN AGRICULTURE AND INDUSTRY

Plant-to-Plant Movement

Being obligate parasites, viruses depend for survival on being able to spread from one susceptible individual to another fairly frequently.

I. INTRODUCTION

Because viruses cannot penetrate the intact plant cuticle and the cellulose cell wall (Figure 12.1B), plants have a barrier to infection. This problem is overcome either by avoiding the need to penetrate the intact outer surface (e.g., in seed transmission or by vegetative propagation) or by some method involving penetration through a wound in the surface layers, such as in mechanical inoculation and transmission by insects. There is considerable specificity in the mechanism by which any one virus is naturally transmitted.

II. TRANSMISSION VIA PLANT MATERIAL

A. Mechanical Transmission

Mechanical inoculation involves the introduction of infective virus or viral RNA into a wound on the plant's surface. When virus establishes itself successfully in the cell, infection occurs. This form of transmission occurs naturally with a few viruses such as *Tobacco mosaic virus* (TMV) and *Potato virus X* (PVX) that are very stable and reach high concentrations in the plant. TMV can readily contaminate hands,

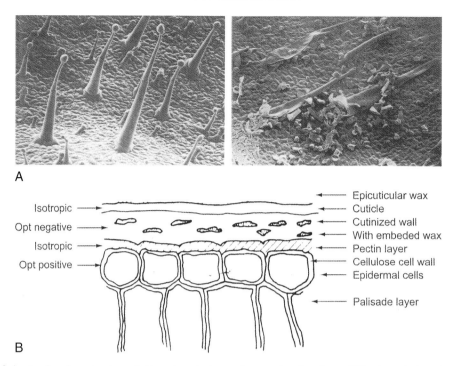

FIGURE 12.1 Leaf surface structure. A. Scanning electron micrographs of surface of *Nicotiana glutinosa* leaf before (left panel) and after (right panel) mechanical inoculation. Note leaf hairs in left panel that are broken in right panel. [From Hull (2002).] B. Diagrammatic representation of cross-section of the upper part of a leaf showing the barriers to virus infection. [From Eglinton and Hamilton (1967; *Science* **156**, 1322–1335), kindly used with permission of Dr. B.E. Juniper.]

clothing, and implements and can be spread by workers and, for instance, birds in tobacco and tomato crops. TMV may be spread mechanically by tobacco smokers because the virus is commonly present in processed tobacco leaf. For example, a survey showed that all 37 brands of cigarette and 60 out of 64 smoking tobaccos contained infectious TMV.

Mechanical transmission is of great importance for many aspects of experimental plant virology, particularly for the assay of viruses, often by local lesion production; in the propagation of viruses for purification; in host range studies; in diagnosis; and in the study of the early events in the interaction between a virus and susceptible cells. Mechanical inoculation is usually done by grinding up infected leaf tissue in a buffer—usually a phosphate buffer

that contains additives that control nucleases and polyphenols—incorporating an abrasive such as celite or carborundum, and then rubbing the extract gently on the leaves of the recipient plant. The gentle application wounds the leaf surface without causing cell death (Figure 12.1A).

B. Seed Transmission

About one-seventh of the known plant viruses are transmitted through the seed of at least one of their infected host plants. Seed transmission provides a very effective means of introducing virus into a crop at an early stage, giving randomised foci of primary infection throughout the planting. Thus, when some other method of transmission can operate to spread the virus within the growing crop, seed transmission may be of very considerable economic importance. Viruses may persist in seed for long periods so commercial distribution of a seed-borne virus over long distances may occur. Seed transmission rates vary from less than 1 to 100 percent, depending on virus and host.

Two general types of seed transmission can be distinguished. With TMV in tomato, seed transmission is largely due to contamination of the seedling by mechanical means. The external virus can be readily inactivated by certain treatments eliminating all, or almost all, seed-borne infection. In the second and more common type of seed transmission, the virus is found within the tissues of the embryo. The developing embryo can become infected either prior to fertilisation by infection of the gametes (indirect embryo invasion or gametic transmission) or by direct invasion after fertilisation. Generally speaking, for infection of the embryo from the mother plant, the earlier the plant is infected, the higher the percentage of seed that will transmit the virus. Obviously, for indirect embryo invasion by pollen, the infection takes place at pollination.

The direct route of seed infection from the mother plant poses problems in that symplastic connection is severed at meiosis. To infect the embryo, the virus has to reach either the floral meristems, which are beyond the limits of normal long-distance movement in the phloem (see Chapter 9), or the embryo itself. The route of direct embryo infection of peas by *Pea seed-borne mosaic virus* has been examined in detail (Box 12.1).

C. Pollen Transmission

Some viruses are transmitted from plant to plant via pollen. As with seed transmission, two mechanisms appear to operate in pollen transmission: gametic infection of the embryo and direct infection of the mother plant.

D. Vegetative Propagation

Vegetative propagation is an important horticultural practice, but it is also, unfortunately, a very effective method for perpetuating and spreading viruses. Economically important viruses spread systemically through most vegetative parts of the plant. A plant once systemically infected with a virus usually remains infected for its lifetime. Thus, any vegetative parts taken for propagation, such as tubers, bulbs, corms, runners, and cuttings, will normally be infected.

E. Grafting

Grafting is essentially a form of vegetative propagation in which part of one plant (the scion) grows on the roots (the stock) of another individual. Once organic union has been established, the stock and scion become effectively a single plant. Where either the rootstock or the individual from which the scion is taken is infected systemically with a virus, the grafted plant as a whole will become infected if both partners in the graft are susceptible. Grafting may succeed in transmitting a virus where other methods fail.

III. TRANSMISSION BY INVERTEBRATES

Many plant viruses are transmitted from plant to plant in nature by invertebrate vectors, members of the *Insecta* and *Arachnida* classes of the *Arthropoda,* and the *Dorylaimida* order of the *Nematoda.* Box 12.2 shows the orders of the *Insecta* that transmit plant viruses. Six of the orders contain insects that feed by chewing. The *Homoptera* feed by sucking sap from plants and are numerically the most important suborder containing plant virus vectors. Figure 12.2 shows three of the most common vectors of plant viruses: aphids, leafhoppers, and whitefly.

DIRECT EMBRYO INFECTION OF PEAS BY *PEA SEED-BORNE MOSAIC VIRUS* (PSbMV)

The route by which PSbMV reached pea seeds has been studied in detail by comparing a variety (Vedette) in which the virus is seed transmitted with one (Progretta) in which it is not (Figure).

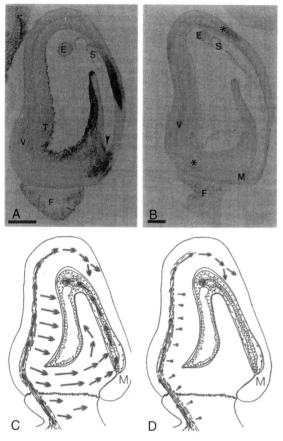

Fig. The pathway to seed transmission of *Pea seed-borne mosaic virus* (PSbMV) in pea. A and B. Analysis of the distribution of PSbMV in longitudinal sections through immature pea seed by immunochemistry using a monoclonal antibody to the virus coat protein shows that a cultivar-virus interaction, which is permissive for seed transmission (e.g., with *Pisum sativum* cv. Vedette in A) results in widespread accumulation of the virus in the testa tissues. In contrast, in the nonpermissive interaction (e.g., with cv. Progretta in B) virus enters the seed through the vascular bundle but is unable to invade the adjacent testa tissues extensively. In both cases there is a gradual reduction in the amount of accumulated virus after invasion such that in cv. Progretta only patches (asterisks) of infected tissue remain detectable. Systematic analyses of the immature seeds of different ages have identified the routes (red arrows) of virus invasion in the two cultivars (illustrated diagrammatically in C for cv. Vedetta and in D for cv. Progretta). The most consistent observation from all these studies is that the virus must reach the micropylar area of the testa for seed transmission to occur, a location providing the closest point of contact (arrowhead in A) between the testa tissues and the embryonic suspensor. In the nonpermissive interaction the virus appears to be blocked (denoted by red squares) in its ability to spread into and/or replicate in the nonvascular testa tissues. E, embryo proper; F, funiculus; M, micropylar region; S, suspensor; T, testa; V, vascular bundle. Bar marker = 500m. [This article was published in *Trends Microbiol.* **4,** A.J. Maule and D. Wang, Seed transmission of plant viruses: A lesson in biological complexity, pp. 153–158, Copyright Elsevier (1996).]

The virus moves through the testa of the immature seed after fertilisation and must reach the micropylar region of the seed for embryo infection to occur. The micropyle is in close contact with the base of the embryonic suspensor, which functions as a conduit for nutrient flow to support growth of the embryo. The suspensor is the route by which the virus invades the embryo itself, but it degrades as part of the seed development programme. This leaves a "window of opportunity" for embryo infection; this window of opportunity is taken by the virus in Vedette but not in Progretta. However, there is no symplastic connection between maternal and embryonic tissue, and it is still unknown how the virus crosses from the maternal testa cells to the embryonic suspensor.

BOX 12.2

VIRUS TRANSMISSION BY INSECTA

Seven of the 29 orders in the living *Insecta* feeding on living green land plants are vectors of plant viruses and are listed here with the approximate number of vector species in parentheses:

1. *Orthoptera*—chewing insects; some feed on green plants (27).
2. *Dermaptera*—chewing insects; a few feed on green plants (1).
3. *Coleoptera*—chewing insects; many feed on green plants; see table.
4. *Lepidoptera*—chewing insects; larvae of many feed on green plants (4).
5. *Diptera*—larvae of a few feed on green plants (2).
6. *Thysanoptera* (thrips)—some are rasping and sucking plant feeders (10).
7. *Hemiptera*—feed by sucking on green plants

   Suborder *Heteroptera*, Families *Myridae,* and *Piesmatidae* ($\sim$ 4)
   Suborder *"Homoptera"*,[a] see Table.

   Distribution of plant virus vectors among selected *Homoptera* and *Coleoptera* families. [From Nault (1997; *Ann. Entomol. Soc. Am.* **90**, 521–541).]

Order, Suborder, Family	Common Name of Insect Group	No. Species Described	No. Vector Species	No. Viruses Transmitted
Homoptera				
Auchenorrhyncha				
Cicadidae	Cicada	3,200	0	0
Membracidae	Treehopper	4,500	1	1
Cercopidae	Spittlebug	3,600	0	0
Cicadellidae	Leafhopper	15,000	49	31
Fulgoroidea	Planthopper	19,000	28	24
Sternorrhyncha				
Psyllidae	Psyllid	2,000	0	0
Aleyrodidae	Whitefly	1,200	3	43
Aphididae	Aphid	4,000	192	275
Pseudococcidae	Mealybug	6,000	19	10
Coleoptera				
Chrysomelidae	Leaf beetle	20,000	48	30
Coccinellidae	Ladybird beetle	3,500	2	7
Cucurlionidae	Weevil	36,000	10	4
Meloidae	Blister beetle	2,100	1	1

[a] "Homoptera" is a widely used generic term for several suborders of the *Hemiptera*.

FIGURE 12.2   Major vectors of plant viruses. A. Aphid (*Aphis glycines*); note the stylet penetrating the leaf (from www.planthealth.info, Purdue University). B. Leafhopper (*Circulifer tenellus*). C. Whitefly (*Bemisia tabaci*).

A. Relationships Between Plant Viruses and Insects

The transmission of viruses from plant to plant by invertebrate animals is of considerable interest from two points of view. First, such vectors provide the main method of spread in the field for many viruses that cause severe economic loss. Second, there is much biological and molecular interest in the relationships between vectors and viruses, especially as some viruses have been shown to multiply in the vector. Such viruses can be regarded as both plant and animal viruses. Even for those that do not multiply in the animal vector, the relationship is usually more than just a simple one involving passive transport of virus on some external surface of the vector ("the flying pin"). Transmission by invertebrate vectors is usually a complex phenomenon involving specific interactions between the virus, the vector, and the host plant, coupled with the effects of environmental conditions. Most of the detailed studies on virus transmission and virus-vector relationships have been made using aphids. Many of the features described following for aphid transmission are applicable to transmission by vectors from other insect orders.

As a general rule, viruses that are transmitted by one type of vector are not transmitted by any of the others. This specificity is not only at the level of vector type, family, genus, or species but can be even at the level of biotype. There are two basic interactions between viruses and their biological vector. They may be taken up internally within the vector, termed *persistent, internally borne* or *circulative,* or they may not pass to the vector's interior, in which case they are termed *nonpersistent, externally borne,* or *noncirculative* (Box 12.3).

BOX 12.3

RELATIONSHIPS BETWEEN PLANT VIRUSES AND THEIR VECTORS

There are two major types of interaction between a virus and its vector: nonpersistent and persistent. Features of the interactions are outlined in the table here, and the pathways that the viruses have with their vector are shown in the figure.

Virus Transmission Group					Transmission Characteristics					
Site in vector	Type of transmission	Virus product interacting with vector	Acquisition time (max. dose)	Retention time (half-life)	Transtadial passage	Virus in vector haemolymph	Latent period	Virus multiplies in vector	Transovarial transmission
Externally-borne	Nonpersistently transmitted stylet-borne	Capsid +/− helper factor	Seconds to minutes	Minutes	No	No	No	No	No
	Nonpersistently transmitted foregut-borne (semipersistent)	Capsid +/− helper factor	Minutes to hours	Hours	No	No	No	No	No
Internally-borne	Persistent circulative		Hours to days	Days to weeks	Yes	Yes	Hours to days	No	No
	Persistent propagative		Hours to days	Weeks to months	Yes	Yes	Weeks	Yes	Often

(continued)

BOX 12.3   (continued)

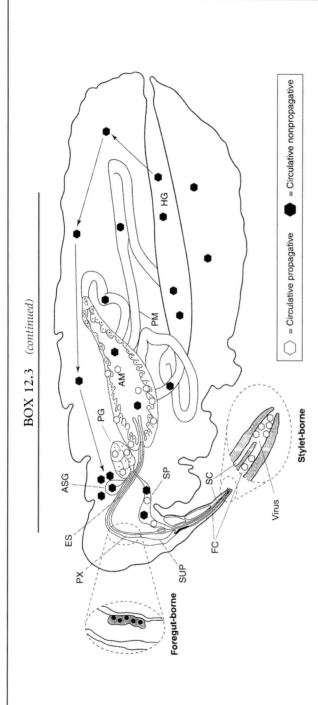

= Circulative propagative   = Circulative nonpropagative

Figure. Mechanism of transmission of plant viruses by arthropods with piercing-sucking mouthparts. The general anatomy of the alimentary system and the salivary system is shown; the areas relevant to virus transmission are labeled. One inset shows a detailed view of the distal end of the mouthparts where the food canal (FC) and the salivary canal (SC) empty into a common space. One current model of transmission of stylet-borne (nonpersistent, noncirculative) viruses suggests that the transmissible virus is retained at the distal tip of the stylets and then released by salivary secretions as the insect salivates during feeding. A second inset shows a detailed view of the foregut-borne (semipersistent, noncirculative) with virus particles attached to the cuticle lining of the foregut, a region that would include the sucking pump (SUP), pharynx (PX), and esophagus (ES). Notice that the virus is embedded in a matrix material attached to the cuticle. The origin or composition of the matrix material is unknown. The circulative nonpropagative viruses pass through the foregut into the anterior midgut (AM), posterior midgut (PM), and then into the hindgut (HG). They do not infect the gut cells but are transported through the posterior midgut and hindgut cells and released into the haemocoel (body cavity). Current information indicates that these viruses specifically associate with the accessory salivary glands (ASG) and are transported across the ASG cells and then released into the salivary canal (SC). The circulative propagative viruses will infect midgut cells and subsequently infect other tissues. These viruses ultimately associate with the principal salivary glands (PG) and possible ASG prior to their release into the SC. SP, salivary pump. [This article was published in *Encyclopedia of virology*, Vol. 1. S.M. Gray and D.M. Rochon (A. Granoff and R. Webster, Eds.), Vectors of plant viruses, pp. 1899–1910, Copyright Elsevier (1999).]

Essentially, there are three stages in the transmission cycle:

1. The *acquisition phase*, in which the vector feeds on the infected plant and acquires sufficient virus for transmission.
2. The *latent period*, in which the vector has acquired sufficient virus but is not able to transmit it. For externally borne viruses, there is little or no latent period.
3. The *retention period* is the length of time during which the vector can transmit the virus to a healthy host.

B. Nonpersistent Transmission by Insects

1. Features of Nonpersistent Transmission

Of the over 300 known aphid-borne viruses, most are nonpersistent. The following virus genera have definite members transmitted in a nonpersistent manner: *Alfamovirus*, *Caulimovirus* by *Myzus persicae*, *Closterovirus*, *Cucumovirus*, *Fabavirus*, *Macluravirus*, and *Potyvirus*. These genera include viruses with helical and isometric particles and with DNA and RNA mono-, bi-, and tripartite genomes. There are no known nonpersistent viruses transmitted by leafhoppers.

Nonpersistently transmitted viruses are acquired rapidly from plants, usually in a matter of seconds. During this time, aphid's stylet does not usually penetrate beyond the epidermal cells, and when it penetrates beyond the epidermis into the mesophyll and vascular tissue, the transmission rate declines rapidly. The initial host-finding behaviour of aphids is short probing, thought to sample the epidermal cells' sap, and fits very well with this mechanism. Since the sampling is especially brief on nonhosts for the aphid, the vectors of nonpersistent viruses are often noncolonisers of that species.

With a nonpersistent virus there is little or no latent period, and aphids begin to lose the ability to infect immediately after the acquisition feed. The rate at which infectivity is lost depends on many factors, including temperature and whether they are held on plants or under some artificial condition.

Different strains of the same virus may differ in the efficiency with which they are transmitted by a particular aphid species. Some strains may not be transmitted by aphids at all. Different strains of the same nonpersistent virus do not usually interfere with each other's transmission, as is sometimes found with propagative viruses.

Aphid species vary widely in the number of different viruses they can transmit. At one extreme, *M. persicae* is known to be able to transmit a large number of nonpersistent viruses, whereas other aphids transmit only one virus. These differences in part reflect the extent to which different aphid species have been tested, but there is no doubt that real differences in versatility occur. Among species that transmit a given virus, one species may be very much more efficient than another. For instance, marked differences were found in the efficiency with which *Potato virus Y* was transmitted by different species even when acquisition feed and test feed times were standardised (Figure 12.3). This can reflect the initial feeding behaviour of different aphid species on the test plant species.

2. Virus-Vector Relationships

As just noted, when they alight on a leaf, aphids may make brief probes into the leaf—often less than 30 seconds. Thus, the initial behaviour of such aphids on reaching a leaf is ideally suited to rapid acquisition of a nonpersistent virus. Sap sampling on a virus-infected plant will contaminate the stylet tips, the food canal, and the foregut. These sites have been favoured for the retention site of virus that will be injected subsequently into a healthy plant following another exploratory probing feed.

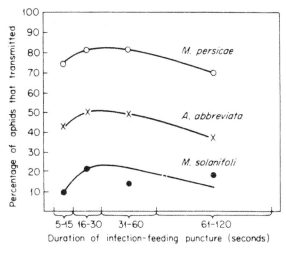

FIGURE 12.3   Relative efficiencies of three aphis species in transmitting *Potato virus Y* after defined acquisition feeding times. [Reprinted with permission from Bradley and Rideout (1953; *Can. J. Zool.* **31**, 333–341).]

The weight of evidence favours the food canal in the maxillae as the site where infective virus is retained during nonpersistent transmission, but it must be remembered that this evidence identifies sites of accumulation but does not give any indication as to whether that virus is transmitted.

There are two phases to the interaction involved in nonpersistent virus transmission: retention of the virus at a specific site and release of the virus. All nonpersistently transmitted viruses have a simple structure of nucleic acid encapsidated in simple icosahedral or rod-shaped particles by one or more coat protein species. Thus, it is the capsid protein that is available for any interactions. Two forms of interaction have been identified in the retention phase: one in which there appears to be direct interaction between the virus capsid and the site of retention in the aphid and the other in which a nonstructural virus-encoded protein is involved. This nonstructural protein is termed a *helper component, helper factor,* or *aphid transmission factor.*

a. Direct Capsid Interaction.   It is thought that *Alfalfa mosaic virus* and *Cucumber mosaic virus* (CMV) transmission involves direct links between the capsid and the binding site within the aphid vector. The prime evidence is that purified virus can be transmitted from artificial feeding systems without the addition of other proteins or factors. Most of the detailed evidence is for CMV. The efficiency of aphid transmission of heterologously assembled particles between the genomes and capsid proteins of a highly aphid-transmitted (HAT) and a poorly aphid-transmitted (PAT) strain of CMV segregated with the source of coat protein. The amino acid differences between the coat proteins of HAT and PAT strains of CMV are associated with both vector transmissibility and virion stability.

The minor coat protein of *Closteroviruses*, which encapsidates only a teminal part of the viral genome to give "rattlesnake" particles (see Figure 5.1), is thought to be involved in aphid transmission.

b. Indirect Interaction Involving Helper Components.   The interactions involving helper components of two groups of viruses, the potyviruses and caulimoviruses, have been studied in detail. The helper components of potyviruses are encoded by the virus and have the following properties in relation to virus transmission:

- The helper factor of one potyvirus may or may not permit the aphid transmission of another potyvirus when tested in an *in vitro* acquisition system. Thus, there is some specificity in the phenomenon.
- The helper component must be acquired by the aphid either during or before virus acquisition. If it is provided after the virus, there is no transmission.
- Potyvirus helper components have MWs in the range of 53 kDa and 58 kDa. They are cleaved from the polyprotein encoded by the virus as a product that, as well as helper

component activity, has various other activities including being a protease. Thus, it is termed HC-Pro.

- Purified helper component can be used to facilitate the transmission of potyviruses by feeding through artificial membranes.
- The biologically active form of helper component appears to be a dimer.

By studying the effects of mutations in the coat protein and helper component on aphid transmission of various potyviruses, a picture appears of some of the molecular interactions involved. The current hypothesis is that the helper component forms a bridge between the virus capsid and the aphid stylet (Figure 12.4A). Close to the N-terminus of the coat protein is the amino acid

FIGURE 12.4   Models for the interactions between viruses, transmission helper factors, and vector. A. Possible interactions between the HC-Pro, the aphid's stylets, and the potyviral coat protein. (1). Position of the virion particles close to the apical section of the food canal. (2). A model assuming an association between two molecules of HC-Pro. Note that one molecule of the HC-Pro is bound to a "receptor" on the stylet, while the second HC-Pro molecule is bound to the coat protein subunit. (3) A model assuming that a dimer is needed to bind to the "receptor" on the stylet. Both HC-Pro molecules are linked to coat protein subunits. (4) A proposed structural binding between the PTK motif of HC-Pro and the DAG motif found on the N-terminus of the coat protein subunit. [This article was published in *Virus-insect-plant interactions*, B. Raccah, H. Huet, and S. Blanc (K.F. Harris, O.P. Smith, and J.E. Duffus, Eds.), Potyviruses, pp. 181–206, Copyright Elsevier (2001).] B. Interaction involved in the aphid transmission of *Cauliflower mosaic virus* (CaMV). The panel on the left is from the figure in Box 12.3. The rest of the figure shows diagrammatically the interactions between a CaMV particle and the aphid's stylet (see text). N and C are the N-terminal and C-terminal regions of P2, respectively.

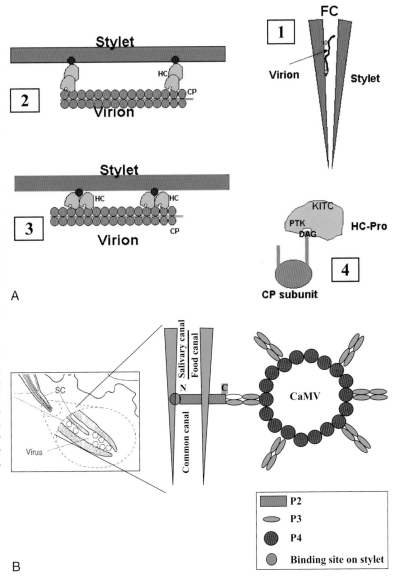

triplet DAG (aspartic acid-alanine-glycine) that is important for transmissibility. Mutagenesis shows that both the DAG sequence itself and the context of the surrounding amino acids affects transmissibility. Biochemical and immunological analyses indicate that this N-terminal region is located on the external surface of the virus particle. Two important regions have been identified in HC-Pro. One, characterised by the amino acid sequence PKT (proline-lysine-threonine), appears to be involved in the interaction with the capsid protein. The other, termed the KITC (lysine-isoleucine-threonine-cysteine) region, appears to be involved with the HC-Pro retention on the aphid's stylets.

Little is known about the mechanisms of release of nonpersistent viruses from the site of binding in the aphid's stylet, but three theories have been proposed:

1. The mechanical transmission theory suggests that the virus is simply inoculated by the stylet.
2. In the ingestion-egestion theory, release is effected by regurgitation and salivation.
3. Since the food and salivary canals of the stylets fuse near the tip of the maxillary stylet, nonpersistently stylet-borne viruses could be released by saliva alone.

Cauliflower mosaic virus (CaMV), and presumably other caulimoviruses, requires a helper component (or aphid transmission factor) when being transmitted by *M. persicae*. The CaMV helper component system has the following properties:

- As with potyviruses, the helper component must be acquired by the aphid either during or before virus acquisition.
- Helper components of other caulimoviruses can complement defects in CaMV helper component.
- The CaMV helper component system involves two noncapsid proteins: the 18 kDa product of ORFII (P2) and the 15 kDa product of ORFIII (P3); (for CaMV gene map, see Profile 4).

- In infected cells, P2 is found in crystalline electrolucent inclusion bodies (see Figure 2.6) and P3 in association with virus particles.
- P2 interacts very strongly with microtubules with binding domains in two regions: one near the N-terminus and the other near the C-terminus.

Thus, the CaMV helper component system is more complex that that of potyviruses. The virus can be transmitted from an *in vitro* acquisition system containing baculovirus-expressed P2 and sap from a plant infected with a P2-defective isolate but not from P2 + purified virus; however, the virus can be transmitted when P3 is added to the purified virus. Secondary structure predictions of P2 suggest two domains: the N-terminus being predominantly β-sheet and the C-terminus predominantly α-helix; the two domains are separated by a predicted random structure. The 61 amino acid C-terminal domain interacts with partially purified virus and with the 30 N-terminal amino acids of P3. Mutations of the N-terminal domain abolish its ability to facilitate transmission but do not affect its ability to bind to semipurified virus. This leads to a model for how the CaMV helper system operates (Figure 12.4B). P3, which forms a tetramer, binds to the virus capsid composed mainly of P4, with the C-terminus of P2 binding to P3. The bridge is completed by the N-terminus of P2 binding to a nonglycosylated protein embedded in the chitin matrix of the common food/salivary duct of the aphid stylet (Figure 12.4B). The role of the microtubule binding activity of P2 is unknown, but it is noted that the microtubule binding domains overlap the P3 and aphid binding domains.

As with nonpersistent viruses, nothing is known about the molecular details of virus release from the vector, but as the virus is retained in the common food/salivary duct at the tip of the aphid's stylet, it is thought that release could be effected by the aphid's saliva.

C. Persistent Transmission by Insects

The main features of persistent transmission are summarised in Box 12.3. Viruses transmitted in this manner are usually transmitted by one or a few species of aphid. Yellowing and leafrolling symptoms are commonly produced by infection with persistently transmitted viruses. Viruses that are internally borne in their aphid vectors may replicate in the vector (propagative) or may not (circulative). For an aphid to become a transmitter by either type of relationship, the virus has to be ingested from the infected plant and reach the salivary glands, usually via the hemolymph, to be egested into the healthy plant. Thus, it has to pass at least two barriers: the gut wall and the wall of the salivary glands.

1. Circulative Viruses

a. Features of Circulative Virus: Vector Interaction.
Circulative viruses are usually phloem-limited, and thus the vector must feed for a longer time to acquire the virus (see Box 12.3). Luteoviruses are the most studied of the circulative (persistent) viruses for which there is no demonstration of replication in the vector.

The minimum acquisition time can be as little as 5 minutes but is usually several hours. This is followed by a latent period of at least 12 hours, after which the virus can be transmitted with an inoculation access time of 10 to 30 minutes. The aphids then remain capable of transmitting for at least several days.

As just noted, persistently transmitted viruses have to cross at least the gut and salivary gland barriers. Particles of *Cereal yellow dwarf virus-RPV* (CYDV-RPV) associate only with the cell membranes of the hindgut of the aphid vector *Rhopalosiphon padi*. It is suggested that the particles entered the hindgut cells by endocytosis into coated pits and coated vesicles and accumulated in tubular vesicles and lysosomes (Figure 12.5A).

Particles are then released into the haemocoel by fusion of the tubular vesicles with the basal plasmalemma. Aphid salivary glands comprise two principal glands and two accessory glands. *Potato leafroll virus* particles have been seen in the basal lamina and plasmalemma invaginations of accessory salivary cells (Figure 12.5B). Particles were also found in tubular vesicles in the cytoplasm near salivary canals and in coated pits connected to the canal membrane. The basal lamina and the basal plasmalemma function as independent barriers to transmission of different luteovirus-aphid combinations. From these studies the route that luteoviruses take across the two barrier tissues in their aphid vector would appear to be by incorporation into coated vesicles and transport across the cell(s). Thus, the main sites of interaction for the virus particles is with the plasma membrane on the gut side of the gut wall cells and with two plasma membranes on the haemocoel side of the salivary gland accessory cells, which suggests a receptor-mediated interaction.

Because purified luteoviruses can be aphid transmitted from *in vitro* acquisition, it is likely that no noncapsid proteins are involved. The capsid comprises the major capsid protein and a minor amount of a larger protein translated via a read-through of the coat protein stop codon (see Chapter 7). Particles containing just the major coat protein without any read-through protein are not transmissible, which led to the widespread assumption that the read-through portion was required for aphid transmissibility. However, there is no clear picture of the luteovirus component of the receptor-mediated recognition.

Several aphid proteins of Mr ranging from 31 to 85 kDa have been shown to interact with purified luteoviruses *in vitro*. Antisera raised against two of these proteins, P31 and P44, react specifically with extracts of accessory salivary glands from vector aphids, suggesting that these proteins might be involved in luteovirus-specific recognition at this site.

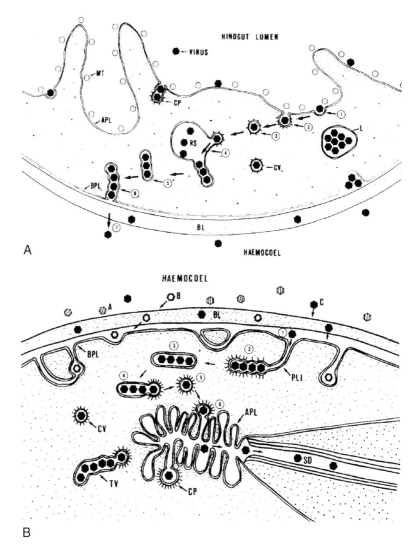

FIGURE 12.5 Models for interactions and transcellular transport of luteoviruses in aphid vectors. A. Transcellular transport through aphid gut epithelium. Visualisation of endocytotic- and exocytotic-associated ultrastructure supports receptor-mediated endocytosis as a mechanism regulating vector-specific luteovirus acquisition. Based on this model, luteoviruses recognised at the gut-cell apical plasmalemma (APL) bind to the membrane (1) initiating virus invagination (2) into coated pits (CP). Coated pits bud off the APL as virus-containing coated vesicles (CV), which transport the virus (3) to larger uncoated vesicles, called receptosomes (RS), which act to concentrate the virus (4). Tubular vesicles containing linear aggregates of virus form on the receptosomes (5) transport the virus to the basal plasmalemma (BPL) and fuse with the BPL, allowing release of the virus from the cell (6). Luteoviruses can then diffuse through the gut basal lamina (BL) and into the haemocoel (7). Eventually, receptosomes (endosomes) mature into lysosomes (L), and any virus particles remaining in the lysosome are probably degraded. MT, microtubules. B. Luteovirus interactions with accessory salivary glands (ASG) determining vector-specific transmission. Luteoviruses in the haemocoel first encounter the extracellular basal lamina (BL) surrounding the ASG. The BL acts as a selective barrier to luteovirus transmission. Depending on the aphid biotype and the specific luteovirus, the virus particles may be prevented from penetrating the BL (A) or may diffuse through (B, C) to the basal plasmalemma (BPL). A second selective barrier occurs at the BPL. Luteovirus particles not recognised at the BPL remain outside the cell in the pericellular space (B). Luteoviruses recognised by putative virus receptors (C) on the BPL (1) are encytosed by coated pits (2) and accumulate into tubular vesicles (TV) in the cytoplasm (3). The TV adjacent to the microvilli-lined canals formed by the apical plasmalemma (APL) bud off coated vesicles (CVs) (4) containing individual virions. The CVs transport the virus (5) to the canals, fuse with the APL (6), forming coated pits (CP), and release the virus into the canal lumen allowing transport of luteovirus out of the aphid along with salivary secretions. PLI, plasmalemma invagination; SD, salivary duct. [From Gildow (1999; in *The Luteoviridae*, H.H. Smith and H. Barker, Eds., pp. 88–112, CAB International, Wallingford, UK).]

Another binding aphid protein, which interacts with many luteoviruses and other viruses, is the 60 kDa symbionin or GroEL from the endosymbiotic bacterium *Buchneria* spp. This protein, found readily in aphid hemolymph, is a member of the molecular chaperone family that is responsible for stabilising the structure of proteins. The interaction of luteoviruses with symbionin is determined by the read-through domain of the minor capsid protein. The luteovirus-symbionin interaction is essential for the retention of the virus in the hemolymph.

b. Dependent Transmission.

As with certain nonpersistent viruses, some persistent viruses require a helper factor—in this case, a virus—to be present in the plant before aphid transmission can occur. For persistent viruses dependent on another virus, it is the presence of the virus itself in a mixed infection that provides the assistance. This type of dependent transmission is due to phenotypic mixing together during replication of the two viruses in the plant, resulting in the encapsidation of the genome of one virus in coat protein subunits of the other virus.

Umbraviruses do not encode a coat protein. For their aphid transmission they associate with a helper luteovirus, which is presumed to supply the coat protein and thus aphid transmission properties. Each definitive Umbravirus species is associated with a specific luteovirus. These systems have the following characteristics:

- Both viruses are transmitted in a circulative nonpropagative manner.
- The dependent virus (umbravirus) is sap transmissible, but the helper is not.
- The dependent virus is only transmitted by aphids from source plants that contain both viruses; in other words, aphids already carrying helper virus cannot transmit the dependent virus from plants infected only with this virus.

- Evidence from a variety of experiments indicates that the dependent virus is transmitted by the aphid vector only when its RNA is packaged in a protein shell made of the helper virus protein.

This phenotypic mixing must take place in doubly-infected plants. *Groundnut rosette virus* depends on its satellite RNA as well as on *Groundnut rosette assistor virus* for transmission by *Aphis craccivora* (see Box 2.1).

2. Propagative Viruses

Propagative viruses can be considered to be viruses of the insect that have become adapted to plants. Two plant virus families, *Reoviridae* and *Rhabdoviridae*, and the *Tenuivirus* and *Marafivirus* genera contain members that replicate in their leafhopper vectors. Such replication usually has little effect on the hoppers. However, from the virus point of view, replication in the vector has two important consequences: Once they acquire virus, the vectors normally remain infective for the rest of their lives, and replication in the vector is often associated with transovarial passage of the virus, thus giving it a means of survival over winter that is quite independent of the host plant. With viruses that replicate in their vectors, there is usually a high degree of specificity between vector and virus or even strains of a virus.

Many of the virus:host interactions resemble those found in animal viruses, with some genes that adapt the virus to either animals or plants. In the plant reoviruses, particular genome segments code for gene products required for replication in the insect but not in the plant. The rhabdo- and tospoviruses need glycoprotein spikes for infecting insects but not for plants. On the other hand, the sc4 gene product in plant rhabdoviruses (see Profile 13 for genome map) that facilitates cell-to-cell spread in plants is not found in animal rhabdoviruses.

Whereas aphids are vectors of many of the persistent circulative viruses, most of the persistent propagative viruses are leafhopper- or planthopper-transmitted. However, several members of the *Rhabdoviridae* replicate in their aphid vector, including *Sowthistle yellow vein virus (SYVV)*.

The latent period of SYVV in the vector is long and depends strongly on temperature. Characteristic bacilliform particles have been observed in the nucleus and cytoplasm of cells in the brain, subesophageal ganglion, salivary glands, ovaries, fat body, mycetome, and muscle. Virus particles appear to be assembled in the nucleus. The virus can be serially transmitted from aphid to aphid by injection of hemolymph, and infection is associated with increased mortality of the aphids. Decreased life span varies with different virus isolates. However, since infected aphids live through the period of maximum larviposition, the intrinsic rate of population growth was hardly affected. The virus is transmitted through the egg of *Hyperomyzus lactucae*, about 1 percent of larvae produced being able to infect plants. Continuous passage of SYVV in the aphid by mechanical inoculation gives rise to isolates that have lost the ability to infect the plant host.

3. Thrip Transmission of Tospoviruses

Transmission of tospoviruses by thrips has several unusual features. Only the first and early second larvae stages can acquire the virus and the competence to acquire decreases with age of the larvae. *Tomato spotted wilt virus* (TSWV) can be acquired or transmitted by first instar nymphs of *Frankliniella occidentalis* in feeding periods of as short as 5 minutes, but the median acquisition access period on infected *Impatiens* plants is more than 100 minutes. The median latent period varies with temperature being 84 hours at 27°C and 171 hours at 20°C. Individuals may retain infectivity for life, but their ability to transmit may be erratic. The virus is not passed through the egg.

As with other internally borne persistently transmitted viruses, tospoviruses have to pass several barriers in the vector, which suggests that there is/are a receptor-mediated mechanism(s). TSWV is enveloped with spikes made up from two virus-coded glycoproteins, extending from the envelope (see Profile 17). Passage of TSWV through plants can result in envelope-deficient isolates. Feeding *F. occidentalis* on plants infected with wild-type and an envelope-deficient isolate showed that the thrips only became infected when they acquired intact virus particles. These and other experiments suggest that the viral glycoproteins bind to a receptor in the vector's midgut. Two proteins from *F. occidentalis* have been shown to bind to TSWV glycoproteins, one of which is localised to the larval thrip midgut and the other present throughout the thrip's body.

A detailed study of the route that TSWV takes through *F. occidentalis* showed that the first infections were found in the midgut (Mg1) region about 24 hours postacquisition (hpa). These infections increased in intensity but remained restricted to the Mg1 epithelium for some time. In late larval stage, it spread to the circular and longitudinal midgut muscle tissues. By the adult stage, the visceral muscle tissues of the midgut and foregut were infected. Infection of the salivary glands was first observed 72 hpa, and at the same time, the ligaments connecting the midgut with the salivary glands became infected. There was no evidence for TSWV in either the haemocoel or the midgut basal lamina. It appeared that the virus reached the salivary glands through the ligaments connecting Mg1 to the salivary glands. This is a different route to that conventionally proposed for persistent viruses, which is movement through the haemocoel from the gut cells to the salivary glands.

D. Virus Transmission by Beetles

Leaf-feeding beetles have chewing mouthparts and do not possess salivary glands. They regurgitate during feeding, which bathes the mouthparts in sap. This regurgitant will contain virus if the beetle has fed on an infected plant. Beetles can acquire virus after a single bite and can infect a healthy plant with one bite. However, beetle transmission is not a purely mechanical process. There is a high degree of specificity between beetle vector and virus, and some very stable viruses, such as TMV, are not transmitted by beetles. The viruses that are transmitted belong to the *Tymovirus, Comovirus, Bromovirus,* and *Sobemovirus* genera.

Sometimes one beetle species will transmit a particular virus with high efficiency, while a related species does so inefficiently. It is suggested that the regurgitant fluid of the beetles contains an inhibitor that prevents the transmission of non-beetle-transmitted viruses but does not affect those that are transmitted. There is good evidence that the inhibitor is an RNase.

E. Nematode Transmission of Viruses

1. Features of Nematode Transmission

Two genera of plant viruses are transmitted by nematodes. Nepoviruses are transmitted by species in the genera *Xiphinema* and *Longidorus,* and tobraviruses are transmitted by species of *Trichodorus* and *Paratrichodorus.* All three tobraviruses are nematode transmitted, but only about one-third of the nepoviruses are transmitted by these vectors. With the exception of *Tobacco ring spot virus* (TRSV), which is reported to also have aphid vectors, none of the viruses in these two genera is known to have invertebrate vectors other than nematodes; some nepoviruses are pollen transmitted.

Nematodes are difficult vectors to deal with experimentally because of their small size and their rather critical requirements with respect to soil moisture content, type of soil, and, to a lesser extent, temperature. To overcome these problems, five criteria have been proposed for establishing the nematode vectoring of viruses.

- Infection of a bait plant must be demonstrated.
- Experiments should be done with hand-picked nematodes.

- Appropriate controls should be included to show unequivocally that the nematode is the vector.
- The nematode should be fully identified.
- The virus should be fully characterised.

A common method for detecting nematode transmission has been to set out suitable "bait" plants (such as cucumber) in a sample of the test soil. These plants are grown for a time to allow any viruliferous nematodes to feed on the roots and transmit the virus and for any transmitted virus to replicate. Extracts from the roots and leaves of the bait plants are then inoculated mechanically to a suitable range of indicator species (see Chapter 13).

2. Virus-Nematode Relationships

The nematode transmission of a virus has been divided into seven discrete but interrelated processes: ingestion, acquisition, adsorption, retention, release, transfer, and establishment. Ingestion is the intake of virus particles from the infected plant, and although it does not require a specific interaction between nematode and virus, it needs a specific interaction between the nematode and plant. In the acquisition phase, the ingested virus particles are retained in an intact state, and specific features on the surface of the particle are recognised by receptor sites in the nematode feeding apparatus leading to adsorption. Once adsorbed, infectious particles can be retained in the nematode for months or even years but not after moulting. Release of the virus particles is thought to occur by a change in pH caused by saliva flow when the nematode commences feeding on a new plant. In the transfer and establishment phases, the virus particles are placed in the healthy plant cell and start replicating and causing infection.

There is specificity in the relationships between nematodes and the viruses they transmit with often an apparent unique association between the virus isolate and the vector species. There are some cases of different virus isolates sharing the same vector species or, conversely, one particular virus isolate being transmitted by several nematode species. There are 13 trichodorid species known to be tobravirus vectors, but only one or two of these transmits each tobravirus. There is a substantial degree of specificity between the nematode vector and the tobravirus serotype. Several nepoviruses are transmitted by more than one vector species, but there can be differences in the observations under laboratory and field conditions.

Once acquired, viruses may persist in transmissible form in starved *Longidorus* for up to 12 weeks, in *Xiphinema* for about a year, and much more than a year in *Trichodorus*. Transmission does not appear to involve replication of the virus in the vector. Plant virus particles have never been observed within nematode cells. Consistent with this is the fact that no evidence has been obtained for virus transmission through eggs of nematode vectors.

Specificity of transmission does not appear to involve the ability to ingest active virus, since both transmitted and nontransmitted viruses have been detected within individuals of the same nematode species. Sites of retention of virus particles within nematodes have been identified by electron microscopy of thin sections. Nepovirus particles are associated with the inner surface of the odontostyle of various *Longidorus* species and with the cuticular lining of the odontophore and esophagus of *Xiphenema* species. Tobravirus particles have been observed absorbing to the cuticular lining of the esophageal lumen.

The genetic determinants for the transmissibility of the nepoviruses *Raspberry ringspot virus* and *Tomato black ring virus* are encoded by RNA2, which expresses, among other proteins, the viral coat protein (see Profile 6 for nepovirus genome organisation).

By making reciprocal pseudo-recombinants between a nematode transmissible and a nontransmissible isolate of the tobravirus *Tobacco rattle virus* (TRV), it was shown that transmissibility segregated with RNA2. As noted in Profile 15, tobravirus RNA2 is variable in size, and, besides encoding the viral coat protein, encodes one to three nonstructural proteins. A recombinant virus, in which the coat protein gene of a nematode nontransmissible isolate of *Pea early browning virus* (PEBV) was replaced with that of a highly nematode transmissible isolate of TRV, was not transmitted by nematodes, which indicated that more than one of the RNA2 genes was involved. Mutations in the 29 kDa and the 23 kDa nonstructural genes of PEBV both abolished nematode transmission without affecting particle formation, as did removal of the C-terminal mobile region of the coat protein. It is suggested that the nonstructural proteins may be transmission helper components analogous to those in some aphid and leafhopper virus transmission systems.

IV. FUNGAL TRANSMISSION OF VIRUSES

Several viruses have been shown to be transmitted by soil-inhabiting fungi or protists. The known vectors are members of the class *Plasmodiophoromycetes* in the division *Myxomycota* or in the class *Chytridiomycetes* in the division *Eumycota*. Both classes include endoparasites of higher plants. Species in the chytrid genus *Olpidium* transmit viruses with isometric particles, while species in two plasmodiophorus genera, *Polymyxa* and *Spongospora*, transmit rod-shaped or filamentous viruses.

The two chytrid vectors, *Olpidium brassicae* and *O. bornavirus*, are characterised by having posteriorly uniflagellate zoospores, while those of the three plasmodiophoral vectors, *Polymyxa graminis, P. betae,* and *Spongospora* subterranean, are biflagellate. All five species are obligate parasites of plant roots and have similar development stages (Figure 12.6). They survive

FIGURE 12.6 Life cycle of a plasmodiophoral fungus. On the left-hand side is the diploid stage in root cells; on the right-hand side is the haploid stage in root hairs. Between are the phases in the soil where plant-to-plant virus transmission can occur. [Reprinted from *Matthews' plant virology*, 4th ed., R. Hull, Transmission 1: by invertebrates, nematodes and fungi, pp. 485–532, Copyright (2002), with permission from Elsevier.]

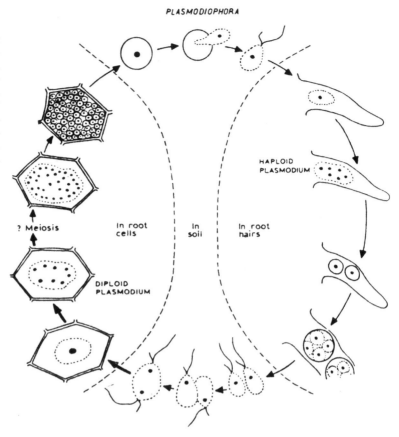

between crops by resting spores that produce zoospores that infect the host. The zoospores form thalli in the host cytoplasm. In the early stages of infection, the cytoplasm of thalli is separated from the host cytoplasm by a membrane, but later the thallus forms a cell wall. The entire thallus is converted into vegetative sporangia or resting spores.

Various degrees of host specificity exist in both the chytrid and plasmodiophoral vectors, with some isolates having a wide host range and others a narrow host range. The wide host range isolates tend to be better virus vectors than are the narrow ones.

Two types of virus-fungal vector relationships have been recognised, termed *in vitro* and *in vivo*.

The *in vitro* virus-vector relationship is found between the isometric viruses of the *Tombusviridae* and two *Olpidium* species. Virions from the soil water adsorb onto the surface of the zoospore membrane and are thought to enter the zoospore cytoplasm when the flagellum is "reeled in." It is unknown how the virus passes from the zoospore cytoplasm to the host cytoplasm, but it is thought that this occurs early in fungal infection of the root. Reciprocal exchange of the coat proteins of *Tomato bushy stunt virus* (not transmitted by *O. bornavirus*) and *Cucumber necrosis virus* (CNV; transmitted by *O. bornavirus*) showed that the coat protein is involved in the uptake of the virus by the zoospore. One amino acid in the coat protein of CNV is important for transmissibility, and binding studies showed that this is associated with recognition of the virus by *O. bornavirus* zoospores.

The model for the *in vivo* virus-vector relationship is demonstrated by the interactions between *Beet necrotic yellow vein virus* (BNYVV) and *Polymyxa betae*. The virus is within the zoospores when they are released from the vegetative sporangia or resting spores and infects the new host when these zoospores establish their own infection of the root. The processes of virus acquisition and release by the zoospores are unknown. The read-through domain from the coat protein (see Profile 2 for genome organisation of BNYVV) is implicated in the fungal transmission. BNYVV RNAs 3 and 4 also have an indirect effect on the transmission, most likely through controlling factors such as spread and accumulation of the virus in the root system.

V. VIRUSES OF OTHER KINGDOMS

In moving to a new host, viruses of vertebrates and invertebrates do not have to cross barriers such as the cuticle and cell wall that face plant viruses. As the respiratory and digestive tracts comprise very large areas of living cells surrounded just by the plasmamembrane, most viruses of vertebrates are transmitted by the respiratory and faecal-oral route. Some, loosely termed arboviruses (e.g., rhabdoviruses, bunyaviruses, and flaviviruses), are transmitted by arthropods in a manner similar to that of propagative persistent plant viruses, the usual vectors being blood-sucking insects such as mosquitoes. From the insects' point of view, the vectors of arboviruses and the analogous plant viruses are the alternate host, the vertebrate, and plant. There is mother-to-child vertical transmission (analogous to seed transmission) of some viruses of vertebrates and invertebrates. Bacterial viruses spread either through cell division or through the surrounding liquid medium.

VI. SUMMARY

- Plant viruses must cross two barriers—the cuticle and the cell wall—before they can infect a plant; this is done by mechanical damage.
- The plant virus can be introduced either from plant material or by a biological vector.
- Introduction from plant material can be by mechanical damage (e.g., breaking leaf hairs), through seed or pollen, or by grafting or vegetative propagation.
- Biological vectors are invertebrates (arthropods or nematodes) and fungi and protests.
- Each plant virus is usually transmitted in nature by just one of the preceding methods.
- Plant viruses have very specific and intricate interactions with their biological vectors.
- There are two basic interactions with insect vectors: nonpersistent or stylet borne, in which the virus interacts with the insect's mouth parts, and persistent or circulative, in which the virus passes though the insect's gut wall into the haemocoel and then into the salivary glands, from where it is injected into the plant.
- The interactions involved the virus coat and in some cases additional virus gene products.
- Interactions with nematode and fungal vectors are also detailed and involve the virus coat protein and sometimes additional viral factors.

Further Reading

Blanc, S. (2008). Vector transmission of plant viruses. *Encyclopedia of Virology*, Vol. 5, 274–281.

Brown, D.J.F., Robertson, W.M., and Trudgill, D.L. (1995). Transmission of viruses by plant nematodes. *Annu. Rev. Phytopathol.* **33**, 223–249.

Brown, J.K. and Czosnek, H.K. (2002). Whitefly transmission of plant viruses. *Adv. Bot. Res.* **36**, 65–100.

Campbell, R.N. (1996). Fungal transmission of plant viruses. *Annu. Rev. Phytopathol.* **34**, 87–108.

Gray, S. and Gildow, F.E. (2003). Luteovirus-aphid interactions. *Annu. Rev. Phytopathol.* **41**, 539–566.

Hogenhaout, S.A., Ammar, E.D., Whitfield, A.E., and Redinbaugh, M.G. (2008). Insect vector interactions with persistently transmitted plant viruses. *Annu. Rev. Phytopathol.* **46**, 327–359.

Hull, R. (2002). *Matthews' Plant Virology*. Academic Press, San Diego.

Maule, A.J. and Wang, D. (1996). Seed transmission of plant viruses: A lesson in biological complexity. *Trends Microbiol.* **4**, 153–158.

Ng., J.C.K. and Falk, B.W. (2006). Virus-vector interactions mediating nonpersistent and semipersistent transmission of plant viruses. *Annu. Rev. Phytopathol.* **44**, 183–212.

Ng., J.C.K. and Perry, K.L. (2004). Transmission of plant viruses by aphid vectors. *Molec. Plant Pathol.* **5**, 505–511.

Nault, L.R. (1997). Arthropod transmission of plant viruses: A new synthesis. *Ann. Entemol. Soc. Am.* **90**, 521–541.

Reisen, W.K. (2008). Vector transmission of animal viruses. *Encyclopedia of Virology*, Vol. 5, 268–273.

Plant Viruses in the Field: Diagnosis, Epidemiology, and Ecology

Plant viruses are a significant agricultural problem and cause major losses. In any disease situation it is important to know what virus is causing the problem, where it comes from, and how it spreads before developing control measures.

I. DIAGNOSIS

A. Introduction

There are three basic situations in which the techniques for recognising and identifying a virus are needed: diagnosis of a viral infection in the field to determine whether it is caused by a known or unknown virus; detection of a known virus, usually in an epidemiological or quarantine situation; and assay for a known virus—for instance, when purifying or manipulating it.

When appraising the relative merits of different methods, the following important factors must be considered:

- What question is being addressed? Is one just determining if the plant is virus infected, what virus is infecting the plant, or what strain of the virus is infecting the plant?

- Sensitivity: How small an amount of virus can be measured or detected?
- Accuracy and reproducibility
- Numbers of samples that can be processed in a given time by one operator
- Cost and sophistication of the apparatus and materials needed
- The degree of operator training required
- Adaptability to field conditions

It must always be remembered that diseased plants in the field may be infected by more than one virus. Thus, an early step in diagnosis for an unknown disease must be to determine whether more than one virus is involved. The methods involved in assay, detection, and diagnosis can be placed in four groups according to the properties of the virus on which they depend: biological activities, physical properties of the virus particle, properties of viral proteins, and properties of the viral nucleic acid.

B. Methods Involving Biology of the Virus

Biological methods for the assay, detection, and diagnosis of viruses are much more time-consuming than most other methods now available. Nevertheless, they remain very important. For diagnosis, in most circumstances only inoculation to an appropriate host species can determine whether a particular virus isolate causes severe or mild disease. However, this group of methods does have a major drawback, especially when used on a new virus or in quarantine situations. It raises the possibility that the production of infected plants might be a source of infection for local crops even in the face of the strictest containment conditions.

1. Indicator Hosts

Disease symptoms on plants in the field are frequently inadequate on their own to give a positive identification. This is particularly so when several viruses cause similar symptoms,

as do yellowing diseases in beet; when a single virus, such as *Cucumber mosaic virus*, is very variable in the symptoms it causes; or when both of these factors are relevant in a single host. Thus, since the early days of plant virology, searches have been made for suitable species or varieties of host plant that will give clear, characteristic, and consistent symptoms for the virus or viruses being studied, usually under glasshouse conditions. Many good indicator species have been found in the genera *Nicotiana*, *Solanum*, *Chenopodium*, *Cucumis*, *Phaseolus*, *Vicia*, and *Brassica*. Certain plant species such as *Chenopodium amaranticolor*, *C. quinoa*, and *Nicotiana benthamiana* react to a wide range of viruses.

Testing large numbers of samples using indicator hosts requires glasshouse facilities that may be occupied for weeks or longer. However, the actual manipulations involved in mechanical inoculation may take less time per sample than many other methods of testing, and many attempts have been made to streamline the procedures for sap extraction and inoculation. As just noted, care must be taken to keep inoculated test plants under suitable containment to minimise escape of the virus(es) to the outside environment.

2. Host Range

Host range is an important criterion in diagnosis. However, there are various factors that must be considered:

- In many of the reported host range studies, only positive results have been recorded.
- Absence of symptoms following inoculation of a test plant has not always been followed by back inoculation to an indicator species to test for masked infection.
- The manner of inoculation may well affect the results. Mechanical inoculation has almost always been used in extensive host range studies because of its convenience, but many plants contain inhibitors of infection

that prevent mechanical inoculation to the species, or from it, or both.

- In studying large numbers of species, it is usually practicable to make tests only under one set of conditions, but it is known that a given species may vary widely in susceptibility to a virus, depending on the conditions of growth.
- Quite closely related strains of a virus may differ in the range of plants they will infect. Host range data may apply only to the virus strain studied.

3. Methods of Transmission

The different methods of virus transmission, discussed in Chapter 12, may be useful diagnostic criteria. Their usefulness may depend on the particular circumstances. For example, a virus with an icosahedral particle that is transmitted through the seed and by nematodes is very probably a nepovirus. On the other hand, such a virus transmitted mechanically and by the aphid *Myzus persicae* might belong to any one of several groups.

4. Cytological Effects

Cytological effects (see Chapter 2) detectable by light microscopy can sometimes be used effectively to supplement macroscopic symptoms in diagnosis. The light microscope has a number of advantages when the inclusions are large enough to be easily observed:

- It is a readily available instrument.
- Specimen preparation techniques can be simple.
- There is a wide field of view, thus allowing many cells to be readily examined.
- A variety of cytochemical procedures are available.

5. Mixed Infections

Simultaneous infection with two or even more viruses is not uncommon. Such mixed infections may complicate a diagnosis based

on biological properties alone, especially if the host response is variable. However, several possible differences in biological properties may be used to separate the viruses:

- If one virus is confined to the inoculated leaf in a particular host, while the other moves systemically
- If a host can be found that only one of the viruses infects
- If the two viruses cause distinctive local lesions in a single host
- If the two viruses have different methods of transmission—for example, by different species of invertebrate vectors

On the other hand, difficulties may arise in sorting out certain diseases in the field that result from a more or less stable association between two or more viruses. For instance, groundnut rosette disease is caused by two viruses and a satellite (see Box 2.1).

C. Methods That Depend on Physical Properties of the Virus Particle

1. Stability and Physicochemical Properties

A virus has certain independently measurable properties of the particle that depend on its detailed composition and architecture. These properties can be useful for identification and as criteria for establishing relationships. The most commonly measured properties are density, sedimentation coefficient and diffusion coefficient, and ultraviolet absorption spectrum (Box 13.1).

2. Electron Microscopy

Knowledge of the size, shape, and any surface features of the virus particle is a basic requirement for virus identification. Electron microscopy can provide this information quickly and, in general, reliably. For the examination of virus particles in crude extracts or purified preparations, a negative-staining procedure is now used

BOX 13.1

ULTRAVIOLET SPECTRA OF VIRUSES

The ultraviolet absorption spectrum of a virus is a combination of the absorption spectra of the nucleic acid and the coat protein (Figure A). Nucleic acids have maximum absorption at about 260 nm and a trough at about 230 nm, whereas the absorption spectrum for proteins peaks at about 280 nm and troughs at about 250 nm. The specific absorption for nucleic acids (20–25 OD units/mg/ml/ at 260 nm) is much greater than that for proteins (about 1 OD unit/mg/ml/ at 280 nm). The combination gives an absorption spectrum for a virus that peaks at about 260 nm and troughs at about 230–240 nm (Figures A, B).

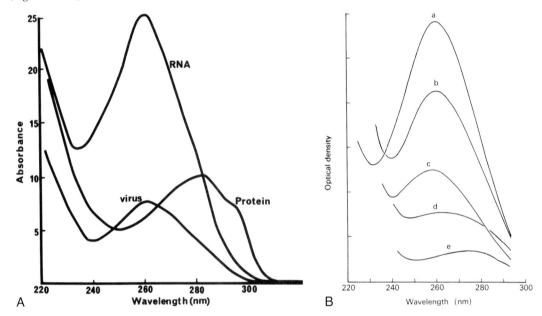

Fig. Ultraviolet absorption spectra. A. Comparative absorption spectra of 1 mg/ml RNA, a virus containing 20% RNA (1 mg/ml) and a typical protein (10 mg/ml); note the shoulder on the protein spectrum at 290 nm due to absorbance by tryptophan. [From Hull (1985; in *Virology, a practical approach,* B.W.J. Mahy, Ed., pp. 1–14, IRL Press, Oxford).] B. Absorption spectra of a suspension of (a) *Turnip yellow mosaic virus* (TYMV) RNA; (b) TYMV nucleoprotein particles (33% RNA); (c) *Cucumber mosaic virus* particles (19% RNA); (d) *Tobacco mosaic virus* particles (5% RNA); and (e) TYMV protein particles (top component). All suspensions contain the same weight of material except (a), which contained half as much as the others. [From Gibbs and Harrison (1976; *Plant virology, the principles,* Edward Arnold, London).]

Thus, this property depends mainly on the ratio of nucleic acid to protein in the virus; the amino acid composition of the protein can also influence the spectrum. Many unrelated viruses with similar compositions have similar absorption spectra. Ultraviolet absorption provides a useful assay for purified virus preparations, provided other criteria have eliminated the possibility of contamination with nonviral nucleic acids or proteins, especially ribosomes. Measurements of A_{260} may be unreliable for rod-shaped viruses that vary substantially in their degree of aggregation from one preparation to another, thus leading to changes in the amount of light scattering for a given virus concentration.

almost universally. Of particular use for rapid results is the epidermal strip technique, in which the broken surface of a strip of epidermis is removed from the leaf is wiped through negative stain on the electron microscope grid. Commonly used negative stains are sodium phosphotungstate, ammonium molybdate, or uranyl acetate, depending on the stability of the virus to these stains.

Approximate particle dimensions can be determined. This is particularly useful for rod-shaped viruses for which particle length distributions can be obtained. Surface features may be seen best in isolated particles on the grid. Depending on size and morphology, a virus may be tentatively assigned to a particular taxonomic group. However, some small icosahedral viruses cannot be distinguished from members of unrelated groups on morphology alone.

For many viruses, examination of thin sections by electron microscopy is a valuable procedure for detecting virus within cells and tissues, but this has its limitations such as differentiating virus particles from normal cell constituents. The large, enveloped viruses, the plant reoviruses, and the rod-shaped viruses can usually be readily distinguished because their appearance in thin sections generally differs from that of any normal structures.

D. Methods That Depend on Properties of Viral Proteins

Some of the most important and widely used methods for assay, detection, and diagnosis depend on the surface properties of viral proteins. For most plant viruses, this means the protein or proteins that make up the viral coat. Different procedures may use the protein in the intact virus, the protein subunits from disrupted virus, or proteins expressed from cloned cDNA or DNA in a system such as *E. coli* or insect cells. More recently nonstructural proteins coded for by a virus have been used in diagnosis.

1. Serology

Serological procedures are based on the interaction between a protein or proteins (termed the antigen) in the pathogen with antibodies raised against them in a vertebrate. The components of a serological reaction are shown in Box 13.2.

2. Types of Antisera

There are two basic types of antisera: polyclonal, which contain antibodies to all the available epitopes on the antigen, and monoclonal (Mab), which contain antibodies to one epitope. There is much discussion as to which is the best for diagnosis, but this will depend on what question the diagnostician is addressing. Monoclonal antisera are much more specific than polyclonal antisera and can be used to differentiate strains of many pathogens. On the other hand, specificity can be a disadvantage, and a variant of the pathogen may not be detected.

The bringing together of two technologies, the production of single-chain antibodies, and the display of recombinant proteins on the surface of bacteriophage (phage display) have resulted in the ability to produce a large range of molecules with the properties of single-type antibodies similar to Mabs without involving injection of animals.

3. Methods for Detecting Antibody-Virus Combination

A wide variety of methods has been developed for demonstrating and estimating combination between antibodies and antigens. The most widely used are the enzyme-linked immunoabsorbent assay (ELISA), immunoabsorbent electron microscopy (ISEM), and "dot blots" that employ either polyclonal or monoclonal antibodies.

a. ELISA Procedures. Many variations of the basic procedure have been described (Box 13.3),

BOX 13.2

COMPONENTS OF A SEROLOGICAL REACTION

1. Antibodies: An antibody is a molecule that binds to a known antigen. Antibodies are secreted by B lymphocytes. Structurally they are composed of one or more copies of a characteristic unit that can be visualised as forming a Y shape (Fig.).

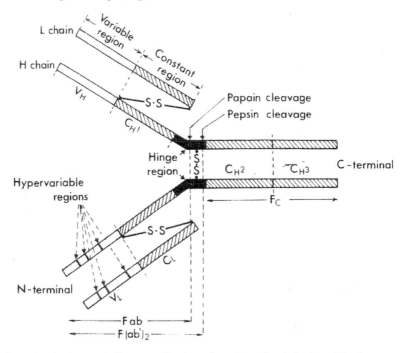

Fig. IgG antibody molecule. Arrows indicate specific sites where the molecule is cleaved by the enzymes papain and pepsin to give the Fab, Fc, and (FabN)₂ fragments. L, light chains; H, heavy chains. The four polypeptide molecules are joined by disulfide bridges. The two antigen-combining sites are made up from the variable regions of the L and H chains. [This article was published in *Serology and immunochemistry of plant viruses*, M.H.V. Van Regenmortel, Copyright Elsevier (1982), Academic Press, New York.]

Each Y contains four polypeptides: two identical copies of the heavy chain and two identical copies of the light chain joined by disulfide bonds. Antibodies are divided into five classes—IgG, IgM, IgA, IgE, and IgD—based on the number of Y-like units and the type of heavy-chain polypeptide they contain.

IgG molecules have three protein domains. Two of the domains, forming the arms of the Y, are identical and are termed the Fab domain. They each contain an antigen-binding site at the end, making the IgG molecule bivalent. The third domain, the Fc domain, forms the stem of the Y. The three domains may be separated from each other by cleavage with the protease papain. The Fc region binds protein A, a protein obtained from the cell wall of *Staphylococcus aureus*, with very high affinity. This property is used in several serological procedures.

(continued)

BOX 13.2 *(continued)*

The N-terminal regions of both the light and heavy chains in the arms of the Y-shaped IgG molecule comprise very heterogeneous sequences. This is known as the variable (V) region. The C-terminal region of the light chains and the rest of the heavy chains form the constant (C) region. The V regions of one heavy and one light chain combine to form one antigen binding site. The heterogeneity of the V regions provides the structural basis for the large repertoire of binding sites used in an effective immune response.

2. Antigens: Antigens are usually fairly large molecules or particles consisting of, or containing, protein or polysaccharides that are foreign to the vertebrate species into which they are introduced. Most have a molecular weight greater than 10 kDa, although smaller peptides can elicit antibody production. There are two aspects to the activity of an antigen. First, the antigen can stimulate the animal to produce antibody proteins that will react specifically with the antigen. This aspect is known as the immunogenicity of the antigen. Second, the antigen must be able to combine with the specific antibody produced. This is generally referred to as the antigenicity of the molecule.

Large molecules are usually more effective immunogens than small ones. Thus, plant viruses containing protein macromolecules are often very effective in stimulating specific antibody production; the subunits of a viral protein coat are much less efficient.

It is specific regions on antigens, termed epitopes, which induce and interact with specific antibodies. Epitopes can be composed of amino acids, carbohydrates, lipids, nucleic acids, and a wide range of synthetic organic chemicals. About 7–15 amino acids at the surface of a protein may be involved in an antigenic site.

There are several different grouping of epitopes:

The *continuous epitope* is a linear stretch of amino acids.

The *discontinuous epitope* is formed from a group of spatially adjacent surface residues brought together by the folding of the polypeptide chain or from the juxtaposition of residues from two or more separate peptide chains.

Cryptotopes are hidden epitopes revealed only on dissociation or denaturation of the antigen.

Neotopes are formed by the juxtaposition of adjacent polypeptides (e.g., adjacent viral coat protein subunits).

Metatopes are epitopes present on both the dissociated and polymerised forms of the antigen.

Neutralisation epitopes are specifically recognised by antibody molecules able to neutralise the infectivity of a virus.

The epitope type giving rise to a monoclonal antibody (MAb) and the relative proportions of different epitopes recognised by a polyclonal antiserum can affect the outcome of certain serological tests. For instance, an antibody to a cryptotope is unlikely to recognise an antigen in DAS-ELISA but is likely to in a Western blot.

3. Interaction Between Antibodies and Antigens. The interaction between an epitope and antibody is affected by both affinity and avidity. Affinity is a measure of the strength of the binding of an epitope to an antibody. As this binding is a reversible bimolecular interaction, affinity describes the amount of antibody-antigen complex that will be found at equilibrium. High-affinity antibodies perform better in all immunochemical techniques, not only because of their higher capacity but also because of thestability of the complex. Avidity is a measure of the overall stability of the complex between antibodies and

(continued)

BOX 13.2   *(continued)*

---

antigens and is governed by three factors: the intrinsic affinity of the antibody for the epitope, the valency of the antibody and antigen, and the geometric arrangement of the interacting components.

Titre is a relative measure of the concentration of a specific antibody in an antiserum. It is often used to define the dilution end point of the antiserum for detection of an antigen. Thus, as the sensitivities of various serological tests differ, the apparent titre is applicable only to the test under discussion.

The serological differentiation index (SDI) is a measure of the serological cross-reactivity of two antigens. It is the number of two-fold dilution steps separating the homologous and heterologous titres. The homologous titre is that of the antiserum with respect to the antigen used for immunising the animal, while the heterologous titre is that with respect to another related antigen.

BOX 13.3

ELISA PROCEDURES

Direct Double-Antibody Sandwich Method

The principle of the direct double-antibody sandwich procedure is summarised in the Figure (A). The technique is widely used but suffers two limitations:

1. It may be very strain specific. For discrimination between virus strains, this can be a useful feature, but for routine diagnostic tests, it means that different viral serotypes may escape detection. This high specificity is almost certainly due to the fact that the coupling of the enzyme to the antibody interferes with weaker combining reactions with strains that are not closely related.
2. It requires a different antivirus enzyme-antibody complex to be prepared for each virus to be tested.

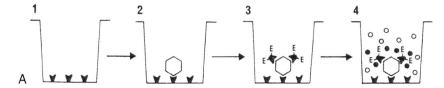

Fig. Immune detection of viruses. A. Principle of the ELISA technique for plant viruses (direct double-antibody sandwich method). (1) The gamma globulin fraction from an antiserum is allowed to coat the surface of wells in a polystyrene microtitre plate. The plates are then washed. (2) The test sample containing virus is added and combination with the fixed antibody is allowed to occur. (3) After washing again, enzyme-labeled specific antibody is allowed to combine with any virus attached to the fixed antibody. For instance, alkaline phosphatase is linked to the antibody with glutaraldehyde.) (4) The plate is again washed and enzyme substrate is added. The colourless substrate *p*-nitrophenyl phosphate (open circle) gives rise to a yellow product with alkaline phosphatase (filled circle), which can be observed visually in field applications or measured at 405 nm using an automated spectrophotometer. [Modified from Clark and Adams (1977; *J. Gen. Virol.* **34**, 475–483).]

(continued)

BOX 13.3 *(continued)*

Indirect Double-Antibody Sandwich Methods

In the indirect procedure, the enzyme used in the final detection and assay step is conjugated to an antiglobulin antibody. For example, if the virus antibodies were raised in a rabbit, a chicken antirabbit globulin might be used. Thus, one conjugated globulin preparation can be used to assay bound rabbit antibody for a range of viruses.

Many variations of these procedures are possible (Figure (B)).

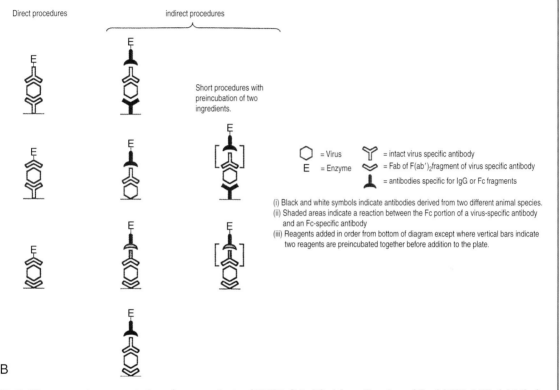

Direct procedures

indirect procedures

Short procedures with preincubation of two ingredients.

○ = Virus

E = Enzyme

Y = intact virus specific antibody

= Fab of F(ab')₂ fragment of virus specific antibody

= antibodies specific for IgG or Fc fragments

(i) Black and white symbols indicate antibodies derived from two different animal species.
(ii) Shaded areas indicate a reaction between the Fc portion of a virus-specific antibody and an Fc-specific antibody
(iii) Reagents added in order from bottom of diagram except where vertical bars indicate two reagents are preincubated together before addition to the plate.

B

Fig.B. Diagrammatic representation of some variants of ELISA. [Modified from Koenig and Paul (1982; *J. Virol. Methods* **5**, 113–125).]

The direct double-antibody sandwich method is the most convenient for the routine detection of plant viruses in situations where strain specificity and very low virus concentrations cause no problems.

Dot ELISA

Several procedures have been developed that use nitrocellulose or nylon membranes as the solid substrate for ELISA tests. For the final colour development, a substrate is added that the enzyme linked to the IgG converts to an insoluble coloured material. An example of a dot blot assay is given in Figure (C).

(continued)

BOX 13.3 *(continued)*

Fig.C. The sensitivity of a dot immunobinding assay. V = purified *Strawberry pseudo mild yellow edge virus;* I = crude sap from infected leaves; and H = crude sap from healthy leaves. The conjugated enzyme was alkaline phosphatase and the substrate for colour development was Fast Red TR salt. [From Yoshikawa *et al.* (1986; *Ann. Phytopathol. Soc. Jpn.* **52**, 728–731).]

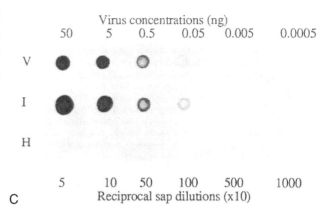

The main advantages of the method are speed (three hours to complete a test), low cost, and the small amount of reagents required. Dot blot immunoassays may be particularly useful for routine detection of virus in seeds or seed samples, especially for laboratories where an inexpensive and simple test is needed. However, endogenous insect enzymes may interfere with tests using insect extracts.

Immuno-Tissue Printing

As an extension to dot ELISA, immuno-tissue printing involves applying the cut surface of a leaf or stem (or any other plant organ) to a nitrocellulose membrane and revealing the presence of an antigen (say viral coat protein) by immunoprobing.

Tissue printing has several advantages:

1. It gives detailed information on the tissue distribution of a virus.
2. Extraction of sap from leaves in which a virus has limited tissue distribution leads to the dilution of the virus with sap from uninfected cells. Since this technique samples the contents of each cell on the cut surface individually there is increased sensitivity.
3. The technique is easily applicable to field sampling; tissue blots can be made in the field, and there is no need to collect leaf samples for sap extraction in the laboratory.

with the objective of optimising the tests for particular purposes. The method is very economical in the use of reactants and readily adapted to quantitative measurement. It can be applied to viruses of various morphological types in both purified preparations and crude extracts. It is particularly convenient when large numbers of tests are needed. It is very sensitive, detecting concentrations as low as 1–10 ng/ml.

As with other diagnostic tests, it is necessary to define and optimise the time of sampling and the tissue to be sampled to achieve a reliable routine detection procedure. The main factor limiting the number of tests that can be done using ELISA procedures is the preparation of tissue extracts.

When many hundreds of field samples have to be processed, it is often necessary to store them for a time before ELISA tests are carried out.

Storage conditions may be critical for reliable results. For this reason, conditions need to be optimised for each virus and host.

b. Serologically Specific Electron Microscopy.

A combination of electron microscopy and serology can provide rapid diagnosis and reveal features of a virus. The method offers a diagnostic procedure based on two properties of the virus: serological reactivity with the antiserum used and particle morphology. Various terms have been used to describe the process: serologically specific electron microscopy (SSEM), immunosorbent electron microscopy (ISEM), solid-phase immune electron microscopy, and electron microscope serology. Basically there are two approaches. Virus particles can be trapped on an electron microscope grid previously coated with an antiserum. The particles can be then negatively stained and reacted with another antiserum that decorates particles of the virus against which the antiserum is directed (Figure 13.1A). The other method is, after being adsorbed onto the EM grid, the virus particles are coated with a virus-specific antibody to which gold particles have been attached (Figure 13.1B).

This method has many advantages:

1. The result is usually clear in the form of virus particles of a particular morphology, and thus false positive results are rare.
2. Sensitivity may be of the same order as with ELISA procedures and may be 1,000 times more sensitive for the detection of some viruses than conventional EM.
3. When the support film is coated with an antibody, much less host background material is bound to the grid.
4. Antisera can be used without fractionation or conjugation, low-titre sera can be satisfactory, and only small volumes are required.
5. Very small volumes (0.1 µl) of virus extract may be sufficient.
6. Antibodies against host components are not a problem inasmuch as they do not bind to virus.
7. One antiserum may detect a range of serological variants (on the other hand, the use of monoclonal antibodies may greatly increase the specificity of the test).
8. Results may be obtained within 30 minutes.

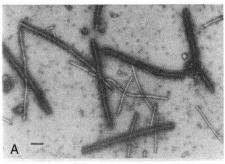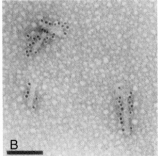

FIGURE 13.1   Immune specific electron microscopy (ISEM). A. A natural mixture of two potyviruses from a perennial cucurbit, *Bryonia cretica*. The antiserum used has decorated particles of only one of the viruses in the mixture. One particle near the centre is longer than normal and decorated for only part of its length. This particle probably arose by end-to-end aggregation between a particle of each of the two viruses. Bar = 100 nm. [From Milne (1991; in *Electron microscopy of plant pathogens*, K. Mendgen and D.E. Lesemann, Eds., pp. 87–102, Springer-Verlag, New York).] B. Gold particle labeling of virus particles. *Rice tungro bacilliform virus* (RTBV) particles were treated first with anti-RTBV protease rabbit antiserum and then with gold-labeled goat antirabbit serum. Bar = 200 nm. [This article was published in *Virology*, **205**, J.M. Hay, F. Grieco, A. Druka, M. Pinner, S.-C. Lee, and R. Hull, Detection of *rice tungro bacilliform virus* gene products *in vivo*, pp. 430–437, Copyright Elsevier (1994).]

9. When decoration is used, unrelated undecorated virus particles on the grid are readily detected (Figure 13.1A).

10. Different proteins on a particle can be decorated (e.g., the two coat proteins on closteroviruses; see Figure 5.1).

11. Prepared grids may be sent to a distant laboratory for application of virus extracts and returned to a base laboratory for further steps and EM examination.

The following are some disadvantages of the procedure:

1. It will not detect virus structures too small to be resolved in the EM (e.g., coat protein monomers).

2. Sometimes the method works inconsistently or not at all for reasons that are not well understood.

3. It involves the use of expensive EM equipment, which requires skilled technical work and is labour-intensive. For these reasons it cannot compete with, say, ELISA for large-scale routine testing.

4. When quantitative results are required, particle counting is laborious, variability of particle numbers per grid square may be high, and control grids are required.

c. Electrophoretic Procedures. Electrophoresis in a suitable substrate separates proteins according to size and net electric charge at the pH used. Polyacrylamide gel electrophoresis in a medium containing sodium dodecyl sulfate (SDS-PAGE) is commonly used. The position and amount of the proteins can then be visualised by a nonspecific procedure such as staining or by a specific procedure such as immunoassay (termed *Western blotting*).

d. Dot Blots. Protein dot blots resemble Western blotting, except that the constituent proteins in the sample are not separated. The samples are spotted onto a solid matrix such as nylon or nitrocellulose and probed with an antiserum to which a reporter group has been attached (see Figure C in Box 13.3). Reporter groups can be either enzymes that give a colour reaction with a specific chemical or a fluorescent compound. Unlike ELISA, the colour product must be insoluble.

Similar to dot blots, there are commercially available kits that use the same technique as pregnancy testing kits, in which the antibody is immobilised on a strip and the sample is diffused along the strip.

E. Methods That Involve Properties of the Viral Nucleic Acid

General properties of a viral nucleic acid, such as whether it is DNA or RNA, double- (ds) or single-stranded (ss), or if it consists of one or more pieces, are fundamental for allocating an unknown virus to a particular family or group. However, with the exception of dsRNA, these properties are usually of little use for routine diagnosis, detection, or assay. The ability to make DNA copies (cDNA) of parts or all of a plant viral RNA genome has opened up many new possibilities. The nucleotide sequence of the DNA copy can be determined, but this is far too time-consuming to be considered as a diagnostic procedure, except in special circumstances.

Basically, the four approaches to using nucleic acids for detection and diagnosis of viruses are the type and molecular sizes of the virion-associated nucleic acids; the cleavage pattern of viral DNA or cDNA; hybridisation between nucleic acids; and the polymerase chain reaction.

1. Type and Size of Nucleic Acid

Double-stranded RNAs are associated with plant RNA viruses in two ways: The plant reoviruses, and cryptoviruses have genomes that consist of dsRNA pieces, and in tissues infected with ssRNA viruses, a ds form of the genome RNA accumulates that is twice the size of the genomic RNA. This is known as the replicative form (Figure 8.1). These dsRNAs have been used for diagnosis following characterisation of the ds forms by PAGE.

2. Cleavage Patterns of DNA

Cleaving cDNAs of RNA genomes and the genomes of DNA viruses at specific sites with restriction enzymes and determining the sizes of the fragments by PAGE are possible procedures for distinguishing viruses in a particular group. For instance, isolates of *Cauliflower mosaic virus* could be distinguished on the restriction endonuclease patterns of the virion DNA.

3. Hybridisation Procedures

These procedures depend on the fact that ss nucleic acid molecules of opposite polarity and with sufficient similarity in their nucleotide sequence will hybridise to form a ds molecule. The theory concerning nucleic acid hybridisation is complex. The Watson and Crick model for the structure of dsDNA showed that the two strands were held together by hydrogen bonds between specific (complementary) bases—namely, adenine and thymidine, cytosine and guanine. This interaction of base-pairing is the basis of all molecular hybridisation. Essentially, there are two stages, disruption of the base-pairing (termed melting or denaturing the nucleic acid) and reinstatement of base-pairing (termed reassociation, renaturation, or hybridisation). Among the factors that affect denaturation are temperature, nucleic acid composition, type and concentration of salt in the buffer, the pH of the buffer, presence of organic solvents, and base-pair mismatch. Factors that affect reassociation include temperature, salt concentration, base-pair mismatch, the length of the nucleic acid fragments, the type of nucleic acid (DNA and/or RNA), the concentration of nucleic acid, and the presence of various anionic polymers.

For diagnosis of virus by hybridisation, the unknown nucleic acid is termed the target and the known nucleic acid is termed the probe. In many of the systems the target is immobilised on a solid matrix and the probe applied in liquid. The probe comprises sequences complementary to the target sequences to which a reporter system, either radioactive or nonradioactive, is attached to reveal when hybridisation has taken place. There are three basic types of nonradioactive reporter: those that directly modify bases in the probe DNA, those that attach precursors (e.g., horseradish peroxidase or alkaline phosphatase) to the probe DNA or RNA, and those that incorporate labeled precursors (e.g., biotinylated nucleotides) into the probe. The detection of the reporter group is usually by an enzyme that gives a coloured product or a luminescent compound on reaction with a substrate. The coloured product has to be insoluble (unlike the coloured product of the ELISA reaction).

Excess probe hybridised to the target immobilized on a solid matrix has to be removed before detection. This is done by washing at selected temperatures and salt concentrations. At this stage relatedness of the probe to the target can be ascertained by using different "stringencies" that relate to the effect of hybridisation and/or wash conditions on the interaction between complementary nucleic acids that may be incompletely matched. The use of different stringencies is one of the more powerful tools of the hybridisation technology. There are various hybridisation formats (Box 13.4).

4. Dot Blots

Dot blots, as just described for proteins, are widely used for detecting nucleic acids using labeled probes (see Box 13.4).

5. Polymerase Chain Reaction

DNA fragments of interest can be enzymatically amplified *in vitro* by the polymerase chain reaction (PCR). The technique involves the hybridisation of synthetic complementary oligonucleotide primers to the target sequence

BOX 13.4

HYBRIDISATION FORMATS

Southern Blotting

See standard molecular biology text books.

Dot Blot Hybridisation

Dot blot hybridisation is now the most commonly used procedure for testing of large numbers of samples. These are the main steps in dot blot hybridisation:

1. A small amount of sap is extracted from the plant under test.
2. The viral nucleic acid is denatured by heating or, if it is DNA, by alkali treatment.
3. A spot of the extract is applied to a membrane.
4. The membrane is baked or exposed to ultraviolet light to bind the nucleic acid firmly to it.
5. Nonspecific binding sites on the membrane are blocked by incubation in a prehybridisation solution containing a protein, usually bovine serum albumin or non-fat dried milk, and small ss fragments of an unrelated DNA, together with salt and other ingredients.
6. Hybridisation of a labeled probe nucleic acid to the test nucleic acid bound to the substrate.
7. Washing off excess (unhybridised) probe and estimation of the amount of probe bound by a method appropriate to the kind of label used for the probe. The prehybridisation step (about two hours) and the hybridisation step (overnight) are carried out in heat-sealable plastic bags in a water bath at about 65°C.

The technique is now widely used in plant virology. For instance, a dot blot technique has been successfully applied to screening large numbers of potato plants in a programme of breeding potatoes for resistance to several viruses.

A nonradioactive dot blot system using minimal equipment was developed for the routine diagnosis of a range of insect-transmitted viruses. The method has been used to assess relationships between tombusviruses, but some unexpected cross-hybridisations were observed. A sensitive nonradioactive procedure has been developed for detecting *Bamboo mosaic virus* and its associated satellite RNA in meristem-tip cultured plants.

Tissue Print Hybridisation

This is similar to immuno-tissue printing as described in Box 13.3 but using labeled nucleic acid probes. Probes can be specifically designed for detecting (+) or (–) strands or specific parts of the genome. Tissue print hybridisation is of especial use for viroids that do not have proteins that can be detected immunologically. Viroids have been detected by molecular hybridisation of imprinted membranes. An adaptation of the procedure, termed *squash-blot* (or *swat blot*), has been designed to assay *Maize streak virus* in single leafhopper vectors squashed directly onto the nitrocellulose filter.

In Situ Hybridisation

In situ hybridisation can give information of the distribution of the target nucleic acid within a cell. The loci at which *Banana streak virus* (BSV) is integrated into the banana chromosomes were identified by *in situ* hybridisation. As an extension of this procedure, fibre stretch hybridisation, in which denatured chromosomal DNA is stretched out on slides before hybridisation, gives information on the detailed structure of the integrated locus; this is described for BSV in Box 8.9.

and synthesis of multiple copies of complementary DNA of the sequence between the primers using heat-stable DNA polymerase. The process goes through a series of amplification cycles, each consisting of melting the ds template DNA molecules in the presence of the oligonucleotide primers and the four deoxyribonucleotide triphosphates at high temperature (melting), hybridisation of the primers with the complementary sequences in the template DNAs at lower temperature (annealing), and extension of the primers with DNA polymerase (DNA synthesis). During each cycle, the sequence between the primers is doubled so that after n cycles, a 2^n amplification should be obtained. Usually the reaction is of 30–50 cycles.

As PCR is based on DNA, it is not directly applicable to most plant viruses that have RNA genomes. However, a cDNA can be made to the desired region of the RNA genome using a primer and reverse transcriptase and this used as the initial template. This procedure, now widely used, is termed *RT-PCR*. A further refinement is to couple PCR with the capture of the virus particles by immobilised antibodies, termed immune capture PCR (*IC-PCR* or *IC-RT-PCR*).

PCR (and RT-PCR) has proved to be a very powerful tool for virus detection and diagnosis. It can be used to directly produce a DNA product of predicted size that can be confirmed by gel electrophoresis. The choice of primers can be used to distinguish between strains of a virus or, with primers containing a variety of nucleotides at specific positions (degenerate primers), be used for more generic determinations. Strains can also be distinguished by amplifying a region that has differences in restriction endonuclease sites. PCR is widely used in producing probes for hybridisation by incorporating reporter nucleotides in the reaction. To illustrate the widespread applicability to the detection and diagnosis of plant viruses, we will look at some examples; these are by

no means a comprehensive collection of the uses to which this technique can be applied.

As described in Box 8.9, *Banana streak virus* (BSV) sequences are frequently found integrated in the host genome. This precludes the use of straightforward PCR for diagnosis of episomal infections of this virus, but IC-PCR can be used in which the episomal virus particles are captured and the host chromosomal DNA (containing the integrant) is removed before PCR. Using an IC-RT-PCR method in a single closed tube, *Tomato spotted wilt virus* can be detected in leaf and root tissue, and *Citrus tristeza virus* and *Plum pox virus* can be detected in plant tissue and single aphid samples.

In an RFLP analysis of 10 *Potato virus Y* isolates representative of four symptomatic groups, the whole genomes were each amplified in two fragments by RT-PCR. Using seven restriction enzymes, three RFLP groups were determined in the 5′ fragment and two in the 3′ fragment that correlated with the biological characters. One group of isolates appeared to have resulted from a recombination event.

Degenerate primers to highly conserved regions have been used to detect whitefly-transmitted geminiviruses. Strains of the leafhopper-transmitted geminivirus *Wheat dwarf virus* can be differentiated by using universal and strain-specific promoters. A highly sensitive procedure for the early detection of *Beet necrotic yellow vein virus* (BNYVV) in plant, soil, and vector samples involves PCR and digoxigenin labeling. PCR is being widely used in the detection and identification of phytoplasmas in both plant and insect samples.

6. DNA Microarray

The principle of microarrays or DNA chips is the hybridisation of fluorescently labeled target sequences to probe sequences spotted onto a solid surface, usually a glass microscope slide. Total RNA from the infected plant is converted to DNA by RT-PCR, and the cDNA is labeled by reaction with a fluorescent dye. Probes to many different viruses or variants can be spotted onto the glass slide, and hybridisation can reveal joint infections with more than one virus.

F. Decision Making on Diagnosis

The application of biological, physical, and molecular techniques has given a large "tool-bag" for detection and diagnosis of plant viruses. This emphasizes the point made in the introduction to this chapter that it is important to identify the question to be addressed. If one wishes to determine if a plant is virus infected—say, for quarantine purposes—one does not necessarily need a sophisticated technique that identifies a virus strain. On the other hand, if one is studying the durability of a potential resistance gene (or transgene), it is very useful to have an understanding of the range of variation of the virus. Thus, one has to select the best technique for what is wanted. In making this decision various points have to be taken into account, including the following.

The sensitivity required. There is much discussion about the relative sensitivity of detection procedures. However, the sensitivity of many of the serological and nucleic acid–based tests is adequate for most purposes, so one should use the system that is most convenient. For ease and speed of operation, and low cost, dot blots based on either an immunological test or nucleic acid hybridisation have a lot to offer. There are field kits available for testing for, say, potato viruses that are based on the same technology as home pregnancy kits. However, for each virus situation it is advisable to compare tests to see which is the most appropriate.

The number of samples. Where large numbers of samples have to be handled, the following factors will be important in choosing a test procedure: specificity; sensitivity; ease and speed of operation; and cost of equipment, consumable supplies, and labour.

The material being sampled. In many cases, and especially in trees, the distribution of virus is not uniform. Thus, one has to be careful with the taking of samples, and it is advisable to take several samples from each tree.

The reliability of the technique. A technique can be unreliable in two ways. False negatives can result from sampling part of the plant not containing virus, inhibition of the reaction by a plant constituent, or limitation in the materials being used (e.g., mismatch in a primer for PCR). False positives can be due to plant constituents, especially when testing new plant species.

The equipment and expertise available. The routine reliable detection of most viruses usually does not require expensive equipment. There are some exceptions, such as the detection of BSV described earlier in this chapter, which requires either ISEM or IC-PCR. However, it is important to have a good, reliable supply of consumables and means for storing them without deterioration. Similarly, most of the basic techniques are relatively easy to learn, but it is important that they are learned properly so potential sources of error can be recognised.

There are international guidelines (FAO/IPGRI *Technical Guidelines for the Safe Movement of Germplasm*) that have been drawn up by expert panels to assist with international germplasm movement. These detail the current ideas on the safest and simplest tests for ensuring that plant propagules do not contain the viruses that are known to infect that species. It should always be remembered that one cannot, of course, test for unknown viruses.

II. EPIDEMIOLOGY AND ECOLOGY

To survive, a plant virus must have (1) one or more host plant species in which it can multiply, (2) an effective means of spreading to and infecting fresh individual host plants, and (3) a supply of suitable healthy host plants to which it can spread. The actual situation that exists for any given virus in a particular locality, or on the global scale, will be the result of complex

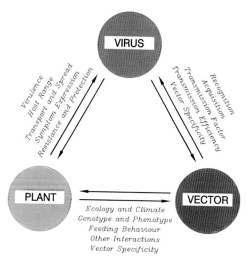

FIGURE 13.2 Some of the interactions involved in the natural infection of a plant by a virus. [This article was published in *Semin. Vinol.* **2**, The movement of viruses within plants R. Hull, pp. 89–95, Copyright Elsevier (1997).]

BOX 13.5

ECOLOGY AND EPIDEMIOLOGY

Ecology describes the factors influencing the behaviour of a virus in a given physical situation. These factors include host range, tissue tropism, pathogenesis, and host responses. It is a fundamental concept based on the relational properties of the virus and is not a property of the environment.

Epidemiology is the study of the determinants, dynamics, and distribution of viral diseases in host populations. It includes a dimensional aspect of the factors determining the spread of a virus into a given situation. The two terms often interlink, but each has a specific meaning.

interactions between many physical and biological factors, the major ones of which are shown in Figure 13.2.

An understanding of the ecology and epidemiology of a virus in a particular crop and locality is essential for the development of appropriate methods for the control of the disease it causes. As with most other obligate parasites, the dominant ecological factors to be considered are usually the way viruses spread from plant to plant and the ways that other factors influence such spread. It is important to distinguish between the terms *epidemiology* and *ecology,* which in many papers have been used interchangeably (Box 13.5). In this chapter, we consider them both together.

A. Epidemiology of Viruses in Agriculture

Taking plant viruses as a whole, the flying, sap-sucking groups of insect vectors, particularly the aphids, are by far the most important agents of spread. There are two stages in the infection of a crop: primary infection and

subsequent secondary spread. These are affected by biological factors such as host range, susceptibility of the crop, and behaviour of the vector, by cultural factors such as time of planting and size of field and by physical factors such as weather and high level air currents. Although these factors all interlink and often present a complex situation, there are several basic points that can be made.

1. Primary Infections

The source of primary infections varies depending on whether the crop is grown on a seasonal basis or whether there is overlapping cropping. If grown on a seasonal basis, the virus must have an overwintering (or oversummering) host. This can be a wild species, a vegetative organ such as a potato or a seed for a seed-transmitted virus, or a vector in which the virus persists (e.g., a nematode).

The initial spread into the crop by an insect vector depends on the virus-vector relationship and on the feeding behaviour of the insect on that crop. As noted in Chapter 12, crop plants are often infected with nonpersistent viruses by aphids that do not colonise the plant. This is because the aphids feed for very short times when seeking a host but for sufficiently long to infect the plant. When an aphid finds a suitable plant, it settles for a longer feed and thus is more likely to transmit a persistent virus, such as a luteovirus. Aphid behaviour is also affected by the growth status of the plant, as they seek the contrasting colours between the plant and the background soil. Thus, when crop leaves have met within and between rows, aphids tend to land initially on plants at the edge of the crop, thus giving the "edge effect" of primary infections. Frequently, primary infection of insect-borne viruses usually shows as scattered patches over the field (Figure 13.3).

Primary infections with nematode- and fungal-transmitted viruses reflect the distribution of the vector in the soil. Because of ploughing and other cultural practice, the infection patches are often elongate (Figure 13.4).

If there are no overwintering hosts in that region, viruses can be brought in by insect vectors carried on high, level air currents from distant regions. Under appropriate climatic conditions, a continuous long-distance journey

FIGURE 13.3   Aerial view of a sugar beet field with patches of plants infected by yellowing viruses and gradients of infection spreading from the patches. (Courtesy of IACR Broom's Barn Research Centre.)

may not be uncommon. Geostrophic airstreams are air movements at altitudes of about 1,000 m or more moving along the isobars of a relatively large-scale weather system. Such airstreams have almost certainly led to the mass transport of winged aphids from Australia to New Zealand, a distance over sea of about 2,000 km. *Lettuce necrotic yellows virus* and several vector species of aphids have probably been introduced in this way. Not only the viruses but also potential aphid vectors are still moving into new continents. Thus, *Metopalophium dirhodum*, a vector *for Barley yellow dwarf virus* (BYDV), was first recorded in Australia in 1985 and was almost certainly a fairly recent arrival. Leafhopper vectors may also travel long distances. For example, large numbers of *Macrosteles fascifrons* may be blown each spring from an overwintering region about 300 km north of the Gulf of Mexico through the midwestern United States and into the prairie provinces of Canada (Figure 13.5). BYDV and cereal aphids follow similar routes.

Commonly, in tropical countries there is overlap between the old maturing crop and the newly planted one; the old crop is a source of primary infection of the new crop. This is thought to be the cause of the great increase of outbreaks of rice tungro disease in Southeast Asia associated with the "green revolution." An analogous situation is where there is a "rolling front" of crop planting over a large region associated with movement of seasonal rainfall. The spread of groundnut rosette disease in sub-Saharan Africa is thought to be associated with the moving seasonal rainfall in the intertropic convergence meteorological zone.

The way a particular crop is grown and cultivated in a particular locality and the ways land is used through the year in the area may have a marked effect on the incidence of a virus disease in the crop. Many diverse situations arise, some examples of which will illustrate the kinds of factors involved.

FIGURE 13.4 Relationship between density of nematode infestation and outbreak of virus disease. Left: Population contours of *Xiphinema diversicaudatum*. Right: Outbreak of *Strawberry latent ring spot virus* in a 10-year-old raspberry plantation. [From Taylor and Thomas (1968; *Ann. Appl. Biol.* **62**, 147–157, Wiley-Blackwell).]

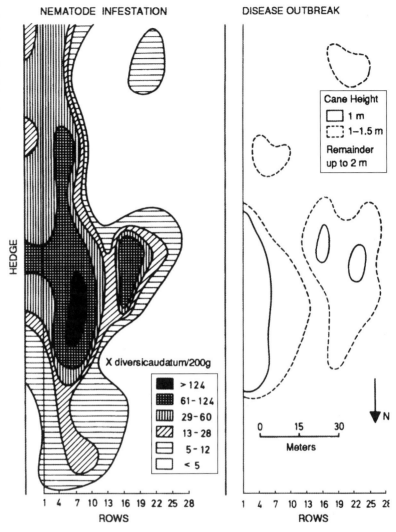

Planting date. A clear-cut example of the way time of sowing seed affects virus incidence occurs with the winter wheat crop in southern Alberta. In a normal season, the percentage infection with streak mosaic is markedly dependent on the time at which the seed is sown. For sowing dates earlier than September, the spring-sown crop, carrying disease, overlaps with the autumn-sown crop. High air temperatures may cause a dramatic reduction in some aphid vector populations. If the planting date for the autumn crop is delayed until such conditions prevail, much less virus spread may take place in a crop.

Crop rotation. The kind of crop rotation practiced may have a marked effect on the incidence of viruses that can survive the winter in weeds or volunteer plants. With certain crops, volunteer plants that can carry viruses may survive in high numbers for considerable periods.

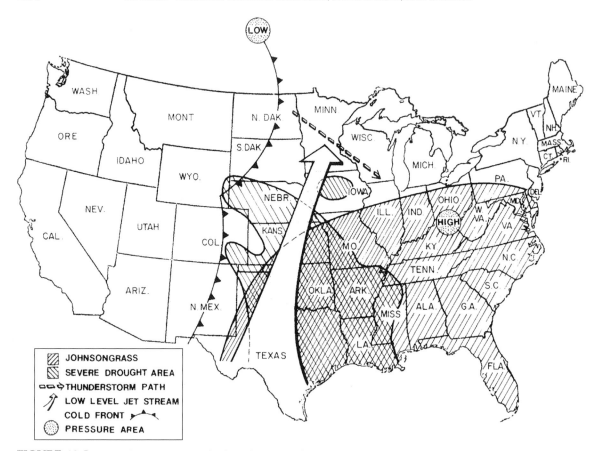

FIGURE 13.5 Hypothesis to account for long-distance aphid movement and a severe outbreak of *Maize dwarf mosaic virus* (MDMV) in corn in Minnesota in 1977. The midcontinental and southern distributions of Johnsongrass, the major wild host of MDMV, are shown relative to a major drought in May and June 1977. The large arrow indicates the path of low-level jet winds (up to 80 km/hr) on 2 July 1977. The smaller dashed arrow indicates the path of a cold front that caused the low-level jet wind to become diffuse and triggered thunderstorm activity moving through Minnesota and into Wisconsin. [From Zeyen *et al.* (1987; *Ann. Appl. Biol.* **111**, 325–336, Wiley-Blackwell).]

Soil cultivation. Soil cultivation practices may affect the spread and survival of viruses in the soil or in plant remains. As just noted, nematode and fungal vectors may be spread by movement of soil during cultivation. The physical movements of soil tillage and harvesting operations are the major means of spread of BNYVV.

Nurseries and production fields as sources of infection. Nurseries, especially where they have

been used for some years, may act for themselves as important sources of virus infection.

2. Secondary Spread

Many of the factors associated with primary infection control the secondary spread of viruses. Thus, nonpersistent viruses are mainly spread by the aphids that are not seen on the crop—those passing through on host-seeking flights. Much of the spread of persistent viruses is by aphids colonising the crop, even walking from plant to plant.

The cultivation of a single crop, or at least a very dominant crop, over a wide area continuously for many years may lead to major epidemics of virus disease, especially if an airborne vector is involved. Soil-borne vectors can also be important from this point of view—for example, with *Grape vine fan leaf virus* in vineyards, where the vines are cropped for many years. Monocropping may also lead to a buildup of crop debris and the proliferation of weeds that become associated with the particular crop.

Weather conditions such as rainfall, temperature, and wind can play a major role in the secondary development of a virus disease in the field. This is mostly by affecting the population size and behaviour of the vector but also can impact upon the crop development and its susceptibility to virus infection. This can lead to wide annual variation in the incidence of a virus disease in an annual crop, as shown in Figure 13.6. Overall, the factors involved in both primary infection and secondary spread are extremely complex and can be illustrated by the epidemiology of luteoviruses (Box 13.6).

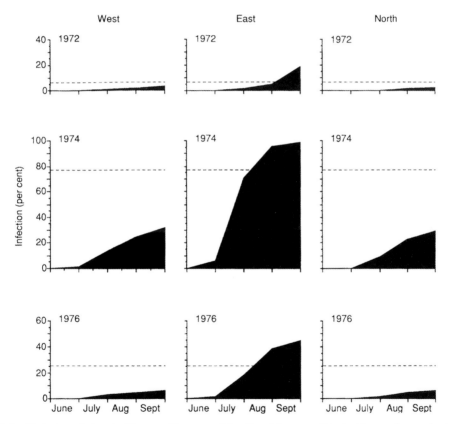

FIGURE 13.6 Variation of virus incidence with year and location. Mean monthly incidence of sugar beet yellows disease in different regions of England in years with greatly differing amounts of spread. [Data supplied by G.D. Heathcote (Broom's Barn Experimental Station) for representative crops sown annually in March or April in eastern England, in wetter areas in western England, and in cooler areas in northern England.] The dashed lines indicate the mean incidence of yellows at the end of September as calculated for the entire English crop. [This article was published in *Ann. Appl. Biol.* **111**, R.J. Zeyen, E.L. Stromberg, and E.L. Kuehnast, Long-range aphid transport hypothesis for *maize dwarf mosaic virus*: History and distribution in Minnesota, pp. 325–336. Copyright Elsevier (1987).]

BOX 13.6

COMPLEXITIES OF EPIDEMIOLOGY

The factors and interactions involved in the ecology and epidemiology of viruses in crops are extremely complex. This can be illustrated by the processes involved in the spread of luteoviruses within and between crops (Fig.). The luteoviruses are among the most studied of virus groups as far as epidemiology is concerned but the complexities shown in the Figure are just as applicable to other crop situations. There are virtually no data on the epidemiology of viruses in noncrop natural systems.

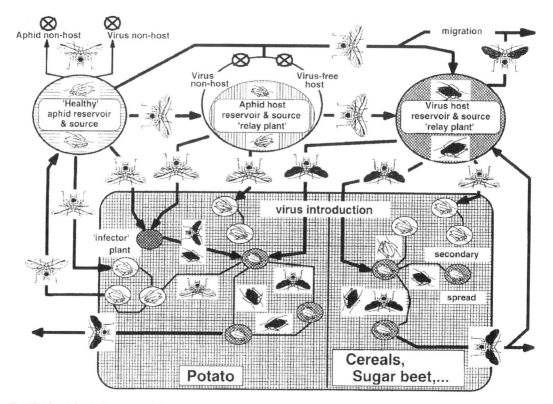

Fig. Epidemiological process of luteovirus spread showing the various routes of infection and the relevant differences between potato and other crops. [From Robert (1999; in *The Luteoviridae*, H.G. Smith and H. Barker, Eds., pp. 221–231. CAB International, Wallingford, UK).]

B. Plant Viruses in the Natural Environment

There is not much information on the impact that virus infections might have on plants in the natural ecosystem. As noted in Chapter 4, a virus that kills its host plant with a rapidly developing systemic disease is much less likely to survive than one that causes only a mild or moderate disease that allows the host plant to survive and reproduce effectively. There is probably a natural selection in the field against strains that cause rapid death of the host plant, and thus viruses in the wild plants are unlikely to show many overt symptoms (see Table 4.1). Even so, there is some evidence that virus infection can reduce the competitive and reproductive ability of a plant species, thus impacting upon the community structure. For instance, in California, BYDV infection reduced native grass growth, survival, and fecundity, thus affecting its fitness in competition with introduced grasses.

C. Emergence of New Viruses

As well as distributing viruses around the world, humans have moved crop species to new countries, often with disastrous consequences as far as virus infection is concerned. Plant species that were relatively virus-free in their native land may become infected with viruses that have long been present in the countries to which they were moved for commercial purposes. Thus, cacao was transferred from the Amazonian jungle to West Africa late in the late 1800s, and since then major commercial production has developed there. The swollen shoot disease caused by *Cacao swollen shoot virus*, first reported in cacao in 1936, was very probably transmitted from natural West African tree hosts of the virus by the mealybug vectors that are indigenous to the region.

Another example of international trade leading to the spread of an important disease is exemplified by the spread of BNYVV, which causes the important rhizomania disease in sugar beet (Figure 13.7).

Important new diseases are also arising from interactions between the genomes of two viruses by either recombination or by synergism induced by suppressors of gene silencing (See Box 11.3).

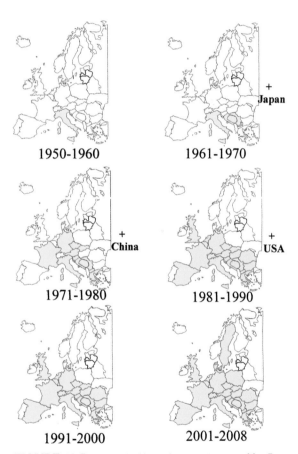

FIGURE 13.7 Spread of beet rhizomania caused by *Beet necrotic yellow vein virus* through Europe and worldwide. [Data courtesy of Dr. M. Stevens (Broom's Barn Research Centre).]

III. VIRUSES OF OTHER KINGDOMS

Diagnostics used for detecting viruses of vertebrates and invertebrates are very similar to those described previously for plant viruses. The epidemiology of vertebrate and invertebrate viruses differs significantly from that of plant viruses. The basic difference is that vertebrates and invertebrates can move around and plants (usually) cannot. Therefore, much of the natural spread of plant viruses is by invertebrate vectors; these are involved in some animal disease conditions, such as bluetongue virus disease of cattle and sheep that is spread by midges. In the natural situation, it is the movement of infected vertebrates into the close proximity to healthy ones that is an important factor in the development of epidemics. Humans play a major role in the spread of crop and farm animal viruses and in the worldwide distribution of viruses.

IV. SUMMARY

- Four major properties of viruses are used in diagnostics, biological properties, physical properties, protein properties, and nucleic acid properties. Most modern diagnostics are based on the latter two properties.
- It is important to identify the reason for the diagnosis before selecting the technique(s) to be used.
- Diagnostic techniques using the virus protein properties focus mainly on serological techniques targeted at the virus coat protein.

- Diagnostics using the viral nucleic acid properties focus mainly on hybridisation techniques and the polymerase chain reaction.
- Invertebrate vectors are among the most important factors in the epidemiology of plant viruses. They introduce primary infections (as does seed transmission) and disseminate the virus in secondary spread.
- Other factors involved in epidemiology include agronomic practices and weather conditions.

Further Reading

Agindotan, B. and Perry, K.L. (2007). Microarray detection of plant RNA viruses using randomly primed and amplified complementary DNAs from infected plants. *Phytopathology* **97**, 119–127.

Cooper, I. and Jones, R.A.C. (2006). Wild plants and viruses: Under-investigated ecosystems. *Adv. Virus Res.* **67**, 1–47.

Fargette, D., Konaté, G., Fauquet, C., Muller, E., Peterschmitt, M., and Thresh, J.M. (2006). Molecular ecology and emergence of tropical plant viruses. *Annu. Rev. Phytopathol.* **44**, 235–260.

Foster, G.D., Johansen, I.E., Hong, Y., and Nagy, P.D. (2008). *Plant Virology Protocols: From viral sequence to protein function*, 2nd ed. Humana Press, Totowa, N.J.

Hull, R. (2002). *Matthews' plant virology*. Academic Press, San Diego.

Jeger, M.J., Seal, S.E., and Van den Bosch, F. (2006). Evolutionary epidemiology of plant virus disease. *Adv. Virus Res.* **67**, 163–203.

Koenig, R., Lesemann, D.-E., Adam, G., and Winter, S. (2008). [Diagnostic techniques] for plant viruses. *Encyclopedia of Virology*, Vol. 2, 18–28.

Reynolds, D.R., Chapman, J.W., and Harrington, R. (2006). The migration of insect vectors of plant and animal viruses. *Adv. Virus Res.* **67**, 453–517.

Vainiopää, R. and Leinikki, P. (2008). [Diagnostic techniques]: Molecular and serological approaches. *Encyclopedia of Virology*, Vol. 2, 29–36.

14

Conventional Control

A range of control measures are used to mitigate the considerable losses that plant viruses cause to crops.

I. INTRODUCTION

The use of fungicidal chemicals to protect crop plants from infection or minimise invasion is an important method for the control of many fungal diseases. No such direct method for the control of virus diseases is yet available. Most of the procedures that can be used effectively involve measures designed to reduce sources of infection inside and outside the crop, to limit spread by vectors, and to minimise the effect of infection on yield. Generally speaking, such measures offer no permanent solution to a virus disease problem in a particular area. Control of virus disease is usually a running battle in which organisation of control procedures, care by individual growers, and cooperation among them is necessary year after year. The few exceptions are where a source of resistance to a particular virus has been found in, or successfully incorporated into, an agriculturally useful cultivar. This is becoming of increasing importance with the development of transgenic protection of plants against viruses that are discussed in Chapter 15. Even with conventional and transgenic resistance, protection may not be permanent when new strains of the virus arise that can cause disease in a previously resistant cultivar.

Correct identification of the virus or viruses infecting a particular crop is essential for effective control measures to be applied. Of major importance in designing a strategy for controlling

a virus in a specific crop is an understanding of the epidemiology of that virus that was discussed in Chapter 13. This enables disease outbreaks to be forecasted (Box 14.1).

The three major approaches to conventional control of plant viruses are the removal or avoidance of sources of infection, protecting plants from systemic infection, and deployment of resistance.

BOX 14.1

DISEASE FORECASTING

An understanding of the epidemiology and ecology of some major crop virus diseases has led to procedures for forecasting potential epidemics. This is very useful in implementing control measures. There are two main approaches to forecasting: monitoring the progress of a disease and developing mathematical models.

Monitoring Virus Disease Progress

Many large-scale farmers routinely monitor their crops and apply control measures at an appropriate time. However, as virus diseases take several days or even weeks to show symptoms after infection, the application of control measures based on symptom appearance can be too late. It also depends on the correct diagnosis of the disease and knowledge of how it is spread.

Mathematical Modelling

There are an increasing number of mathematical models directed at forecasting the outcome of the spread of a disease into an agronomic situation. Basically, the two types of models are prediction models, to predict a possible epidemic, and simulation models, to understand the factors that give rise to and control a given

situation. A model is developed to answer specific questions, and there is no general model to predict the potential and outcome of all potential viral epidemics. In developing a model, as many factors as can be predicted are taken into account. These include, knowledge on the virus, its vector, virus-vector interaction, type of crop, the cropping system, and various environmental factors that can impact on these biological factors. A good model enables one to make strategic management decisions on whether the problem is going to be significant and, if so, when and how to deal with it.

The efficiency of prediction from even a good model is only limited by the amount and reliability of the data fed into it. The data must be obtained from various sources and collated. An example is a model for predicting virus yellows disease of sugar beet in the United Kingdom that is based on the preceding winter weather (especially the number of frost days) and the dates when the aphid vectors begin their spring migration and region in which the beet is being grown (eastern, western, and northern regions). This model has been refined to allow for the numbers of migrating *Myzus persicae*, the major vector.

II. AVOIDING INFECTION

A. Removal of Sources of Infection

It is obvious that there will not be a virus problem if the crop is free of virus when planted and when there is no source of infection near enough to allow it to spread into the crop. Sources of infection are discussed in Chapter 13.

To eliminate these sources, it may be worthwhile to remove infected plants (rogue) from a crop. If the spread is occurring rapidly from sources outside the crop, roguing the crop will have no beneficial effect. In certain situations, roguing may increase disease incidence by disturbing vectors on infected plants. In many crops, newly infected plants may be acting as sources of virus for further vector infection before they show visible signs of disease.

Most of the successful eradication schemes have been on tree crops. The following are among the factors that dictate success:

- Relatively small numbers of infected trees and infection foci
- Low rate of natural spread
- Good data on extent and distribution of infection
- Rapid, reliable, and inexpensive diagnostic procedure for the virus and resources for rapid and extensive surveys and tree removal.

Two examples of roguing and eradication schemes are given in Box 14.2.

B. Virus-Free Seed

Where a virus is transmitted through the seed, such transmission may be an important source of infection, since it introduces the virus into the crop at a very early stage, allowing infection to be spread to other plants while they are still young. In addition, seed transmission introduces scattered foci of infection throughout the crop. Where seed infection is the main or only source of virus, and where the crop can be grown in reasonable isolation from outside sources of infection, virus-free seed may provide a very effective means for control of a disease.

Lettuce mosaic virus is a good example of controlling a virus problem through clean seed. Crops grown from virus-free seed in California had a much lower percentage of mosaic at harvest than adjacent plots grown from standard commercial seed. To obtain effective control by the use of virus-free or low-virus seed, a certification scheme is necessary, with seed plants being grown in appropriate isolation.

C. Virus-Free Vegetative Stocks

For many vegetatively propagated plants, the main source of virus is chronic infection in the plant itself. With such crops, one of the most successful forms of control has involved the development of virus-free clones—that is, clones free of the particular virus under consideration. Two problems are involved. First, a virus-free line of the desired variety with good horticultural characteristics must be found. When the variety is 100 percent infected, attempts must be made to free a plant or part of a plant from the virus. Second, having obtained a virus-free clone, a foundation stock or "mother" line must be maintained virus free, while other material is grown up on a sufficiently large scale under conditions where reinfection with the virus is minimal or does not take place. These stocks are checked that they are "virus free" (e.g., below a set level of detected virus) and are then used for commercial planting.

As a plant is usually infected with a virus for life, various techniques have to be used to free them of virus, including heat therapy in which the plant is kept at a temperature usually in the range of 35 to 40°C for periods of weeks, meristem tip culture, taking advantage of the fact

BOX 14.2

ROGUING AND ERADICATION CONTROL OF PLANT VIRUSES

Banana Bunchy Top Virus (BBTV)

One of the most successful examples of disease control by roguing of infected crop plants has been the reduction in incidence of BBTV in bananas in eastern Australia. Legislation to enforce destruction of diseased plants and abandoned plantations was enacted in the late 1920s. Within about 10 years, the campaign was effective to the point where bunchy top disease was no longer a limiting factor in production. The success of the scheme was attributed to the absence of virus reservoirs other than bananas, together with a small number of wild bananas; knowledge that the primary source of the virus was planting material and that spread was by aphids; cultivation of the crop in small, discrete plantations rather than as a scattered subsistence crop; strict enforcement of strong government legislation; and the cooperation of most farmers.

Cocoa Swollen Shoot Virus (CSSV)

Cocoa swollen shoot disease (CSSD) is caused by CSSV. The disease was discovered in Ghana in 1936 and is one of the most devastating scourges of cocoa, in the 1940s threatening to wipe out the

cocoa industry in what is now Ghana. A massive nationwide eradication campaign began in 1946 after it had been shown that the swollen shoot and dieback disease was caused by a virus that is spread from tree to tree by several species of mealybugs (*Pseudococcidae*). The eradication campaign has continued to the present time, but there have been serious interruptions and discontinuities. Thus, financial resources and personnel who could have otherwise been used to improve the standard of husbandry, raise cocoa production, and make improvements of other crops in the agricultural sector have been diverted into eradicating the disease by cutting out diseased and at times neighbouring "contact" trees. By the 1980s, more than 190 million trees had been removed, but despite massive expenditure, swollen shoot was more prevalent in Ghana than ever before. Among the main problems were the financial and logistic problems in mounting and sustaining such a large and complex eradication programme, lack of cooperation from farmers who were reluctant to lose several years' production, and lack of detailed epidemiological information. Since then the main approach to control is by trying to find and deploy resistance to the virus.

that most viruses do not invade the plant meristem (see Chapter 9), and chemotherapy, by treating select plants with antiviral compounds such as an analogue of guanosine (ribavirin, also called virazole) in combination with *in vitro* tissue culture. Such techniques are only used on elite material of high-cost crops such as soft fruit and flowers. It is very important to include long-term virus testing into the

programme for producing virus-free mother plants and also to maintain the nuclear stock in a virus-free environment.

D. Modified Agronomic Practices

Virus infection can be reduced by modifying agronomic practices such as breaking the infection where one major susceptible annual crop or group of related crops is grown in an area and where these are the main hosts for a virus in that area by ensuring that there is a period when none of the crop is grown. A good example of this is the control of planting date of the winter wheat crops in Alberta to avoid overlap with the previous spring- or winter-sown crop (Figure 14.1). This procedure, together with elimination of volunteer wheat and barley plants and grass hosts of *Wheat streak mosaic virus* before the new winter crop emerges, can give good control in most seasons.

Other approaches include changing planting dates to avoid young plants being exposed to major migrations of the insect vector and using close-spaced planting to reduce the attractiveness to flying aphids.

E. Quarantine Regulations

Most agriculturally advanced countries have regulations controlling the entry of plant material to prevent the introduction of diseases and pests not already present. Many countries now have regulations aimed at excluding specific viruses and their vectors, sometimes from specific countries or areas. The setting up of quarantine regulations and providing effective means for administering them is a complex problem. Economic and political factors frequently have to be considered. Quarantine measures may be well worthwhile with certain viruses, such as those transmitted through seed, or in dormant vegetative parts such as fruit trees and bud wood.

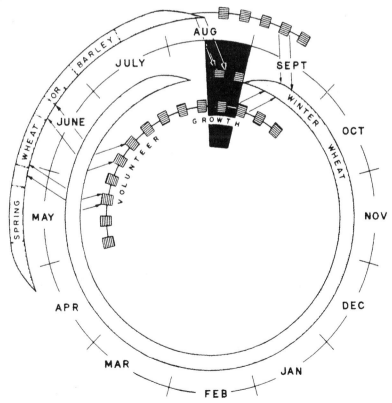

FIGURE 14.1 Wheat streak mosaic disease cycle. Preventing the infection of winter wheat in the autumn is the key to controlling this disease in southern Alberta. Dark area period during which effective control can normally be achieved; broken hatched bands, problems presented by volunteer seedlings, early-seeded winter wheat, and/or late-maturing spring wheat or barley; arrows, transfer of virus by windblown mites. (Diagram courtesy of T.G. Atkinson.)

The value of quarantine regulations will depend to a significant degree on the previous history of plant movements in a region. For example, active exchanges of ornamental plants between the countries of Europe have been going on for a long period, leading to an already fairly uniform geographical distribution of viruses infecting this type of plant. On the other hand, the European Plant Protection Organisation found it worthwhile to set up quarantine regulations against fruit tree viruses not already recorded in Europe.

In spite of many countries having regulations designed to prevent the entry of damaging viruses, they can spread internationally very rapidly. A good example is the rhizomania disease of sugar beet, shown in Figure 13.7.

III. STOPPING THE VECTOR

As described in Chapter 12, plant viruses are usually transmitted by arthropod vectors, but some are transmitted by fungal vectors, and others, particularly *Tobacco mosaic virus* (TMV)

and *Tomato mosaic virus* (ToMV), may be transmitted mechanically (by "human vectors"). Once TMV or ToMV enters a crop like tobacco or tomato, it is very difficult to prevent its spread during cultivation and particularly during such processes as tying-up of plants. Control measures consist of treatment of implements and washing of the hands. Workers' clothing may become heavily contaminated with TMV and thus spread the virus by contact.

A. Air-Borne Vectors

Before control of virus spread by air-borne vectors can be attempted, it is necessary to identify the vector. This information has sometimes been difficult to obtain. Not uncommonly, it is an occasional visitor rather than a regular coloniser that is the main or even the only vector of a virus. Furthermore, some aphid species are more efficient vectors than others. For instance, the brown citrus aphid (*Toxoptera citricida*) is a much more efficient vector of *Citrus tristeza virus* (CTV) than is the melon aphid (*Aphis gossypii*; see Figure 14.2).

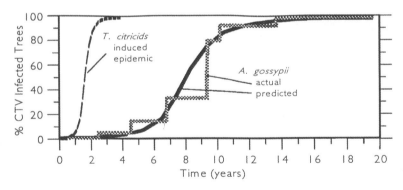

FIGURE 14.2 Comparative increase of *Citrus tristeza virus* infection in field situations when vectored by the brown citrus aphid (*Toxoptera citricida*), an efficient vector, and the melon aphid (*Aphis gossypii*), a less-efficient vector. Data for the brown citrus aphid were taken from test plots in Costa Rica and the Dominican Republic. Data for the melon aphid were taken from surveys and experimental plots in Spain, Florida, and California. Initial infection levels were less than 1%. Note "stairstep" progression in infection with the melon aphid, which is believed to correspond to periodic heavy aphid migrations. [From Garnsey *et al.* (1998; in *Plant virus disease control*, A. Hadidi, R.K. Khetarpal, and H. Koganezawa, Eds., pp. 639–658, APS Press, St. Paul, MN).]

1. Insecticides

The application of insecticides is currently one of the main ways of controlling insect pests of plants. To prevent an insect from causing direct damage to a crop, it is necessary only to reduce the population below a damaging level. Control of insect vectors to prevent infection by viruses is a much more difficult problem, as relatively few winged individuals may cause substantial spread of virus. Contact insecticides would be expected to be of little use unless they were applied very frequently. Persistent insecticides, especially those that move systemically through the plant, offer more hope for virus control. Viruses are often brought into crops by winged aphids, and these may infect a plant during their first feeding, before any insecticide can kill them. When the virus is nonpersistent, the incoming aphid, when feeding rapidly, loses infectivity anyway, so killing it with insecticide will not make much difference to infection of the crop from the outside. On the other hand, an aphid bringing in a persistent virus is normally able to infect many plants, so killing it on the first plant will reduce spread.

As far as subsequent spread within the crop is concerned, similar factors should operate. Spread of a virus that is nonpersistent should not be reduced as much by insecticide treatment as a persistent virus where the insect requires a fairly long feed on an infected plant. Thus, spread of the persistent *Potato leaf roll virus* (PLRV) in potato crops was substantially reduced by appropriate application of insecticides, but spread of the nonpersistent *Potato virus Y* (PVY) was not.

As discussed in Box 14.1, disease forecasting data can be an important factor in the economic use of insecticides. Sometimes a long-term programme of insecticide use aimed primarily at one group of viruses will help in the control of another virus. Thus, the well-timed use of insecticides in beet crops in England, aimed mainly at reducing or delaying the incidence of yellows diseases (*Beet yellows virus* and *Beet mild yellows virus*), has also been a major factor in the decline in the importance of *Beet mosaic virus* in this crop. A warning scheme to spray against the vectors of beet yellows viruses was initiated in the United Kingdom in 1959 and is based on monitoring populations of aphids in crops from May until early July.

As well as the problems just described, there may be other adverse biological and economic consequences related to the use of insecticides, including development of resistance by the target insect to the insecticide, resurgence of the pest once the insecticide activity has worn off, and possible effects on humans and other animals in the food chain.

2. Insect Deterrents

The application of various chemicals or materials can deter aphids from landing on or feeding on crop plants. Spraying mineral oils on plants affects the feeding behaviour of aphids and leafhoppers and can give some protection against nonpersistent viruses. Derivatives prepared from the pheromone (E)-ß-farnesene and related compounds have been shown to interfere with the transmission of PVY by *Myzus persicae* in glasshouse experiments. Laying aluminium strips on the ground between crop rows repels aphids coming into the crop through reflecting UV light.

3. Agronomic Techniques

A tall cover crop will sometimes protect an undersown crop from insect-borne viruses. For example, cucurbits are sometimes grown intermixed with maize. It is thought that incoming aphids land on the barrier crops, feed briefly, and either stay there or fly away.

A major approach is the use of crops bred for resistance to insect pests. Sources of resistance have been found against most of the airborne vector groups. The basis for resistance to the vectors is not always clearly understood, but some factors have been defined. In general terms, there are two kinds of resistance relevant to the control of vectors. First, *nonpreference* involves an adverse effect on vector behaviour, resulting in decreased colonisation, and second, *antibiosis* involves an adverse effect on vector growth, reproduction, and survival after colonisation has occurred. These two factors may not always be readily distinguished. Some specific mechanisms for resistance are sticky material exuded by glandular trichomes, such as those in tomato; heavy leaf pubescence in soybean; A-type hairs on *Solanum betrhaultii* that when ruptured, entrap aphids with their contents and B-type hairs on the same host, which entangle aphids, making them struggle more and so rupture more A-type hairs; inability of the vector to find the phloem in *Agropyron* species; and interference with the ability of the vector to locate the host plant. For example, in cucurbits with silvery leaves, there was a delay of several weeks in the development of 100 percent infection in the field with *Cucumber mosaic virus* (CMV) and *Clover yellow vein virus*. This effect may be due to aphids visiting plants with silvery leaves less frequently because of their different light-reflecting properties.

There may be various limitations on the use of vector-resistant cultivars: Sometimes such resistance provides no protection against viruses. For example, resistance to aphid infestation in cowpea did not provide any protection against *Cowpea aphid-borne mosaic virus*. If a particular virus has several vector species, or if the crop is subject to infection with several viruses, breeding effective resistance against all the possibilities may not be practicable, unless a nonspecific mechanism is used (e.g., tomentose leaves). Perhaps the most serious problem is the potential for new vector biotypes to emerge following widespread cultivation of a resistant cultivar, as may happen following the use of insecticides (Box 14.3).

BOX 14.3

INSECT BIOTYPES OVERCOMING PLANT RESISTANCE

This breakdown of plant resistance to insects is well illustrated by the history of the rice brown planthopper (*Nilaparvata lugens*) (BPH). With the advent of high-yielding rice varieties in Southeast Asia in the 1960s and 1970s, the rice BPH and *Rice grassy stunt virus*, which it transmits, became serious problems. Cultivars containing a dominant gene (*Bph1*) for resistance to the hopper were released about 1974. Within about three years, resistance-breaking populations of the hopper emerged. A new, recessive resistance gene (*bph2*) was exploited in cultivars released between 1975 and 1983. They were grown successfully for a few years until a new hopper biotype emerged that overcame the resistance. A study of the adaptation of three colonies of *N. lugens* to rice cultivars containing different resistance genes showed that the *bhp1* and *bhp3* resistance genes were overcome more readily by colonies that had been exposed for about 10 years to those genes. However, rice cultivar IR64 which contained *bph1* and some minor resistance genes showed a greater durability of resistance than other cultivars. DNA markers to BPH resistance genes are being used in breeding programmes.

B. Soil-Borne Vectors

Most work on the control of viruses transmitted by nematodes and fungi has centred on the use of soil sterilization with chemicals. However, several factors make general and long-term success unlikely:

- Huge volumes of soil may have to be treated.
- A mortality of 99.99 percent still leaves many viable vectors.
- The use of some of the chemicals involved has been banned in certain countries, and such bans are likely to be extended.

In any event, chemical control can be justified economically only for high-return crops or crops that can remain in the ground for many years. However, some recent advances in nematode control procedures may be applicable to the control of viruses that they transmit and may be adaptable to the control of fungus-transmitted viruses.

1. Nematodes

There are four basic strategies for nematode control:

1. Exclusion or avoidance usually by quarantine.
2. Reduction of the initial population density by cultural approaches such as use of clean planting stock or crop rotation with a break crop of a species that is not a host for the target nematode, by chemical nematicides, by biological tactics such as introducing biological agents antagonistic to nematodes and organic amendments, or by the use of nematode-resistant crop varieties that will reduce nematode populations.
3. Suppression of nematode reproduction by chemicals, organic amendments, and certain natural and transgenic resistance traits.
4. Restriction of the current or future crop damage by nematode resistance. However, tolerant cultivars will reduce crop damage due to nematode feeding but will not reduce the chances of virus infection.

2. Fungi

Fungus-transmitted viruses are important in two agronomic situations: nutrient or aquatic systems, and fields. The major control measures are the use of three types of chemicals: surfactants, heavy metals, and sometimes fungicides for use in nutrient or aquatic systems; soil amendments and fungicides to control the fungal vector in the soil; and soil partial sterilants or disinfectants to reduce the active and resting spore stages of fungal vectors in the field. In general, attempts to control infection with viruses having fungal vectors by application of chemicals to the soil have usually not been successful.

IV. PROTECTING THE PLANT

Even if sources of infection are available and the vectors are active, a third kind of control measure is available: protecting inoculated plants from developing systemic disease. There are essentially three approaches that have been used to protect plants, using a mild strain of the virus (termed cross-protection or mild strain protection), the use of chemicals, and genetic protection (conventional resistance and transgenic resistance).

A. Protection by a Plant Pathogen

Inoculation of plants with either a mild virus strain or with satellite RNA (termed cross-protection) has been used to protect against severe virus strains. The phenomenon of cross-protection is described in Chapter 10. Infection of a plant with a strain of virus causing only mild disease symptoms (the protecting strain; also known as the mild, attenuated, hypovirulent, or avirulent strain) may protect it from infection with severe strains (the challenging strain). Thus, plants might be purposely infected with a mild strain as a protective measure against severe disease.

Although such a procedure could be worthwhile as an expedient in very difficult situations, it is not to be recommended as a general practice for the following reasons:

- So-called mild strains often reduce yield by about 5 to 10 percent.
- The infected crop may act as a reservoir of virus from which other more sensitive species or varieties can become infected.
- The dominant mild strain of virus may change to a more severe type in some plants.
- Serious disease may result from mixed infection when an unrelated virus is introduced into the crop.
- For annual crops, introduction of a mild strain is a labour-intensive procedure.
- The genome of the mild strain may recombine with that of another virus, leading to the production of a new virus.

In spite of these difficulties, the procedure has been used successfully, at least for a time, with some crops. A suitable mild isolate should have the following properties:

- It should induce milder symptoms in all the cultivated hosts than isolates commonly encountered and should not alter the marketable properties of the crop products.
- It should give fully systemic infections and invade most, if not all, tissues.
- It should be genetically stable and not give rise to severe forms.
- It should not be easily disseminated by vectors to limit unintentional spread to other crops.

- It should provide protection against the widest possible range of strains of the challenging virus.
- The protective inoculum should be easy and inexpensive to produce, check for purity, provide to farmers, and apply to the target crops.

Mild protecting strains are produced from naturally occurring variants, from random mutagenesis, or from directed mutagenesis of severe strains. The control of CTV provides the most successful example for the use of cross-protection (Box 14.4).

Satellite viruses and RNAs are described in Chapter 3, and, as far as potential biocontrol agents, fall into three categories: those that enhance the helper virus symptoms, those that have no effect, and those that reduce the helper virus symptoms. The last one has potential as a control agent. Most of the work has focussed on the satellites of CMV with some successes in field application, especially against the necrogenic satCMV (CARNA5) described in Chapter 3.

However, there has been concern over the durability of using satellites as biocontrol agents. There is a wide range of necrogenic and other virulent strains of satCMV. Passage of a benign satellite of CMV through *Nicotiana tabacum* led to the satellite rapidly mutating to a pathogenic form and mutations of a single or a few bases can change a nonnecrogenic variant to a necrogenic one (Figure 14.3). Necrogenic variants of the CMV satellite have a greater virulence than nonnecrogenic variants, but, as they depress the accumulation of the helper virus more than do nonnecrogenic variants, the necrogenic variants are not so efficiently aphid transmitted.

B. Antiviral Chemicals

Considerable effort has gone into a search for inhibitors of virus infection and multiplication that could be used to give direct protection to a crop against virus infection in the way that

BOX 14.4

CONTROL OF *CITRUS TRISTEZA VIRUS* BY CROSS-PROTECTION

Worldwide, *Citrus tristeza virus* (CTV) is the most important virus in citrus orchards. In the 1920s, after its introduction to South America from South Africa, the virus virtually destroyed the citrus industry in many parts of Argentina, Brazil, and Uruguay. The application of cross-protection by inoculation with mild CTV isolates in Brazil proved to be successful particularly with Pera oranges, with more than 8 million trees being planted in Brazil by 1980. Protection continues in most individual plants through successive clonal generations. However, in an eight-year assessment of the ability of four mild isolates to suppress severe CTV isolates in Valencia sweet orange on sour orange rootstock in Florida, about 75 percent of the mild-strain protected trees had severe symptoms compared with about 85 percent of the unprotected trees. The use of the same isolates gave better protection of Ruby Red grapefruit on sour orange rootstock. Thus, there are differences in the responses of the scion:rootstock combination, but it is also important to have a compatible mild strain. The search for improved attenuated strains of the virus continues, and the technique is being adopted in other countries.

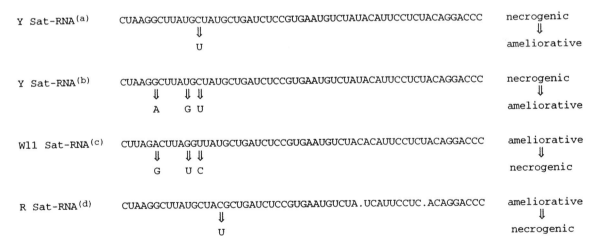

FIGURE 14.3 Alignment of the 55 3′-terminal residues of the CMV satellite RNA variants mutated from a necrogenic form towards a nonnecrogenic one or vice versa. The arrows indicate the positions found to be determinant for necrogenicity. [From Jacquemond and Tepfer (1998; in *Plant virus disease control*, A. Hadidi, R.K. Khetarpal, and H. Koganezawa, Eds., pp. 94–120, APS Press, St. Paul, MN).]

fungicides protect against fungi. There has been no successful control on a commercial scale by the application of antiviral chemicals due to these major difficulties:

- An effective compound must inhibit virus infection and multiplication without damaging the plant. This is a major problem, as virus replication is so intimately bound up with cell processes and any compound blocking virus replication is likely to have damaging effects on the host.
- An effective antiviral compound would need to move systemically through the plant if it is to prevent virus infection by invertebrate vectors.
- A compound acting systemically would need to retain its activity for a reasonable period. Frequent protective treatments would be impracticable. Many compounds that have some antiviral activity are inactivated in the plant after a time.
- For most crops and viruses, the compound would need to be able to be produced on a large scale at an economic price. This might not apply to certain relatively small-scale, high-value crops, such as greenhouse orchids.
- For use with many crops, the compound would have to pass food and drug regulations. Many of the compounds that have been used experimentally would not be approved under such regulations.

Because of these difficulties, there are only a few cases of the use of chemicals to produce virus-free stock plants.

V. CONVENTIONAL RESISTANCE TO PLANT VIRUSES

A. Introduction

When one considers the advantages and disadvantages of the control measures just described, it is obvious that the use of crop plants that are resistant to viruses is likely to be the most promising approach. Thus, for many years plant breeders have been attempting to produce virus-resistant varieties. There are two sources of resistance gene: natural ones from sexually compatible species and nonconventional ones introduced by genetic modification; the latter is discussed in Chapter 15.

There are three types of genes that plant breeders consider for control of plant viruses: those conferring immunity, those conferring field resistance, and those conferring tolerance (Box 14.5). Some molecular aspects of these virus:plant interactions are discussed in Chapter 10.

The following points concerning the effects of host genes on the plant's response to infection emerge from many different studies:

- Both dominant and recessive Mendelian genes may have effects. However, while most genes known to affect host responses are inherited in a Mendelian manner, cytoplasmically transferred factors may sometimes be involved.
- There may or may not be a gene dose effect.
- Genes at different loci may have similar effects.
- The genetic background of the host may affect the activity of a resistance gene.
- Genes may have their effect with all strains of a virus or with only some.
- Some genes influence the response to more than one virus.
- Plant age and environmental conditions may interact strongly with host genotype to produce the final response.
- Route of infection may affect the host response. Systemic necrosis may develop following introduction of a virus by grafting into a high-resistant host that does not allow systemic spread of the same virus following mechanical inoculation.
- Resistance originally thought to be to the virus may be really to the vector.

BOX 14.5

TYPES OF RESISTANCE TO A PLANT VIRUS

There are three main types of resistance and immunity to a particular virus:

1. *Immunity* involves every individual of the species; little is known about the basis for immunity, but it is related to the question of the host range of viruses discussed in Chapter 2.
2. *Cultivar resistance* describes the situation where one or more cultivars or breeding lines within a species show resistance, whereas others do not.

3. *Acquired or induced resistance* is present where resistance is conferred on otherwise susceptible individual plants following inoculation with a virus.

Some authors have considered that immunity and cultivar resistance are based on quite different underlying mechanisms. However, studies with a bacterial pathogen in which only one pathogen gene was used show that for this class of pathogen at least the two phenomena have the same basis.

B. Genetics of Resistance to Viruses

Resistance to viruses in many crop virus combinations is controlled by a single dominant gene (Table 14.1). However, this may merely reflect the fact that most resistant cultivars were developed in breeding programmes aimed at the introduction of a single resistance gene. Furthermore, incomplete dominance may

TABLE 14.1 Summary of Number of Virus Resistance Genes Reported.

Resistance Gene	Monogenic	Oligo- or Polygenic
Dominant	81	10
Recessive	43	20
Incompletely dominant	15	6
(Nature unknown)	—	4
Total number of resistance genes	139	40

From Khetarpal *et al.* (1998; in *Plant virus disease control*, A. Hadidi, R.K. Khetarpal, and H. Koganezawa, Eds., pp. 14–32, APS Press, St. Paul, MN).

be a reflection of gene dosage or be due to environmental factors. Some specific examples of dominant and recessive genes for resistance were shown in Table 10.2. The current reports about the mechanisms of resistance by either total immunity or hypersensitive response were also discussed in Chapter 10.

In several plant species, the resistance virus resistance genes are clustered to specific loci on the chromosomes; in this, virus resistance resembles that for fungi. For instance, in *Pisum sativum*, the resistances to the lentil strain of *Pea seed-borne mosaic virus*, *Bean yellow mosaic virus*, *Watermelon mosaic virus-2*, *Clitoria yellow vein virus*, and *Bean common mosaic virus* NL-8 strain (all potyviruses) are controlled by tightly linked recessive genes on chromosome 2.

C. Tolerance

The classic example of genetically controlled tolerance is the Ambalema tobacco variety. TMV infects and multiplies through the plant, but in the field, infected plants remain almost normal in appearance. This tolerance is due to a pair of independently segregating recessive genes, r_{m1} and r_{m2}, and perhaps to others, as well with minor effects. On the other hand, tolerance to *Barley yellow dwarf virus* is controlled by a single dominant gene in barley, with different alleles giving different degrees of tolerance.

D. Use of Conventional Resistance for Control

A review of the consideration in a breeding programme for resistance to an important virus—that causing rhizomania of sugar beet—is given in Scholten and Lange (2000). Many of the aspects that they discuss are applicable to breeding programmes for resistance to other viruses. In this section, we examine the application of conventional resistance to the control of viruses.

1. *Immunity*

Although many searches have been made, true immunity against viruses and viroids, which can be incorporated into useful crop cultivars, is a rather uncommon phenomenon.

2. *Field Resistance*

Where suitable genes can be introduced into agriculturally satisfactory cultivars, breeding for resistance to a virus provides one of the best solutions to the problem of virus disease. However, there are two major problems. It has proved difficult to find resistance genes in species that are sexually compatible with the crop species. There have been widespread searches in wild species for such genes, and techniques for wide crosses, such as embryo rescue, have been used. Chemical and radiation mutagenesis of the crop plant has also been used to provide useful resistance

The second problem has been the durability of resistance. How long can the gene be deployed successfully before a resistance breaking (virulent) strain of the virus emerges? Of 87 host-virus combinations for which resistance genes have been found, more than 75

percent of those tested were overcome by virulent virus isolates. Fewer than 10 percent of the resistance genes have remained effective when tested against a wide range of virus isolates over a long period. However, some of the virulent isolates were found only in laboratory tests rather than field outbreaks.

The costs of a breeding programme must be weighed against the possible gains in crop yield. Many factors are involved, such as the seriousness of the virus disease in relation to other yield-limiting factors; the "quality" of the available resistance genes—for example, resistance genes against CMV are usually "weak" and short-lived, which may be due, at least in part, to the many strains of CMV that exist in the field; the importance of the crop (compare, for instance, a minor ornamental species with a staple food crop such as rice); and crop quality. Good virus resistance that gives increased yields may be accompanied by poorer quality in the product, as happened with some TMV-resistant tobacco cultivars.

The difficulties in finding suitable breeding material are compounded when there are strains of not just one but several viruses to consider. Cowpeas in tropical Africa are infected to a significant extent by at least seven different viruses. In such circumstances, a breeding programme may utilise any form of genetic protection that can be found. Sources of resistance, hypersensitivity, or tolerance have been found for five of the viruses. However, several of these viruses have different strains or isolates that may break resistance to other isolates. There is, of course, the further problem of combining these factors with multiple resistance to fungal and bacterial diseases. For example, genetic resistance to TMV, cyst nematodes, root-knot nematodes, and wildfire from *Nicotiana repanda* has been incorporated into *N. tabacum*.

3. *Tolerance*

Where no source of genetic resistance can be found in the host plant, a search for tolerant varieties or races is sometimes made. However, tolerance is not nearly as satisfactory a solution as genetic resistance for several reasons:

- The infected tolerant plants may act as a reservoir of infection for other hosts. Thus, it is bad practice to grow tolerant and sensitive varieties together under conditions where spread of virus may be rapid.
- Large numbers of virus-infected plants may come into cultivation. The genetic constitution of host or virus may change to give a breakdown in the tolerant reaction.
- The deployment of tolerant varieties removes the incentive to find immunity to the virus until the tolerance breaks down in an "out of sight, out of mind" attitude.
- Virus infection may increase susceptibility to a fungal disease (see Chapter 10).

However, tolerant varieties may yield very much better than standard varieties where virus infection causes severe crop losses and where large reservoirs of virus exist under conditions where they cannot be eradicated. Thus, tolerance has, in fact, been widely used. Cultivars of wheat and oats commonly grown in the midwestern United States have probably been selected for tolerance to BYDV in an incidental manner because of the prevalence of the virus.

VI. STRATEGIES FOR CONTROL

Three kinds of situations are of particular importance: annual crops of staple foods such as grains and sugar beet that are either grown on a large scale or are subsistence crops and that under certain seasonal conditions may be subject to epidemics of viral disease; perennial crops, mainly fruit trees with a big investment in time and land, where spread of a virus disease, such as citrus tristeza or plum pox, may be particularly damaging; and high-value cash crops such as tobacco, tomato, cucurbits,

peppers, and a number of ornamental plants that are subject to widespread virus infections.

With almost all crops affected by viruses, an integrated and continuing programme of control measures is necessary to reduce crop losses to acceptable levels. Such programmes will usually need to include elements of all three kinds of control measure just discussed. In developing strategies for the integrated approach, it is essential to have a full understanding of the disease, its epidemiology and ecology, and the pathogen, its genetic makeup and functioning and its potential for variation.

VII. VIRUSES OF OTHER KINGDOMS

Some of the ways of controlling animal viruses, such as avoidance of infected individuals, are the same as those described previously for plant viruses. However, plants do not have an innate immune system, so control by immunisation is not a viable approach. As we will see in Chapter 15, there is an analogous approach in plants to immunisation in that they can be transformed to activate the RNA silencing defence system. Also, they can be transformed to produce antibodies that have been shown to mitigate some viruses.

Although there are examples of genetic resistance to viruses infecting vertebrates, more effort is put into control by immunisation and chemoprophylaxis. There are some examples of breeding virus resistance genes into invertebrates (e.g., shrimps).

VIII. SUMMARY

- There are four basic approaches to controlling plant virus diseases: avoiding infection, stopping the vector, protecting the plant, and breeding for resistance.

- The first two approaches involve agronomic practices such as using clean planting material, changing the planting time, and using insecticides against vectors.
- Insecticides are better at preventing the spread of viruses with a persistent interaction with their vector than those with a nonpersistent interaction.
- Plants can be protected by inoculating them with a mild strain of the virus (cross-protection). This is only viable with high-cost perennial crops.
- Breeding for resistance is considered to be the best approach but has the difficulties of sources of resistance genes in sexually compatible species and the durability of resistance.

Further Reading

Brown. J.K. (2008). Plant resistance to viruses. *Encyclopedia of Virology*, Vol. 4, 164–169.

Caranta, C. and Dogimont, C. (2008). Natural resistance associated with recessive genes. *Encyclopedia of Virology*, Vol. 4, 170–176.

Hull, R. (2002). *Matthew's plant virology.* Academic Press, San Diego.

Jones, R.A.C. (2006). Control of plant virus diseases. *Adv. Virus Res.* **67**, 205–244.

Lecoq, H., Moury, B., Desbiez, C., Palloix, A., and Pitrat, M. (2003). Durable virus resistance in plants through conventional approaches: A challenge. *Virus Res.* **100**, 31–39.

Moffett, P. and Klessig, D.F. (2008). Natural resistance associated with dominant genes. *Encyclopedia of Virology*, Vol. 4, pp. 170–176, Elsevier Academic Press, Oxford.

Scholten, O.E. and Lange, W. (2000). Breeding for resistance to rhizomania in sugar beet: A review. *Euphytica*, **112**, 219–231.

Ziebell, H. (2008). Mechanics of cross-protection. *CAB Review: Perspectives in Agriculture, Veterinary Science, Nutrition and Natural Resources* **3**, No. 049, 1–13 (www.cababstractsplus.org/cabreviews).

15

Transgenic Plants and Viruses

Genetic modification technology has opened up many possibilities for protecting plants against virus infection and for using plant viruses as gene vectors.

I. TRANSGENIC PROTECTION AGAINST PLANT VIRUSES

A. Introduction

It is now possible to introduce almost any foreign gene into a plant and obtain expression of that gene. In principle, this should make it possible to transfer genes for resistance or immunity to a particular virus, across species, genus, and family boundaries. Furthermore, genes can be designed to interfere with directly, or induce the host to interfere with, the virus infection cycle. Several approaches to producing transgenic plants resistant to virus infection are being actively explored.

There are essentially three sources of transgenes for protecting plants against viruses: natural resistance genes; genes that are derived from viral sequences, giving what is termed *Pathogen-derived resistance* (PDR); and genes from various other sources that interfere with the target virus.

B. Natural Resistance Genes

Molecular aspects of genes found in plant species that confer resistance to various viruses are discussed in Chapter 10. When a resistance gene has been identified, it can be isolated and transferred to another plant species. The *Rx1* gene that gives extreme resistance to *Potato virus X* (PVX) has been isolated from potato and transformed into *Nicotiana benthamiana* and *N. tabacum,* where it provides resistance to the virus. Similarly, the *N* gene, found naturally in *N. glutinosa,* and that confers hypersensitive resistance to TMV, gives resistance to TMV when transferred to tomato. Much of the specificity of *R* genes is determined by the leucine repeat regions (see Box 10.2); this region can be manipulated *in vitro* to give yet further sources of resistance.

II. PATHOGEN-DERIVED RESISTANCE

Since the mid-1980s, PDR has attracted major interest and is the main method by which transgenic protection is being produced against viruses in plants. The concept of PDR is explained in Box 15.1. In this rapidly expanding subject, there are various terminological problems. The main one is whether to term this phenomenon *resistance* or *protection,* since the reactions of various forms of transgenic protection give a great range of responses, varying from delay in symptom production for just a few days to complete immunity. This book uses *protection* wherever possible, but in situations where it has been used widely (such as pathogen-derived resistance), *resistance* is used.

Currently, there are two basic molecular mechanisms by with PDR is thought to operate. In some systems the expression of an unmodified or a modified viral gene product interferes with the viral infection cycle (called *protein-based protection*). The second mechanism does not involve the expression of a protein product (called *nucleic acid–based protection*).

A. Protein-Based Protection

As noted in Box 15.1, the first demonstration of PDR involved the expression of TMV coat protein. Since then, there have been many examples of the use of this coat protein–mediated protection; the phenomenon is often referred to as "coat protein–mediated resistance." The expression of other viral gene products also gives protection to a greater or lesser extent against the target virus.

1. Transgenic Plants Expressing a Viral Coat Protein

The sequences encoding viral coat proteins are the most widely used for conferring protection in plants because this gene was used in the first example of this approach and because coat protein genes are relatively easy to identify and to clone. These are some of the properties of the protection given by TMV coat protein:

- The higher the amount of virus inoculum, the lower the protection afforded by TMV coat protein.
- There is a positive correlation between the level of protections and the sequence similarity between the transgene coat protein and that of the challenge virus. For instance, TMV coat protein gives better protection against *Tomato mosaic virus,* which has 82 percent sequence identity than against *Ribgrass mosaic virus,* which has 45 percent identity.
- TMV coat protein does not protect against inoculation with viral RNA.

The evidence suggests that the resistance conferred by the coat protein is an early event in the virus infection cycle, possibly by affecting cotranslational disassembly of the incoming virus particles (see Box 7.1 for an explanation of cotranslational disassembly).

2. Other Viral Proteins

Among other virus-encoded proteins that have been explored experimentally to give

BOX 15.1

PATHOGEN-DERIVED RESISTANCE

The ideas leading up to the concept of pathogen-derived resistance for plant viruses are encapsulated as a general concept in a paper by Sandford and Johnson (1985). They suggested that the transgenic expression of pathogen sequences might interfere with the pathogen itself terming this concept *parasite-derived resistance*. Since then, several names have been used for this approach including *nonconventional protection, transgenic resistance,* and *engineered virus resistance*, but the generally accepted term is now *pathogen-derived resistance (PDR)*.

The basic idea arising out of Sanford and Johnson's concept is that, if one understands the molecular interactions involved in the functioning of a pathogen, mechanisms can be devised for interfering with them. Although this concept applies to all pathogens and invertebrate pests, it has mainly been used against plant viruses because of their relatively simple genomes. In developing the concept it was recognised that the interactions of interest occur at all stages of the virus infection cycle and that they can potentially be interfered with in various ways—for example, by blocking the interaction or by decoying one or more of the molecules involved in the interaction. This then led to the idea of the overall strategy as being one of attacking specific viral "targets" with specific molecular "'bullets." However, in practice, much of the

development of this approach was done without detailed knowledge of the precise molecular mechanisms involved, and analysis of these results has thrown light on several new mechanisms. Perhaps the most important is the gene silencing phenomenon described in Chapter 11.

The first demonstration of PDR against plant viruses was by Powell-Abel *et al.* (1986), who showed that the expression of *Tobacco mosaic virus* (TMV) coat protein in tobacco plants protected those plants against TMV. They showed that transgenic plants expressing TMV coat protein either escaped infection following inoculation or developed systemic disease symptoms significantly later than plants not expressing the gene. Plants that showed no systemic disease did not accumulate TMV in uninoculated leaves. Transgenic plants produced only 10 to 20 percent as many local lesions as controls when inoculated with a strain of TMV causing local lesions. The idea that transgenic plants resist initial infection rather than subsequent replication was suggested by results obtained using transgenic Xanthi nc tobacco plants, in which fewer local lesions were produced than on control plants. However, the lesions that did develop were just as big as on control leaves, indicating that once infection was initiated, there was no further block in the infection cycle.

protection are both complete or modified cell-to-cell movement proteins and replicase proteins. The rationale behind using modified proteins is that they would block an essential stage of the infection cycle. Varying results have been obtained, with some constructs giving reasonable protection. However, follow-up experiments have shown that the protection could be

bypassed by another virus complementing the defect.

B. Nucleic Acid–Based Protection

Three potential forms of protection based on the expression of viral RNA sequences have been recognised: that induced by the viral RNA

BOX 15.2

UNEXPECTED RESULTS WITH COAT PROTEIN PROTECTION

In the early days of developing pathogen-derived resistance (see Box 15.1), it was thought that a viral protein had to be expressed to confer the protection. However, some unexpected results were found, especially with controls designed not to express the protein (this shows the importance of controls!). For instance, there was no correlation between the protection and the expression of *Potato virus Y* (PVY) or *Potato leaf roll virus* coat proteins in potato. The untranslatable coat protein gene of *Tobacco etch virus, Tomato spotted wilt virus,* and PVY gave higher levels of protection than either full-length or truncated translatable constructs. This led to the realisation that the protection was RNA-mediated and was an important step in the understanding of the RNA silencing defence system.

---

sequence, that induced by the expression of satellite RNAs, and that in which ribozymes are targeted to viral genomes. Early in the development of coat protein–mediated protection, there were some unexpected observations (Box 15.2). These and other observations suggested that the protection, at least in these cases, was mediated by nucleic acid rather than by protein.

1. RNA-Mediated Protection

As no promoterless transgenes have been shown to confer protection, it must be assumed that either the RNA transcript or a protein that is encoded give the protection. In a plant that is transformed with a construct that does not give a protein, any protection is obviously due to the RNA. When a plant is transformed with a construct designed to produce a viral protein, it can often be difficult to distinguish between protection due to the expression of the protein itself or due to the RNA transcript. However, there are various features of the protection that tend to be characteristic for RNA-mediated protection:

- There is no direct correlation between RNA expression levels and the level of protection.
- Usually, no transgene-encoded protein can be detected, and the steady state of the transcript in inoculated plants is often in low amounts.
- The protection is usually narrow and against strains of the virus that have very similar sequences to that of the transgene.
- Unlike coat protein–mediated protection, the protection is not overcome by inoculating RNA.
- Also, unlike coat protein–mediated protection, RNA-mediated protection is not dose dependent and operates at high levels of inoculum.
- The insert in the host genome often comprises multiple copies of the transgene, particularly with direct repeats of coding regions.
- Copies of the transgene may be truncated and/or in an antisense orientation.
- Transgene sequences and sometimes their promoter(s) may be methylated.

2. Molecular Basis of RNA-Mediated Protection

When transcript levels have been examined, three general classes of resistance phenotype have been recognised:

1. *Plants that are fully susceptible.* These plants have low to moderate levels of transgene transcription and steady-state RNA.

2. *Plants that become infected and then recover.*
   These have moderate to high levels of
   transgene transcription and steady-state
   RNA in uninfected plants but low-level
   steady-state RNA in recovered tissues.
3. *Plants that are highly resistant.* These plants
   have high levels of transgene expression but
   low steady-state levels.

The recovery phenomenon associated with
low steady states of transgene RNA in recovered tissues and the low steady-state RNA
levels in highly resistant plants, coupled with
the narrow range of protection against viruses
with homologous sequences to the transgene
are strongly indicative of homology-dependent
RNA silencing, which is described in detail in
Chapter11.

Because of the variation in response, large
numbers of independent transformants should
be tested, not only to obtain lines with the best
protection characteristics but also to rule out
the possibility that protection is not given by a
particular construct.

3. Sequences for RNA-Mediated Protection

RNA-mediated protection has been given
by a range of sequences from viral genomes.
In many cases, it has resulted from attempts
to transform plants with the viral genes just
described or from constructs designed to produce antisense RNA, which is intended to
block translation of the viral mRNA.

It is the constructs that are important as
they need to transcribe to give dsRNA, which
initiates the silencing pathway. The hairpin
construct comprising sequences that are transcribed to give (+) strand and (−) strand of
the target (viral) RNA separated by a spacer
(Figure 11.2F) are the most efficient.

4. Ribozymes

As described in Box 3.2, ribozymes are
catalytic RNAs that can cleave at specific sites
in complementary target RNAs. Since the ribozyme has to be complementary to the target

viral sequence, it can be considered to be an
antisense RNA. Thus, any effect can be difficult
to distinguish from that of RNA silencing. Incorporation of a ribozyme into an antisense RNA to
TMV gave no significant advantage over the
antisense RNA itself but constructs directed
against *Plum pox virus* that containing a hammerhead ribozyme gave stronger protection
than the ordinary antisense RNA construct.

5. Relationship Between Natural Cross-Protection and Protection in Transgenic Plants

The mechanism for transgenic protection
against a virus infection, especially coat protein–mediated protection, has been compared
with natural cross-protection or mild-strain
protection (see Chapter 10). There are several
similarities that have been used to support the
idea:

- In both situations, the degree of resistance
  depends on the inoculum concentration,
  with high concentrations reducing the
  observed resistance.
- Both are effective against closely related
  strains of a virus, less against distantly
  related strains, and not at all against
  unrelated viruses.
- In some circumstances, cross-protection
  can be substantially overcome when
  RNA is used as inoculum rather than
  whole virus.

On the other hand, there appear to be some differences between natural cross-protection and
coat protein–induced resistance. When cross-
protection between related strains of a virus is
incomplete, the local lesions produced may be
much smaller than in control leaves. This indicates reduced movement and/or replication of
the superinfecting strain.

It is quite possible that there are several
mechanisms that give cross-protection. One of
them is likely to involve the RNA silencing host
defence system and thus, to resemble RNA-
mediated protection.

6. Transgenic Protection by Satellite and DI Nucleic Acids

The general nature of satellite RNAs is described in Chapter 3, and the ability of some satellite RNAs to attenuate the symptoms of the helper virus was also discussed there. Defective interfering (DI) nucleic acids are described in Chapter 10. They are mutants of viral genomes that are incapable of autonomous replication but contain sequences that enable them to be replicated in the presence of the parent helper virus. In many cases, they are amplified at the expense of the parent virus, ameliorating the symptoms induced by that virus.

Transgenic plants expressing either satellite or DI RNA are less severely diseased when inoculated with the helper virus. However, the use of satellite RNAs in transgenic plants to protect against the effect of virus infection has both advantages and disadvantages. The protection afforded is not affected by the inoculum concentration, as it is with viral coat protein transformants. The losses that do occur in transgenic plants because of slight stunting will affect only the plants that become naturally infected in the field. Furthermore, the resistance may be stronger in transgenic plants than in plants inoculated with the satellite. Inoculation is not needed each season, and the mutation frequency is lower. Nevertheless, there are distinct risks and limitations with the satellite control strategy. The satellite RNA could cause virulent disease in another crop species or could mutate to a form that enhances disease rather than causing attenuation (see Chapter 14). Another risk is the reservoir of virus available to vectors in the protected plants. Last, the satellite approach will be limited to those viruses for which satellite RNAs are known.

C. Other Forms of Transgenic Protection

Various forms of protection against viruses have been shown for a variety of transgenes that are not derived from viruses themselves.

For instance, the ß-1,3-glucanases are proteins believed to be part of the constitutive and induced defence system of plants against fungal infection. Unexpectedly, plants deficient in these enzymes due to expression of an antisense RNA show markedly reduced lesion number and size on inoculation of *Nicotiana tabacum* Havana 425 with TMV and reduced severity and delay of symptoms of TMV in transgenic *N. sylvestris*.

Plants do not have an immune system like that of animals, in which specific antibody proteins are formed in response to an infection. Transgenic plants can produce antibodies (Box 15.3): The expression in plants of a single-chain Fv antibody derived from a panel of monoclonal antibodies against *African cassava mosaic virus* coat protein reduced the infection incidence and delayed symptom development.

In mammalian systems, interferons are effective antiviral molecules. When one of the components of the virus-inhibiting pathway, 2′-5′oligoadenylate synthetase was expressed in potato plants, it gave protection against PVX. The virus concentration in transgenic plants was lower than it was in plants expressing PVX coat protein.

D. Field Releases of Transgenic Plants

1. Potential Risks

There have been concerns expressed about the release and use of plants modified by genetic manipulation. This has led to plants produced by this means being treated in a different manner from those produced by conventional breeding techniques and being subject to specific regulatory structures. The concerns are in two areas: potential risk to human and other animal health and potential risk to the environment. There are strong controls over the use of new sources of human food and animal feed, and, as plant viruses are a normal component of human and animal diet, little problem is

BOX 15.3

ANTIBODIES IN PLANTS

It has long been assumed that plants could not produce antibodies. However, plants transformed with constructs expressing either single gamma (heavy) chains (see Box 13.2 for antibody structure) or single kappa (light) chains and then crossed yield a functional antibody comprising more than 1 percent of the leaf proteins. The

two variable chains of an antibody can be expressed in *E. coli* as a fusion protein by joining them with a flexible linker protein. These can be put into a phage display system, and a very large number of combinations of the two chains can be expressed and then selected. Suitable clones can then be transformed into plants.

---

foreseen here. However, there has been a long-term debate over the use of the *Cauliflower mosaic virus* (CaMV) 35S promoter in transformation constructs (Box 15.4).

A major consideration for the use of plant lines transgenically protected against viruses is the possible environmental risks. The area of virus-protecting transgenes that has attracted specific interest is the use of virus sequences. The basic question is "What is the risk of any

interactions that might arise between a virus or virus-related sequence integrated in the host genome and another virus super-infecting that plant?" Three scenarios are considered: heteroencapsidation, recombination, and synergism.

Heteroencapsidation involves the superinfection of a plant expressing the coat protein of virus A by the unrelated virus B, the expression of the virus A coat protein not protecting the plant against virus B. The risk is that the

BOX 15.4

USE OF THE CAULIFLOWER MOSAIC VIRUS (CaMV) 35S PROMOTER IN TRANSGENIC PLANTS

The CaMV 35S promoter is one of the most widely used promoters for transforming plants and is used in most currently released genetically modified (GM) crop plants. The similarity in replication between CaMV and retroviruses (see Chapter 8) has led to suggestions that this promoter might give rise to some risks. One of the suggested scenarios is that on eating GM crops, the 35S promoter could integrate into the human genome leading to the activation of cancers. This argument has been countered by

pointing out that humans have been eating brassicas infected with CaMV for many hundreds of years and that no problem has been identified. Furthermore, infected plants contain more than 1,000 times as many copies of the promoter as do GM plants, and the human digestive system breaks down plant DNA to small pieces. This illustrates the arguments about GM crops and the use of viral sequences in producing them. For further information, see Hull *et al.* (2000).

coat protein of virus A might encapsidate the genome of virus B, thereby conferring on it other properties, such as different transmission characteristics (Figure 15.1). Heteroencapsidation by transgenically expressed coat protein has been reported for closely related viruses, such as *Cucumber mosaic virus* and *Alfalfa mosaic virus,* and between potyviruses but only occurs

Plant transgenic in coat protein of virus A

Resists virus A

Susceptible to virus B

Transcapsidation of virus B in virus A coat protein

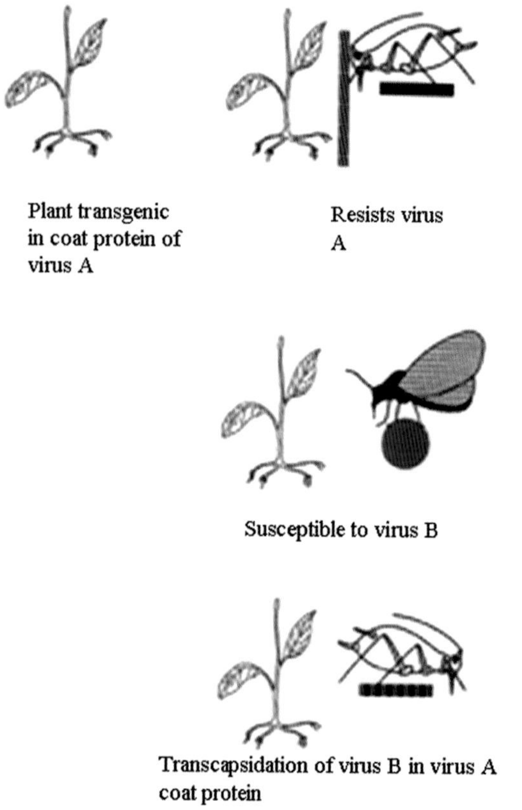

FIGURE 15.1   Hypothetical scheme to illustrate hetero-encapsidation. The plant on the left is transformed with the coat protein of virus A, a rod-shaped virus transmitted by aphids; it is protected against virus A (top right) but not against virus B, an isometric virus transmitted by whitefly (middle right). In the heteroencapsidation situation, the genome of virus B is encapsidated by the coat protein of virus A expressed in the transgenic plant and thus the het-eroencapsidant would be transmissible by aphids (bottom right). As noted in the text, this situation has not been found with unrelated viruses.

between closely related viruses, the particles of which have similar forms of stabilisation.

Recombination between the transgene and superinfecting virus might lead to a new virus. As described in Box 4.5, there are examples of "new" viruses naturally arising from recombination. The question is, would the deployment of plants transgenic in viral sequences lead to any more recombination than occurs in the natural situation?

As described in Chapter 8, synergistic interaction between two unrelated viruses are potentiated by distinct virus sequences. Thus, there is a possibility that the effect of a superinfecting virus could be exacerbated by a viral transgene. However, it appears that synergism is due to interactions between suppressors of RNA silencing (see Box 11.3), so the use of such genes should be avoided in transgenic construct design.

The construct design is important in minimising potential risk. Most of the preceding scenarios depend on a viral protein or large transcript being produced in the transgenic plant. The use of the RNA silencing protection approach leads to the production of relatively short transcripts that are processed to siRNA and thus minimise any potential risk. If a viral protein has to be produced, understanding the molecular interactions involved in the potential risk situations can lead to methods for the "sanitising" of the transgene. For example, as described in Chapter 12, aphid transmission of potyviruses involves an amino acid triplet (asp, ala, gly; DAG) in the coat protein, the mutation of which blocks aphid transmission but does not affect the protection offered by the transgene. Similarly, understanding of the factors involved in recombination will lead to transgene constructs that lessen the possibility of new molecules being formed between the transgene and a superinfecting virus.

In all of these risk assessments, it is important to compare the transgenic situation with the nontransgenic situation. Thus, there is the possibility of the described above potential risks occurring in mixed natural infections between viruses.

2. Field Performance

Testing the field performance of transgenic plants is essentially no different from testing plant lines that have been obtained by traditional breeding. The testing objectives include evaluating the plant appearance, typeness, growth vigour, yield, and quality. Of especial importance is to assess the stability and durability of the protecting transgene under these conditions. Two main factors can affect stability and durability: possible climatic effects on the expression of the transgene and the presence of protection-breaking strains or isolates of the virus that are present in the viral ecosystem but were not recognised in the initial glasshouse tests.

The use of transgenic protection in papaya to *Papaya ring spot virus* has been a great success in controlling the disease in Hawaii and maintaining papaya production there. However, attempts to transfer the technology to Thailand showed the necessity of having the construct targeted to the local strains of the virus.

III. POSSIBLE USES OF PLANT VIRUSES FOR GENE TECHNOLOGY

A. DNA Viruses as Gene Vectors

In the early 1980s, there was considerable interest in the possibility of developing plant viruses as vectors for introducing foreign genes into plants. At first, interest centred on the caulimoviruses, the only plant viruses with dsDNA genomes, because cloned DNA of the viruses was shown to be infectious. Interest later extended to the ssDNA geminiviruses and then to RNA viruses when it became possible to reverse transcribe these into dsDNA, which could produce infectious RNA transcripts.

These are the main potential advantages of a plant virus as a gene vector:

- The virus or infectious nucleic acid could be applied directly to leaves, thus avoiding the need to use transformation technologies and the consequent difficulties in plant regeneration.
- It could replicate to high copy number.
- There would be no "position effects" of insertion into a site in the plant chromosomal DNA.
- The virus could move throughout the plant, thus offering the potential to introduce a gene into an existing perennial crop, such as orchard trees.

Such a virus vector would have to be able to carry a nonviral gene (or genes) in a way that did not interfere with replication or movement of the genomic viral nucleic acid. Ideally, it would also have the inability to spread from plant to plant in the field, providing a natural containment system; induction of very mild or no disease; a broad host range, which would allow one gene vector to be used for many species but would be a potential disadvantage in terms of safety; and maintenance of continuous infection for the lifetime of the host plant.

These are the major general limitations in the use of plant viruses as gene vectors:

- They are not inherited in the DNA of the host plant, and therefore genes introduced by viruses cannot be used in conventional breeding programmes.
- Plants of annual crops would have to be inoculated every season unless there was a very high rate of seed transmission.
- The foreign gene introduced with the viral genome may be lost quite rapidly by recombination or other means with the virus reverting to wild type.
- It would be necessary to use a virus that caused minimal disease in the crop cultivar. The virus used as vector might mutate to

produce significant disease or be transmitted to other crops that were susceptible.
Infection in the field with an unrelated second virus might cause very severe disease.

1. Caulimoviruses

There appear to be several constraints to the use of CaMV as a gene vector. These include the packing capacity of the CaMV particle, the amount of viral DNA that can be removed without affecting the functioning of the genome, and the interactions between different parts of the genome in expression and replication. Removal of nonessential regions of the genome should enable about 1,000 bp of sequence to be inserted, but it is not certain as to whether all this sequence is really nonessential.

2. Geminiviruses

Much attention has been focused on the geminiviruses as potential gene vectors because of their DNA genomes and because the small size of the genomes makes them convenient for *in vitro* manipulations. Nevertheless, this small size may restrict the amount of viral DNA that can be deleted. However, this is counterbalanced by the fact that for some geminiviruses a viable coat protein and encapsidation are not necessary for successful inoculation by mechanical means or for systemic movement through the plant.

There are other potential difficulties. Recombination can occur to give parental-type molecules. Most geminiviruses are restricted mainly to the phloem and associated cells. However, the wide host range of the geminiviruses (compared with the caulimoviruses) makes them of considerable interest. The fact that some members infect cereal crops would be particularly useful except they are not seed transmitted and are mechanically transmitted only with difficulty. In any event, inoculations on the scale needed for cereal crops would be impractical.

B. RNA Viruses as Gene Vectors

The ability to manipulate RNA virus genomes by means of a cloned cDNA intermediate has opened up the possibility of using RNA as well as DNA viruses as gene vectors. In principle the known high error rate in RNA replication (see Figure 4.5) might place a limitation on the use of RNA viruses. The experimental evidence to date suggests that mutation may not be a major limiting factor, at least in the short term. Viruses with isometric and rod-shaped particles have been studied as potential vectors, but those with rod-shaped particles have better potential because there is less constraint on the amount of nucleic acid that can be inserted. Four basic strategies have been used for RNA virus gene vectors (see Figure 15.2).

In the gene replacement strategy, a viral gene is removed and replaced with the gene of interest. This has been attempted with several RNA viruses with varying success, but because of the integrated and coordinated expression of most viral genomes, it does not appear to be a general viable approach.

Instead of removing a viral gene, in the gene insertion strategy, the gene of interest is

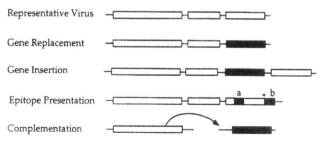

FIGURE 15.2   Comparison of strategies used to express foreign genes (black boxes) from different viruses; white boxes indicate viral genes. The epitope method involves translational fusion (a) of a small sequence inside the coat protein gene or translational read through (b) of an amber stop codon (*) at the 3′ end. [From Scholthof *et al.* (1996; *Ann. Rev. Phytopathol.* **34**, 299–323. Reprinted, with permission, from the *Annual Review of Phytopathology*, Volume 34 © 1996 by Annual Reviews www.annualreviews.org).]

inserted usually attached to a viral subgenomic promoter. This enables the gene of interest to be expressed in a manner similar to downstream viral genes (see Chapter 7 for subgenomic RNAs). Viable gene vector systems have been established for several rod-shaped RNA viruses [e.g., *Tobacco mosaic virus* (TMV), *Potato virus X* (PVX), and *Tobacco rattle virus* (TRV)] whose structure can accommodate the extra nucleic acid from the insert. One of the advantages of TRV is that it expresses foreign proteins efficiently in roots. One problem with this approach is that if the subgenomic promoter for the gene of interest is the same as that for the normal viral gene, there is a likelihood of recombination leading to loss of the insert. For the TMV vector, this was overcome by developing a hybrid vector containing sequences from two tobamoviruses: TMV-U1 and *Odontoglossum ring spot virus* (ORSV). In this vector, the gene of interest is expressed from the TMV coat protein subgenomic promoter and the coat protein from the ORSV promoter.

In the epitope presentation strategy the epitope-encoding sequence is inserted into a location in the viral coat protein gene so it is presented on the surface of the virus particle. This will be described in more detail following.

For the complementation system, the gene of interest is inserted into a virus that is "disarmed" through the removal of an essential gene. This vector is then inoculated to plants transgenic in the "missing" gene, which complements the defect. Infectious clones of defective interfering (DI) RNAs may be used as vectors.

C. Viruses as Sources of Control Elements for Transgenic Plants

Certain plant viral nucleic acid sequences have been found to have useful activity in gene constructs as promoters of DNA and RNA transcription and as enhancers of mRNA translation.

1. DNA Promoters

Transcription of CaMV DNA gives rise to a 19S and a 35S mRNA (Chapter 7). The 19S and 35S promoters are both strong constitutive promoters, but the 35S promoter is much more effective than the 19S in several systems. For example, expression of the a-subunit of ß-conglycinin in petunia plants under control of the 35S promoter was 10–50 times greater than that from the 19S promoter. A variant (enhanced) 35S promoter that contains a tandem duplication of 250 bp of upstream sequences gives about a tenfold increase in transcriptional activity.

Several of the promoters from other DNA viruses, such as badnaviruses and geminiviruses, have been shown to have activity in transformation constructs but are not used as widely as the CaMV 35S promoter.

2. RNA Promoters

The subgenomic promoters for several RNA viruses have now been identified. These are used in the RNA virus vectors just described and may prove useful for gene amplification.

3. Translation Enhancers

Untranslated leader sequences of several viruses have been shown to act as very efficient enhancers of mRNA translational efficiency both *in vitro* and *in vivo* and in prokaryotic and eukaryotic systems (see Chapter 7). For instance, the TMV leader sequence (the Ω sequence) enhances translation of almost every mRNA in every system that has been tested.

D. Viruses for Producing Vaccines

There are two approaches to using plant viruses to produce vaccines and other materials of medical and veterinary interest: expressing the target vaccine gene using a plant viral vector or presenting the epitope on the coat protein of a plant virus (Figure 15.3).

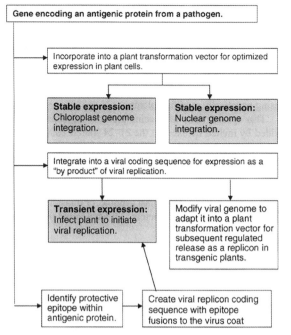

FIGURE 15.3   Plant-derived vaccine research strategies. [This article was published in *Vaccine* 23, C. Arntzen, S. Plotkin, and B. Dodet, Plant-derived vaccines and antibodies: Potential and limitations, pp. 1753–1756, Copyright Elsevier (2005).]

1. Vaccines Using Plant Virus Vectors

Target virus proteins have been expressed from plant RNA virus vectors. For instance, hepatitis B surface antigen and *Norwalk virus* capsid proteins expressed in *Nicotiana benthamiana* and tomato assembled into virus-like particles and conferred immunogenicity in animals when applied orally.

2. Viruses for Presenting Heterologous Peptides

The production of small peptides is required for several reasons, including acting as epitopes for vaccines and biologically active peptides. As described in Box 13.2, epitopes are patches of amino acids that adopt specific conformations. Free peptides can act as epitopes, but the immunogenicity is enhanced by presentation

on the surface of a macromolecular assembly. One approach to presenting the peptide sequence in the correct conformation is to incorporate it into a viral coat protein sequence in such a way that they are exposed on the surface of the virus particle. The virus particle can then be used as a vaccine. There are several advantages to doing this with plant viruses:

- The virus can be produced in large amounts and in less-developed countries where the technology for animal virus vaccine production may be limited.
- Such vaccines may be given orally as part of the normal food supply.
- The virus will not infect humans or animals and thus is completely inactive.
- The system is not subject to contamination by other virulent animal pathogens.

A potential disadvantage is that the high rate of mutation of RNA viruses could result in the deletion or loss of inserted sequences, especially as they would not be under selection pressure; however, experience is indicating that this problem may not be as significant as at first feared. Several plant viruses have been used for the presentation of foreign peptides.

a. Cowpea Mosaic Virus (CPMV).
The structure of the particles of CPMV has been solved to atomic resolution. The capsid comprises two types of protein: the L protein that has two ß-barrel domains and the S protein that has one ß-barrel domain (see Chapter 5 for capsid protein structure). Analysis of the three-dimensional structure suggested that loops between the ß-strands would be suitable for the insertion of sequences to be expressed as epitopes, as these loops are not involved in contacts between protein subunits. The ßB-ßC loop of the S protein is highly exposed and was used for most of the insertions (Figure 15.4); some insertions have been made in other loops.

Early studies on inserting sequences at the ßB-ßC loop site gave guidelines for construction

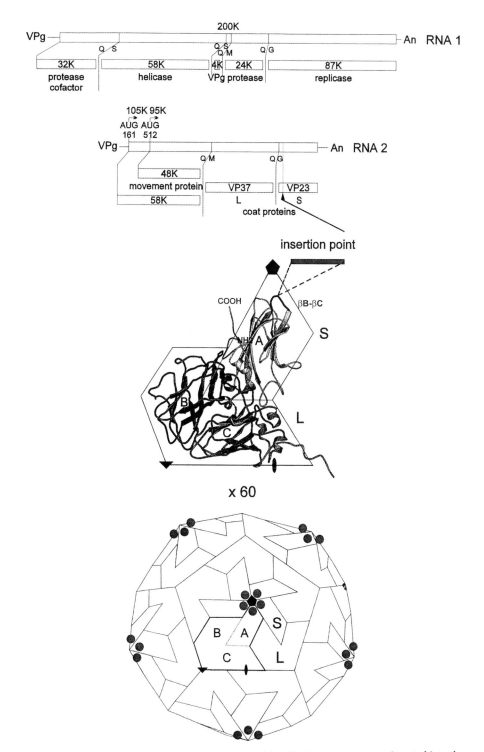

FIGURE 15.4 Generation of chimeric *Cowpea mosaic virus* particles. Foreign sequences are inserted into the gene for the S coat protein on RNA2. Both RNA1 and RNA2 are translated into polyproteins and undergo a cascade of cleavages, whose sites and final products are shown. RNA2 (bearing the heterologous sequence) needs to be coinoculated with RNA1 (unmodified) to initiate infection in cowpea plants. S protein harbouring a foreign peptide in its βB-βC loop of the β-barrel structure (see Chapter 5) and native L protein assemble at 60 copies each into icosahedral virus particles on which the foreign insert is expressed around the 5′-fold axes of symmetry. [From Porta and Lomonossoff (1998; *Rev. Medical Virol.* **8**, 25–41).]

of viable, genetically stable chimeras. These included: (i) foreign sequences should be inserted as additions and not replacements of the CPMV sequence; (ii) sequence duplication should be avoided as this led to loss of insert by recombination; (iii) the precise site of insertion was important for maximising growth of chimeras. Understanding of these guidelines gave a standard procedure for inserting foreign DNA into the ßB-ßC loop of the S protein.

Chimeras with inserts for the sequence for up to 38 amino acids have been successfully made in which the presence of foreign sequences did not significantly affect the ability of the modified virus to replicate. Various epitopes have been inserted, including ones from *Human rhinovirus 14*, *Human immunodeficiency virus type 1*, and *Canine parvovirus*.

b. Tobacco Mosaic Virus (TMV). With the development of infectious cDNA clones to TMV, it was possible to use a self-replicating system in plants. Fusion of a foreign sequence to the C-terminus of the coat protein prevented particle assembly. To overcome this problem, the insert was placed after an amber stop codon at the C-terminus of the coat protein gene (Figure 15.5A) so it could be expressed as a read-through protein. Particles were assembled with about 5 percent of the coat protein subunits expressing the inserted sequence. Replacement of two amino acids on a surface loop near the C-terminus of the coat protein gave particles with 100 percent of the subunits containing the insert (Figure 15.5B), as did inserts into another part of the C-terminal region not involved in particle assembly (Figure 15.5C).

E. Viruses in Functional Genomics of Plants

The gene silencing induced by virus infection of plants is described in Chapter 11.

In virus-induced gene silencing (VIGS) a gene incorporated into a virus vector—say, a TMV-based vector, a PVX-based vector, a TRV-based vector, or, for monocots, a *Barley stripe mosaic virus*–based vector—will generate dsRNA and thus silence a homologous gene in a plant (Figure 15.6).

In selecting a vector for a specific purpose, the virus must be able to infect that host. Among the advantages of this approach to genomics is as the virus construct is inoculated to seedlings or mature plants, it overcomes the problem with insertional mutagenesis (the other main approach) of identifying genes whose disruption is lethal before the plant has developed. With the VIGS approach, the lethality would be apparent from the death of the mature plant that had been inoculated. VIGS is also simple, rapid, and allows for a high-throughput screening. However, as described in Chapter 11, viruses can suppress gene silencing, so the effect on the plant gene may be only temporary. The VIGS phenomenon was initially shown for transgenes and has now been demonstrated for a wide range of plant genes (Figure 15.7).

F. Plant Viruses in Nanotechnology

Plant viruses, such as CPMV, have a number of features that can be exploited for nanoscale biomaterial fabrication. These include the ordered structure of the capsids, the detailed knowledge of the molecular biology of the virus, the ease of inoculation and high yield, and the lack of biological hazards to humans. The structure of the capsid provides regularly spaced attachment units for a nanoscaffold. The approach is similar to that of presentation of epitopes and involves decorating surface residues such as lysine, tyrosine, and carboxylate groups with functional compounds to form the nanoscaffold.

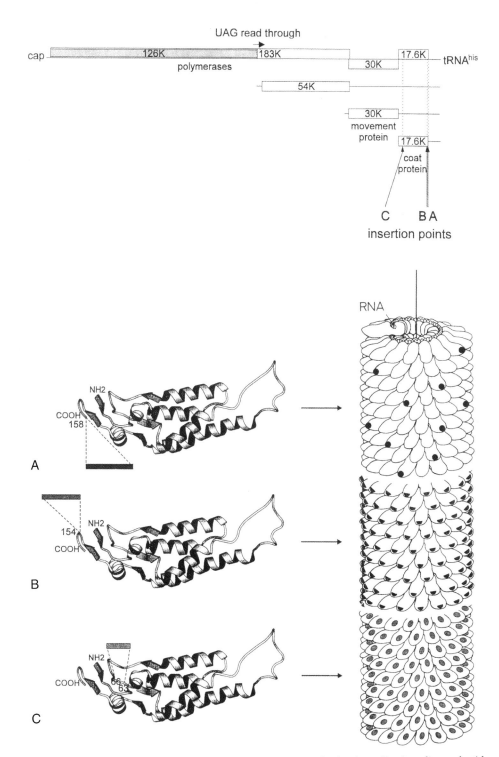

FIGURE 15.5 Production of chimeric *Tobacco mosaic virus* (TMV) particles *in planta*. Foreign oligonucleotide sequences are introduced at one of three positions, labeled A, B, and C, in the gene of TMV coat protein, which is expressed from the most 3' of the viral subgenomic RNAs. *In vitro* transcripts of the altered full-length genomic cDNA are inoculated onto tobacco plants. The resulting recombinant TMV coat proteins are represented as ribbon drawings with the numbers, indicating the insertions site. Upon assembly of these coat proteins, chimeric rod-shaped virions are formed on which the foreign peptides are differentially displayed and distributed; a maximum of 5% of the coat proteins present an insert in position A; all of the coat proteins are modified in positions B and C. [From Porta and Lomonossoff (1998; *Rev. Medical Virol.* **8**, 25–41).]

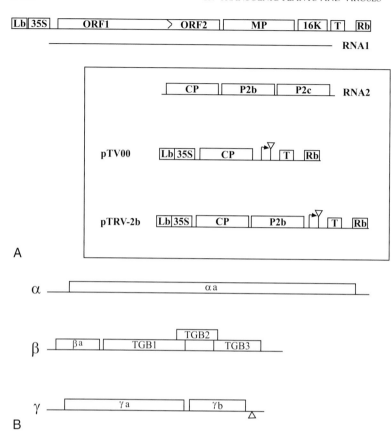

A

B

FIGURE 15.6 Examples of VIGS constructs showing different approaches. A. *Tobacco rattle virus* (see Profile 15 for TRV genome organisation). For the VIGS tests the RNA1 construct and the modified RNA2-construct containing the target insert are inserted into an *Agrobacterium tumefaciens* plasmid, which is then inoculated into the target species. The top diagram shows the genome organisation of a cDNA clone of RNA1; Lb is left border; Rb is right border; 35S, *Cauliflower mosaic virus* 35S promoter. The box contains information on RNA2. The top line shows the genome organisation; the middle line shows the agroinoculation plasmid PTV00 in which ORFs P2b and P2c are removed; the arrow indicates the duplicated coat protein promoter and the triangle, a multiple cloning site for insertion of the VIGS sequence targeting a specific host gene. The bottom line is a construct for agroinoculation to obtain gene silencing in roots. [Data from Valentine *et al.* (2004; *Plant Physiol.* **136**, 3999–4009).] B. *Barley stripe mosaic virus* constructs for VIGS in monocots; (see Profile 1 for BSMV genome organisation). BSMV infection requires three RNA species (a, ß, γ), cDNAs of which are cloned behind a T7 promoter and transcripts are expressed in *Escherichia coli* for inoculation to plants. The top two lines show the genome organisation of RNAs a and ß. The bottom line shows RNA γ with the site modified to enable insertion of the VIGS target sequence. [Data from Halzberg *et al.* (2002; *Plant J.* **30**, 315–327).]

IV. VIRUSES OF OTHER KINGDOMS

Viruses are also being used to modify vertebrates. There is an increasing number of cases of viral vectors (such as adenoviruses and retroviruses) being used to deliver therapeutic genes into the host genome for gene therapy. Unlike plant viruses, the animal virus, together with the gene of interest, integrates into the host genome. Promoter sequences, such as those from *Simian virus 40* and *Human herpesvirus 5* (cytomegalovirus), have been used widely in constructs for transforming animals.

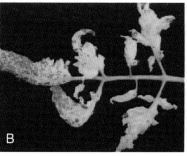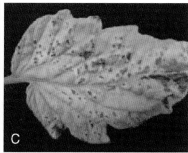

FIGURE 15.7 Some examples of VIGS experiments. A. Tomato leaf infected with *Tobacco rattle virus* (TRV); B. *PDS* gene silenced in tomato using a TRV construct; photobleached tissue indicates regions of silencing; C. TRV:Prf silenced tomato leaf treated with an avirulent *Pseudomonas syringae* strain; visible disease symptoms are the result of loss of resistance because of the suppression of the *Prf* gene. [From Burch-Smith *et al.* (2004; *The Plant J.* **39**, 734–746, Wiley-Blackwell).]

V. SUMMARY

- Genetic modification technology has opened the possibility of conferring protection in plants against viruses.
- The most commonly used sequences for protecting plants are viral sequences, either coding for a viral protein which interferes with the replication cycle of the target virus, or a noncoding sequence, which primes the RNA silencing defence system.
- Plants containing viral transgenes have been released to the field but are subject to strong regulations.
- Plant viruses also provide sequences such as promoters and translation enhancers for constructs used for introducing other genes into plants.
- The coat proteins of plant viruses have been engineered to express epitopes against animal viruses or nanoscaffolds for nanotechnology.
- In virus-induced gene silencing (VIGS), plant viral genomes are engineered to contain sequences that would silence (through RNA silencing) target host sequences and are of use in functional genomics.

References

Abel, P.P., Nelson, R.S., De, B., Hoffmann, N., Rogers, S.G., Fraley, R.T., and Beachy, R.N. (1996). Delay of disease development in transgenic plants that express the tobacco mosaic virus coat protein gene. *Science* **232**, 738–743.

Hull, R., Covey, S.N., and Dale, P. (2000). Genetically modified plants and the 35S promoter: Assessing the risks and enhancing the debate. *Microb. Ecol. Health Dis.* **12**, 1–5.

Sanford, J.C., and Johnston, S.A. (1985). The concept of parasite-derived resistance. *J. Theor. Bio.* **113**, 395–405.

Further Reading

Dietzgen, R.G. and Mitter, N. (2006). Transgenic gene silencing strategies for virus control. *Austr. Plant Pathol.* **35**, 605–618.

Fuchs, M. (2008). Engineering resistance. *Encyclopedia of Virology*, Vol. 4, 156–163.

Gleba, Y., Marillonnet, S., and Klimyuk, V. (2008). Plant virus vectors (gene expression systems). *Encyclopedia of Virology*, Vol. 4, 229–237.

Padmanabhan, M.S. and Dinesh-Kumar, S.P. (2008). Virus-induced gene silencing (VIGS). *Encyclopedia of Virology*, Vol. 5, 375–379.

Rybicki, E.P. (2008). Vaccine production in plants. *Encyclopedia of Virology*, Vol. 5, 221–225.

Steinmetz, N.F., Lomonossoff, G.P., and Evans, D.J. (2006). Cowpea mosaic virus for material fabrication: Addressable carboxylate groups on a programmable nanoscaffold. *Langmuir* **22**, 3488–3490.

Young, M. Willits, D., Uchida, M., and Douglas, T. (2008). Plant viruses as biotemplates for materials and their use in nanotechnology. *Annu. Rev. Phytopathol.* **46**, 361–384.

Appendix: Profiles

These profiles give brief details of selected plant viruses that either have great economic significance and/or have contributed to the understanding of viral functions and that are discussed in this book. Where there are two or more species in the same taxonomic group, they are listed under the appropriate family or genus so comparisons can be made within that group. These profiles are of only 32 of the more than 1,000 plant virus species. Further information on these viruses includes reviews, where available, and the *Association of Applied Biologists Descriptions of Plant Viruses*, which describes much of the biology of each virus (see www.dpvweb.net).

INDEX

1. *Barley stripe mosaic virus*
2. *Beet necrotic yellow vein virus*
3. Family *Bromoviridae* (*Brome mosaic virus*, *Cucumber mosaic virus*, *Alfalfa mosaic virus*)
4. Family *Caulimoviridae* (*Cauliflower mosaic virus*, *Banana streak virus*, *Rice tungro bacilliform virus*)
5. Genus *Closterovirus* (*Citrus tristeza virus*, *Beet yellow virus*)
6. Family *Comoviridae* (*Cowpea mosaic virus*, *Tobacco ringspot virus*)
7. Family *Geminiviridae* (*Tomato yellow leaf curl virus*, *African cassava mosaic virus*, *Abutilon mosaic virus*, *Bean golden mosaic virus*, Cotton leaf curl disease, *Maize streak virus*)
8. Family *Luteoviridae* (*Barley yellow dwarf virus-PAV*, *Potato leafroll virus*)
9. *Potato virus X*
10. Genus *Potyvirus* (*Potato virus Y*, *Papaya ringspot virus*, *Plum pox virus*, *Tobacco etch virus*)
11. *Rice ragged stunt virus*
12. *Rice stripe virus*
13. *Sonchus yellow net virus*
14. *Tobacco mosaic virus*
15. *Tobacco rattle virus*
16. Family *Tombusviridae* (*Tomato bushy stunt virus*, *Carnation mottle virus*)
17. *Tomato spotted wilt virus*
18. *Turnip yellow mosaic virus*

Profile No. 1

BARLEY STRIPE MOSAIC VIRUS (BSMV)

Classification

Type species of the *Hordeivirus* genus.

Symptoms and Host Range

BSMV naturally infects barley (*Hordeum vulgare*) and wheat (*Triticum aestivum*). The symptoms in barley depend on strain and range, from a mild stripe mosaic (Fig. left panel; courtesy of A.O. Jackson) to lethal necrosis; crop losses can be 20% or more. (Fig. right panel shows infection of a susceptible variety compared with that of a resistant variety; courtesy of A.O. Jackson.) It has an artificial host range of more than 250 monocot and dicot species.

Strains

There are several strains causing different symptoms in barley, wheat, and oats.

Geographic Distribution

Worldwide

Transmission

No natural vector known. BSMV is highly seed transmitted in barley and is pollen-borne. It is mechanically transmissible.

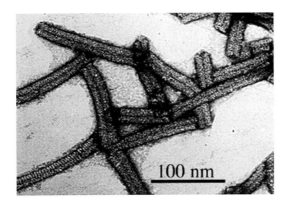

100 nm

Particles

BSMV has tubular rod-shaped particles 22 nm in diameter and of 2–4 lengths (100–150 nm), depending on strain.

Genome and Genome Organisation

The infectious genome is divided between three species of positive sense ssRNA that are designated α, ß, and γ. Their sizes vary between strains, with some containing sequence duplications. Each RNA has a 5' cap, a 3' tRNA-like structure, and a poly(A) sequence internal from the 3' end.

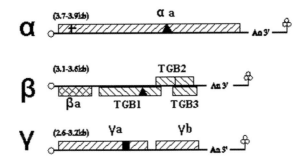

○ represents the 5' cap, ⚘ the 3' t-RNA-like structure, An3' shows a poly(A) sequence, methyl transferase domain, ▲ helicase domain, ■ RdRp domain. RNAα encodes a single replicase subunit protein (αa); RNAß encodes four proteins: ßa, the virus coat protein and TGB1, TGB2, and TGB3 that for the triple gene cell-to-cell movement proteins; RNAγ encodes two protein: γa, which together with αa forms the complete replicase and γb, which affects the pathogenicity of the virus. TGB1, TGB2, TGB3, and γb are expressed from subgenomic RNAs.

Notes

Infection with BSMV can cause mutation in host plants.

Further Information

Atabekov, J.G. and Novikov, V.K. (1989). Barley stripe mosaic virus. *Association of Applied Biologists descriptions of plant viruses*, No. 344.
Bragg, J.N., Lim, H.-S., and Jackson, A.O. (2008). Hordeiviruses. *Encyclopedia of Virology*, Vol. 2, 459–466.

Profile No. 2

BEET NECROTIC YELLOW VEIN VIRUS (BNYVV)

Classification

BNYVV is the type species of the genus *Benyvirus*.

Symptoms and Host Range

BNYVV causes rhizomania disease of sugar beet (*Beta vulgaris*). The most characteristic symptom is root stunting and proliferation of the lateral roots (Fig). In early infections, plants are often stunted, and can wilt and die; in later infections, virus may reach the leaves, causing yellow and necrotic vein symptoms. The host range is very narrow. [Figure courtesy of *Association of Applied Biologists Descriptions of Plant Viruses*, No. 144.]

Strains

There are three major strain groups of BYNVV (A, B, and P-type) based on nucleic acid sequence and RNA composition.

Geographic Distribution

Worldwide in most countries in which sugar beet is grown.

Transmission

BYNVV is transmitted by the plasmodiophorid vector, *Polymyxa betae* (see Chapter 12 for fungus transmission). It is mechanically transmissible to experimental hosts and is not seed transmitted.

Particles

BNYVV is a multicomponent virus with rigid rod-shaped particles; predominant lengths of 390, 265, 105, 90, and 80 nm; and diameters of 20 nm. Each segment of the multipartite genome is encapsidated in separate particle by a single coat protein species of 21 kDa. [Figure courtesy of M. Stephens.]

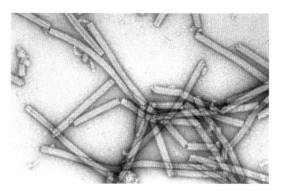

Genome and Genome Organisation

The infectious genome comprises two species of positive sense ssRNA (RNA1 and 2), but in natural infections three other RNA species (RNA 3, 4, and 5) are found that contribute to the biology of the virus. Each of the RNAs has a 5′ cap and is polyadenylated at the 3′ end.

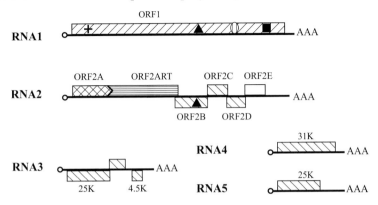

○ 5′ cap; + methyl transferase; ○ protease; ▲ helicase; ■ RdRp; ᴀᴀᴀ the 3′ Poly-A; 〉 = read-through. RNA1 encodes a single polypeptide of 237 kDa that is the replicase; RNA2 has 6 ORFs: ORF2A is the 21 kDa coat protein that reads through into ORF2ART, the read-through protein being involved in virus assembly and fungus transmission, ORFs 2B, 2C, and 2D form the triple block involved in cell-to-cell movement, ORF2E encodes a 14 kDa protein, which is a silencing suppresser; RNA3 has three ORFs, and RNAs 4 and 5 each contain one ORF; ORFs 2B, 2C, 2D, 2E, and the two 3′ ORFs in RNA3 are expressed from subgenomic RNAs.

Notes

Rhizomania is the most severe virus disease of sugar beet, limiting production in many areas.

Further Information

Schirmer, A., Link, D., Cogmat, V., Moury, B., Beuve, M., Meunier, A., Bragard, C., Gilmer, D., and Lemaire, O. (2005). Phylogenetic analysis of isolates of Beet necrotic yellow vein virus collected worldwide. *J. Gen. Virol.* **86**, 2897–2911.

Stevens, M., Lui, H.-Y., and Lemaire, O. (2006). The viruses, in *The sugar beet book* (A.P. Draycott, Ed.), Blackwells Publishing, Oxford, UK, pp. 256–285.

Tamada, T. (2002). Beet necrotic yellow vein virus. *Association of Applied Biologists Descriptions of Plant Viruses*, No. 144.

Profile No. 3

FAMILY *BROMOVIRIDAE*
GENUS *BROMOVIRUS*

Brome mosaic virus (BMV)

Classification

BMV is the type species of the genus *Bromovirus* in the family *Bromoviridae*.

Symptoms and Host Range

Causes mosaic and brown streaks in leaves (Figure). Natural host range limited to family *Poaceae*; experimental host range includes *Nicotiana benthamiana* and several *Chenopodium* spp., in which it produces local lesions. BMV will also replicate in yeast, *Saccharomyces cerevisiae*. [Figure shows symptoms of BMV in barley; courtesy of Wooley and Kao (2004), *Descriptions of Plant Viruses*, No. 405.]

Strains

No strains reported.

Geographic Distribution

Worldwide

Transmission

Natural transmission not determined; *Xiphenema* nematodes shown to transmit experimentally.

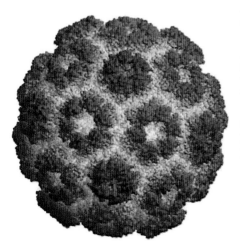

Particles

Isometric 27 nm diameter, T = 3 symmetry. Three particles components, one containing a molecule of RNA1, one containing a molecule of RNA2, and one containing one molecule each of RNA3 and RNA4. [Figure shows surface structure of BMV particle; this article was published in *Encyclopedia of Virology*, Bujarsky (B.W.G. Mahy and M. van Regenmortel, Eds.), Bromoviruses, No. 638. Copyright Elsevier (2008).]

Genome and Genome Organisation

The infectious genome is divided between three species of positive sense ssRNA; RNA1, 3.23 kb; RNA2, 2.87 kb; RNA3, 2.12 kb; the subgenomic RNA4 (1.2 kb) encoding coat protein is also encapsidated. The 5' end of each RNA is capped, the 3' end has a t-RNA-like structure (see Chapter 6).

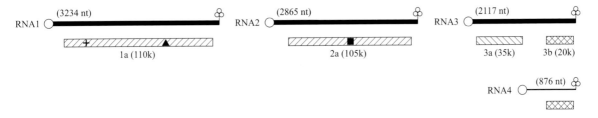

○ represents the 5' cap, ⚲ the 3' t-RNA-like structure, + methyl transferase domain, ▲ helicase domain, ■ RdRp domain. RNAs 1 and 2 encode the replicase; RNA3 is bi-cistronic encoding the-cell-to-cell movement protein and the coat protein that is expressed by subgenomic RNA4.

Notes

BMV has been extensively studied as a model for RNA replication, gene expression, and virion assembly of positive-strand RNA viruses.

Further Information

Bujarski, J.J. (2008). Bromoviruses. *Encyclopedia of Virology*, original 638.

Kao, C.C. and Sivakumaran, K. (2000). *Brome mosaic virus*, good for an RNA virologist's basic needs. *Molec. Plant Pathol.* **1**, 91–97.

Wang, X. and Ahlquist, P. (2008). Brome mosaic virus. *Encyclopedia of Virology*, Vol. 1, 381–385.

Wooley, R.S. and Kao, C.C. (2004). Brome mosaic virus. *Association of Applied Biologists Descriptions of Plant Viruses*, No. 405.

GENUS *CUCUMOVIRUS*

Cucumber mosaic virus (CMV)

Classification

Type member of genus *Cucumovirus* in family *Bromoviridae*.

Symptoms and Host Range

CMV in zucchini

CMV distortion of zucchini

CMV, causing fernleaf in tomato (from www.avdrc.org)

Causes mosaic symptoms in leaves of many dicot species and chlorotic streaks in monocot species, deformation of leaves (fernleaf in tomato) and of fruits, and sometimes severe necrosis. Very broad host range including >1,300 species in >100 monocot and dicot plant families; isolated from natural infections of >500 species.

Strains

Several strains differing in symptoms, host range, transmission, serology, and nucleotide sequence. Nucleotide sequence and serology give two subgroups, I and II; subgroup I divided into IA and IB on nucleotide sequence.

Geographic Distribution

CMV is found worldwide.

Transmission

Transmitted by >80 spp. of aphid (in 33 genera) in the nonpersistent manner (see Chapter 12). Mechanically transmissible and also seed transmitted in some plant spp.

Particles

Isometric 28–30 nm diameter, T = 3 symmetry. Three particles components, one containing a molecule of RNA1, one containing a molecule of RNA2, and one containing one molecule each of RNA3 and RNA4. The Figure shows particles of CMV negatively stained in uranyl acetate, bar marker = 100 nm. [Courtesy of Palukaitis and Garcia-Arenal (2003), *Descriptions of Plant Viruses*, No. 400.]

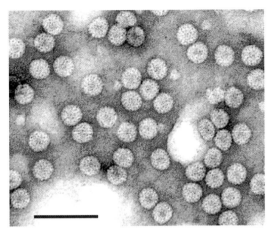

Genome and Genome Organisation

The infectious genome is divided between three species of positive sense ssRNA; RNA1, 3.36 kb; RNA2, 3.05 kb; RNA3, 2.16 kb; the subgenomic RNA4, (1.03 kb) encoding coat protein is also encapsidated. The 5′ end of each RNA is capped, the 3′ end has a t-RNA structure (see Chapter 6).

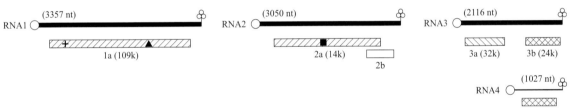

O represents the 5′ cap, ⌓ the 3′ t-RNA structure, + methyl transferase domain, ▲ helicase domain, ■ RdRp domain. RNAs 1 and 2 encode the replicase; RNA3 is bi-cistronic encoding the cell-to-cell movement protein and the coat protein that is expressed by subgenomic RNA4.

Notes

CMV is an economically important virus worldwide. Combined with a satellite RNA, it causes a severe necrotic disease of tomato (see Box 3.3).

Further Information

Garcia-Arenal, F. and Palukaitis, P. (2008). Cucumber mosaic virus. *Encyclopedia of Virology*, Vol. 1, 614–619.

Palukaitis, P. and Garcia-Arenal, F. (2003). Cucumber mosaic virus. *Association of Applied Biologists Descriptions of Plant Viruses*, No. 400.

GENUS *ALFAMOVIRUS*

Alfalfa mosaic virus (AMV)

Classification

AMV is the type and only member of the genus *Alfamovirus* in the family *Bromoviridae*.

Symptoms and Host Range

AMV infections are economically important in alfalfa (*Medicago sativa*), where they can cause mosaics and mottles; often infection is symptomless but can render the plants susceptible to frost. It is found naturally in more than 150 species. The experimental host range is more than 600 species. In tobacco (*Nicotiana tabacum*) it causes necrotic and chlorotic local lesions and a systemic mottle (Figure).

Strains

There are numerous strains of AMV that differ in the symptoms that they produce.

Geographic Distribution

Worldwide

Transmission

AMV is transmitted by aphids in the nonpersistent manner. It is also transmitted mechanically and by seed.

Particles

There are four major classes of particles called bottom component (B), middle component (M), top component *b* (Tb), and top component *a* (Ta). B, M, and Tb are bacilliform and contain genomic RNAs 1, 2, and 3, respectively; Ta component contains two molecules of subgenomic RNA4 and has two forms: bacilliform Ta-b and spheroidal Ta-t particles. The bacilliform particles (Fig) are made up of a single coat protein species (24.3 kDa) and are 19 nm wide and have lengths of 56 nm (B), 43 nm (M), 30 nm (Tb), and 30 nm (Ta-b). The Figure shows a preparation of AMV negatively stained in uranyl acetate; bar marker is 50 nm. (See also Figure 5.8.)

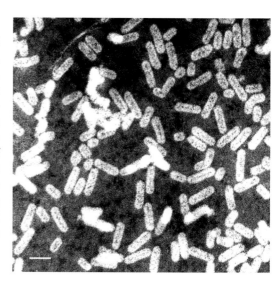

Genome and Genome Organisation

The infectious genome is divided between three species of positive sense ssRNA; RNA1, 3.64 kb; RNA2, 2.59 kb; RNA3, 2.14 kb; the subgenomic RNA4 (0.88 kb) encoding coat protein is also encapsidated. The 5′ end of each RNA is capped, the 3′ end has a t-RNA-like structure (see Chapter 6).

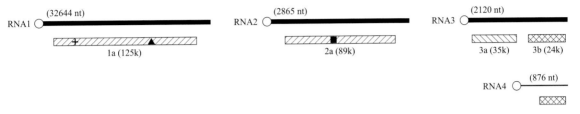

$\bigcirc$ represents the 5′ cap, $\oplus$ the 3′ t-RNA-like structure, + methyl transferase domain, ▲ helicase domain, ■ RdRp domain. RNAs 1 and 2 encode the replicase; RNA3 is bi-cistronic encoding the cell-to-cell movement protein and the coat protein that is expressed by subgenomic RNA4.

Notes

AMV requires its coat protein for replication. This is supplied either by the capsid of the incoming virus or by the presence of RNA4 in RNA preparations.

Further Information

Bol, J.F. (2003). *Alfalfa mosaic virus:* Coat protein-dependent initiation of infection. *Molec. Plant Pathol.* **4**, 1–8.
Bol, J.F. (2008). Alfalfa mosaic virus. *Encyclopedia of Virology*, Vol. 1, 81–86.
Jaspars, E.M.J. and Bos, L. (1980). Alfalfa mosaic virus. *Association of Applied Biologists Description of Plant Viruses*, No. 229.

Profile No. 4

FAMILY *CAULIMOVIRIDAE*
GENUS *CAULIMOVIRUS*

Cauliflower mosaic virus (CaMV)

Classification

Type species of the genus *Caulimovirus* in the family *Caulimoviridae*.

Symptoms and Host Range

CaMV in cauliflower   Mosaic in   Symptoms of CaMV strains in turnip
                       cauliflower leaf   leaves; healthy in centre.

CaMV infection causes a mosaic and mottle disease in many *Cruciferae*. The central picture shows chlorotic vein-banding. The natural host range is limited to the family *Cruciferae*; some strains also infect *Nicotiana clevelandii* and *Datura stramonium* on artificial infection.

Strains

CaMV is a variable virus with many strains that vary in virulence from very severe to very mild (Figure right hand panel).

Geographic Distribution

Worldwide

Transmission

CaMV is transmitted by aphids in the nonpersistent manner using a virus-encoded helper protein (see Chapter 12). The virus is mechanically transmissible but is not seed transmitted.

Particles

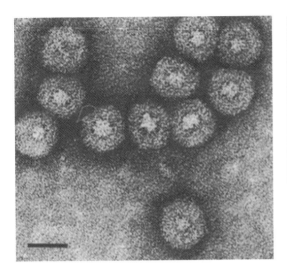
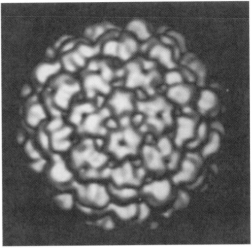

CaMV particles negatively stained in ammonium molybdate. Bar marker = 50 nm.

Reconstruction of CaMV surface structure. From Cheng et al (1992; *Virology* **186**, 655–668).

The virus has isometric particles (T = 7) 50 nm in diameter (see Chapter 5). The Figure panels show negatively stained CaMV preparation and the T = 7 structure.

Genome and Genome Organisation

Each particle contains a single circular molecule of dsDNA of about 8 kbp. Most strains have three discontinuities in the DNA, two in one strand and one in the other; one strain has only one in each strand. Replication is by reverse transcription (see Chapter 8).

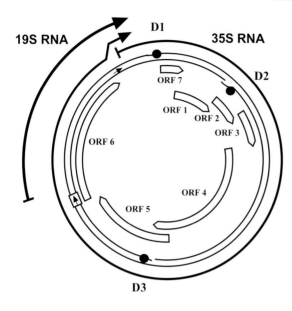

The double complete circles represent the dsDNA genome with the positions of the discontinuities marked D1, D2, and D3. The boxed arrows show the positions of the promoters. Outside the double circles the positions of the 35S and 19S transcripts are shown by the arrowhead marking the 3′ ends. Inside the double circle the arced boxes represent ORFs. ORF product functions: 1, cell-to-cell movement; 2, aphid transmission; 3, DNA-binding protein; 4, coat protein; 5, protease, reverse transcriptase, RNaseH; 6, translational transactivator; 7 unknown.

Notes

CaMV was the first plant virus shown to replicate by reverse transcription. The 35S promoter is widely used in constructs for plant transformation.

Further Information

Haas, M., Bureau, M., Geldereich, A., Yot, P., and Keller, M. (2002). *Cauliflower mosaic virus*: Still in the news. *Molec. Plant Pathol.* **2**, 419–430.

Hohn, T. (2008). [*Caulimoviridae*] Molecular biology. *Encyclopedia of Virology*, Vol. 1, 464–468.

Schoelz, J.E. (2008). [*Caulimoviridae*] General features. *Encyclopedia of Virology*, Vol. 1, 457–463.

Shepherd, R.J. (1981). Cauliflower mosaic virus. *Association of Applied Biologists Descriptions of Plant Viruses*, No. 243.

GENUS *BADNAVIRUS*

Banana streak virus (BSV)

Classification

A species in the genus *Badnavirus* in the family *Caulimoviridae*.

Symptoms and Host Range

Causes chlorotic streaks in leaves, which become necrotic with time; the expression of these leaf symptoms can often be periodic. Also may cause stunting of the plant, constriction of the emerging bunch, and detachment and splitting of outer sheaths of pseudostem. Host range limited to *Musaceae*.

Strains

Very variable with several viruses (BSV-Mys, BSV-GF, BSV-IM, BSV-OL) causing similar symptoms.

Geographic Distribution

BSV probably occurs wherever bananas are grown.

Transmission

Transmitted by mealybugs, most probably in a semipersistent manner. Transmitted in cultivation mainly by planting infected suckers. The virus is not mechanically transmissible.

Particles

Bacilliform particles 30 nm in diameter and 130–150 nm in length; longer particles are found with some strains. The figure shows a BSV preparation negatively stained in uranyl acetate; bar marker = 100 nm.

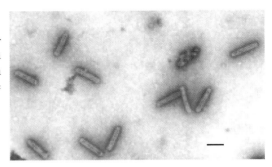

Genome and Genome Organisation

dsDNA circular molecule of about 7.4 kbp with one discontinuity in each strand. Replicated by reverse transcription (see Chapter 8). Viral sequences integrate into the host genome and in some cases can be activated to give episomal infection (see Box 8.9).

The double complete circles represent the dsDNA genome. Outside the double circles the arced boxes represent ORFs. ORF product functions: 1, unknown; 2, unknown; 3, a polyprotein cleaved to give the cell-to-cell movement protein, the coat protein, the aspartate protease and the replicase comprising reverse transcriptase and RNase H.

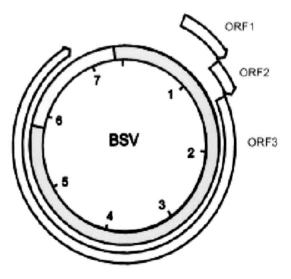

Notes

BSV was the first plant virus shown to have sequences integrated in the host genome that are activatable to give episomal infection.

Further Information

Geering, A.D.W. and Thomas, J.E. (2002). Banana streak virus. *Association of Applied Biologists Descriptions of Plant Viruses*, No. 390.

Harper, G., Hull, R., Lockhart, B., and Olszewski, N. (2002). Viral sequences integrated into plant genomes. *Annu. Rev. Phytopathol.* **40**, 119–136.

GENUS *TUNGROVIRUS*

Rice tungro bacilliform virus (RTBV)

Classification

Type species of the genus *Tungrovirus* in the family *Caulimoviridae*.

Symptoms and Host Range

Causes stunting of the plant, red or orange colouring of the leaves, and reduction in tiller number. When complexed with *Rice tungro spherical virus* (RTSV) the symptoms are more severe. It has a limited host range mainly in the *Poaceae*.

Strains

The genomes of RTBV isolates from Southeast Asia differ from those from the Indian subcontinent in that the latter have a deletion in a noncoding region. There are strains of the Southeast Asian type that differ in response to resistance genes in rice.

Geographic Distribution

RTBV occurs in South East and East Asia and in the Indian subcontinent (see Figure 4.4).

Transmission

Transmitted by leafhoppers only when RTSV has been acquired previously or at the same time. Transmission is in a semipersistent manner by several leafhopper species, the most important being the rice green leafhopper, *Nephotetix virescens*.

Particles

Bacilliform particles 30 nm in diameter and 130–150 nm in length very similar to those of *Banana streak virus* shown previously.

Genome and Genome Organisation

dsDNA circular molecule of about 8.0 kbp with one discontinuity in each strand. Replicated by reverse transcription (see Chapter 8).

The double complete circles represent the dsDNA genome with the discontinuities marked D1 and D2. Inside the double circles the arced boxes represent ORFs. ORF product functions: 1, (P24) unknown; 2, (P12) unknown; 3, (P194) a polyprotein cleaved to give the cell-to-cell movement protein, the coat protein, the aspartate protease and the replicase comprising reverse transcriptase and RNase H; 4, (P48) the transactivator protein. Outside the double circles, the positions of the 35S RNA transcript and the splices mRNA for ORF4 are shown, with arrowheads marking the 3′ ends.

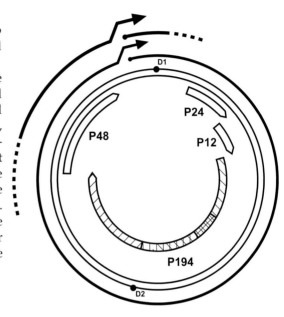

Notes

RTBV complexes with RTSV to give the very severe tungro disease of rice. RTBV has various interesting molecular features such as the non-AUG start codon for ORF1 and the transactivator properties of the product of ORF4 that are described in this book.

Further Information

Hull, R. (2004). Rice tungro bacilliform virus. *Association of Applied Biologists Descriptions of Plant Viruses*, No. 406.

Profile No. 5

GENUS *CLOSTEROVIRUS*

CITRUS TRISTEZA VIRUS (CTV)

Classification

A member of the genus *Closterovirus* in the family *Closteroviridae*.

Symptoms and Host Range

The severity of tristeza or quick decline depends on the virus strain, the citrus species, and the rootstock onto which it has been grafted; it is important on commercial varieties of citrus on sour orange (*C. aurantium*) rootstocks. Severe strains cause rapid wilting of the scion (see left-hand Figure), stem pitting of the trunk, and honeycombing immediately below the bud union; other milder symptoms include vein yellowing (right-hand Figure). The host range is mainly limited to the plant family *Rutaceae*. [Figures from Moreno *et al.* (2008; *Molec. Plant Pathol.* **9**, 251–268).]

Strains

CTV has many strains that show a wide variety of symptoms.

Geographic Distribution

Worldwide wherever citrus is grown.

Transmission

The virus is transmitted by aphids (mainly *Toxoptera citricida* and *Aphis gossypii*) in semipersistent manner.

Particles

Very flexuous rod-shaped particles about 2,000 nm in length and 10 nm in diameter. Figure. Bar = 100 nm [This figure was published in *Virus Taxonomy, 7th Report of the International Committee on the Taxonomy of Viruses*, G.P. Martelli, A.A. Agranovsky, M. Bar-Joseph *et al.* (M.H.V. van Regenmortel, C.M. Fauquet, D.H.L. Bishop *et al.*, Eds.), Closteroviridae, p. 943–964, Copyright Elsevier Academic Press, San Diego (2000).]

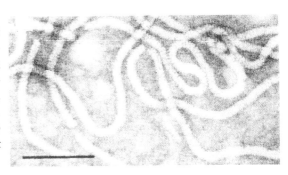

Genome and Genome Organisation

The genome of CTV is (+)-sense ssRNA of 19.3 kb. It has a 5′ cap and a 3′ hydroxyl group and contains 12 ORFs.

19.3 kb

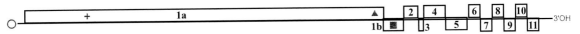

○ represents the 5′ cap, 3′OH the 3′ hydroxyl group, + methyl transferase domain, ▲ helicase domain, ■ RdRp domain. The viral replicase is expressed from ORFs 1a and 1b, 1b reading through from 1a; ORF2 encodes a 33 kDa protein of unknown function; ORF3, a 6 kDa hydrophobic protein; ORF4 a 65 kDa homologue of heat shock protein HSP70; ORF5, a 61 kDa protein involved in virion assembly; ORF6, the 27 kDa modified coat protein; ORF7, the 25 kDa coat protein; ORF8, an 18 kDa protein and ORF9, a 13 kDa protein, both of unknown function; ORF10, a 10 kDa protein and ORF11, a 23 kDa protein, both suppressors of silencing. ORFs 2–11 are expressed from subgenomic RNAs (see Figure 7.5).

Notes

CTV is the most important virus of citrus and can be a serious constraint to production. It has the largest genome of any plant virus.

Further Information

Bar-Joseph, M. and Lee, R.F. (1989). Citrus tristeza virus. *Association of Applied Biologists Descriptions of Plant Viruses*, No. 353.

Bar-Joseph, M. and Dawson, W.O. (2008). Citrus tristeza virus. *Encyclopedia of Virology*, Vol. 1, 520–524.

Dolja, V.V., Kreuze, J. F., and Valkonen, J.P.T. (2006). Comparative and functional genomics of closteroviruses. *Virus Res.* **117**, 38–51.

Moreno, P., Ambrós, S., Albiach-Marti, M.R., Guerri, J., and Pena, L. (2008). Citrus tristeza virus: A pathogen that changed the course of the citrus industry. *Molec. Plant Pathol.* **9**, 251–268, Wiley-Blackwell.

BEET YELLOWS VIRUS (BYV)

Classification

BYV is the type member of the genus *Closterovirus* in the family *Closteroviridae*.

Symptoms and Host Range

BYV causes vein clearing and vein yellowing symptoms (see right-hand Figure) in the young leaves of infected plants, leading to yellowing and thickening (left-hand Figure) of older leaves. It has a moderate host range infecting more than 120 species.

Strains

There are several strains of BYV differing in the symptoms in beet and other hosts.

Geographic Distribution

Worldwide in sugar beet growing areas.

Transmission

The virus is transmitted by aphids (mainly *Myzus persicae* and *Aphis fabae*) in semipersistent manner. It is mechanically transmissible but not seed transmitted.

Particles

As with CTV very flexuous rod-shaped particles about 1,250–1,450 nm long and 19 nm in diameter. They are composed of the genomic RNA and two species of coat protein, a major one (p22) and a minor one (p24) associated with aphid transmission (see Figure 5.1). [Figure courtesy of M. Stevens.]

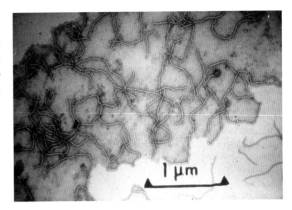

Genome and Genome Organisation

The genome of BYV is (+)-sense ssRNA of 15.5 kb. It has a 5′ cap and a 3′ hydroxyl group and contains 9 ORFs. ○ represents the 5′ cap, 3′OH the 3′ hydroxy, + methyl transferase domain, ▲ helicase domain, ■ RdRp domain.

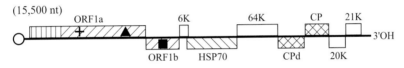

ORFs 1a and 1b encode the replication enzymes; the two coat proteins are expressed from ORFs Cp and CPd; the 6 K, HSP70, and 64 K are involved in cell-to-cell movement together with the two coat proteins; the 20 K and 21 K products are thought to act as silencing suppressors.

Notes

BYV is one of the major viruses of sugar beet.

Further Information

Agronovsky, A.A. and Lesemann, D.E. (2000). Beet yellows virus. *Association of Applied Biologists Descriptions of Plant Viruses*, No 377.

Dolja, V.V., Kreuze, J.F., and Valkonen, J.P.T. (2006). Comparative and functional genomics of closteroviruses. *Virus Research* **117**, 38–51.

Peremyslov, V.V., Andreev, I.A., Prokhnevsky, A.I., Duncan, G.H., Taliansky, M.E., and Dolja, V.V. (2004). Complex molecular architecture of beet yellow virus particles. *Proc. Natl. Acad. Sci.* **101**, 5030–5035.

Profile No. 6

FAMILY COMOVIRIDAE
GENUS COMOVIRUS

Cowpea mosaic virus (CPMV)

Classification

Type species of genus *Comovirus* in family *Comoviridae*.

Symptoms and Host Range

Causes a mosaic, decrease in leaf area, and reduced flower production; yield loss of up to 75%. Limited host range mainly in family *Leguminosae*; experimental infection of *Chenopodium amaranticolor* giving local lesions.

Strains

Several strains differing mainly in symptoms have been described.

Geographic Distribution

Nigeria, Tanzania, Japan, Surinam, Cuba, and the United States.

Transmission

Transmitted by beetles, especially members of family *Chrysomelidae* (see Chapter 12). Mechanically transmissible; seed transmitted in cowpea.

Particles

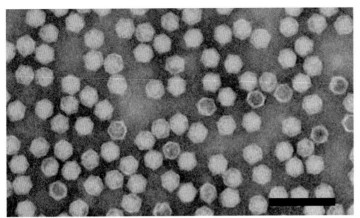

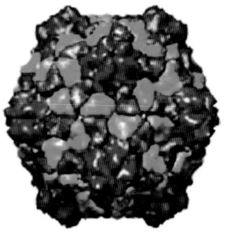

Purified preparation of CPMV negatively stained in 2% phosphotungstic acid, pH 6.8; bar = 100 nm. [Courtesy of *Descriptions Plant Viruses*, No. 378.]

Model showing structure of CPMV. Source www.primidi.com/2006/03/20.html

Isometric particles, 28–30 nm in diameter (left-hand picture) with T = 1 (pseudo T = 3) symmetry (right-hand picture) (See Chapter 5). Particles composed of two coat protein species. Three sedimenting components, B containing B-RNA; M, M-RNA; and T, no RNA (empty) that show in electron micrograph.

Genome and Genome Organisation

Infectious genome divided between two molecules of positive-sense ssRNA with 5′ VPg and 3′ Poly-A (see Chapter 6); M-RNA, 5.90 kb; B-RNA, 3.48 kb.

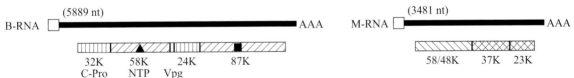

□ represents the 5′ VPg, ᴀᴀᴀ the 3′ Poly-A, ■ the RdRp motif, ▲ the helicase motif. Each of the two RNAs is expressed as a polyprotein that is processed to give: B-RNA; a 32 K and 24 K protease, a 58 K nucleotide-binding protein, the VPg and the 78K replicase protein; M-RNA; the 58/48 K cell-to-cell movement protein and the 37 K and 23 K coat proteins.

Notes

The coat protein of CPMV is modified to make vaccines against animal viruses (see Chapter 15).

Further Information

Pouwels, J., Carette, J.E., van Lent, J., and Wellink, J. (2002). *Cowpea mosaic virus*: Effects on host cell processes. *Molec. Plant Pathol.* **2**, 411–418.

Lomonossoff, G. P. Cowpea mosaic virus. *Encyclopedia of Virology*, Vol. 1, 569–573.

Van Kammen, A., van Lent, J., and Wellink, J. (2001). Cowpea mosaic virus. *Association of Applied Biologists Descriptions of Plant Viruses*, No. 378.

GENUS *NEPOVIRUS*

Tobacco ringspot virus (TRSV)

Classification

TRSV is the type member of the genus *Nepovirus* in the family *Comoviridae*. It belongs to sub-group A of the genus.

Symptoms and Host Range

The virus causes ring and line patterns in the leaves of tobacco (*Nicotiana tabacum*) (Fig.). It also causes a severe disease, bud blight of soybean in which the buds become brown and brittle, and necrotic streaking appears on the stem. TRSV has a wide experimental host range.

Strains

There are many variants of TRSV distinguishable on symptoms.

Geographic Distribution

The virus is endemic in central and eastern North America and is found in crops (especially perennial plants) in various other places worldwide.

Transmission

TRSV is transmitted by nematodes, *Xiphenema* spp. (see Chapter 12 for nematode transmission). It is also mechanically and seed transmitted.

Particles

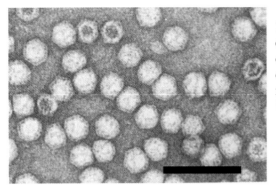

The particles are isometric, 28 nm in diameter, T = 1 (pseudo T = 3) with the two genomic segments encapsidated separately in a capsid of 60 subunits of the 57 kDa coat protein. Figure shows preparation of TRSV negatively stained in uranyl acetate; bar marker = 100 nm. [Courtesy of *Descriptions of Plant Viruses*, No. 309.]

Genome and Genome Organisation

The infectious genome is divided between two species of positive-sense ssRNA (RNA1 and 2), the 5′ end of each having a VPg and the 3′ end being polyadenylated.

The diagram shows the genome organisation of *Beet ringspot virus* as a representative of subgroup A nepoviruses. □ represents the 5′ VPg, ᴀᴀᴀ the 3′ Poly-A, ▪ the RdRp motif, ▲ the helicase motif. Each of the two RNAs is expressed as a polyprotein that is processed to give: B-RNA; P1A and P1C are proteases, P1B is a nucleotide-binding protein, the VPg and P1E the replicase protein; M-RNA; P2A is the cell-to-cell movement protein and P2B the coat protein. The processing pathway is similar to that of CPMV/

Notes

There are three subgroups of nepoviruses (A, B, C), which differ in genome organisation. Nepoviruses often have satellite RNAs associated with them (see Chapter 3).

Further Information

Sanfaçon, H. (2008) Nepovirus. *Encyclopedia of Virology*, Vol. 3, 405–412.
Stace-Smith, R. (1985). Tobacco ringspot virus. *Association of Applied Biologists Descriptions of Plant Viruses*, No. 309.

Profile No. 7

FAMILY GEMINIVIRIDAE
GENUS BEGOMOVIRUS

Tomato yellow leaf curl virus (TYLCV)

Classification

A species in the genus *Begomovirus* in the family *Geminiviridae*. TYLCV is a single component begomovirus (see ACMV following).

Symptoms and Host Range

Infected plants are stunted with small chlorotic leaves that have upward curling of the margins [Figure courtesy of Gafni (2003), *Molec. Plant Pathol.* **4**, 9–15.] The virus has a wide natural host range including *Phaseolus vulgaris* (common bean) and petunia.

Strains

There are numerous variants of this virus that, depending on sequence differences, are called strains of TYLCV (e.g., Israel, Iran, etc.) or different viruses (e.g., *Tomato yellow leaf curl Sardinia virus*, *Tomato yellow leaf curl Sudan virus*).

Geographic Distribution

Worldwide

Transmission

Transmitted in a circulative nonpropagative manner by the whitefly *Bemisia tabaci* (see Chapter 12). Not transmitted by seed or mechanically.

Particles

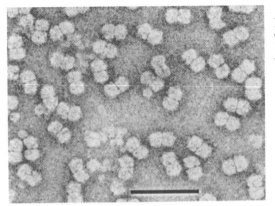

The particles are geminate, 20 × 30 nm. The Figure shows negatively stained particles of TYLCV; bar = 100 nm. [Courtesy of Gafni (2003), *Molec. Plant Pathol.* **4**, 9–15.]

Genome and Genome Organisation

The genome is a covalently closed circle of ssDNA, 2.79 kb. The diagram shows the genome organisation; the circle represents the ssDNA; the black box the common region; the arced boxes within the circle are the ORFs, the function being: VI, the coat protein; V2, cell-to-cell movement protein; C1, the Rep protein; C2, the transcriptional activator protein; C3, the replication enhancer protein; C4, unknown function.

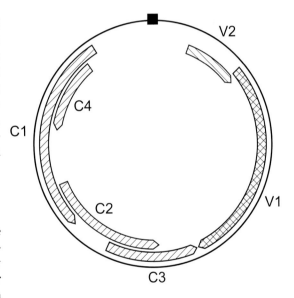

Notes

TYLCV is widespread and causes a severe disease of tomatoes. There is much recombination between its variants giving rise to new strains and viruses (see Chapter 4). Some other begomoviruses can cause similar symptoms in tomato.

Further Information

Czosnek, H. (1999). Tomato yellow leaf curl virus. *Association of Applied Biologists Descriptions of Plant Viruses*, No. 368.

Czosnek, H. (2008). Tomato yellow leaf curl virus. *Encyclopedia of Virology*, Vol. 5, 138–144.

Gafni, Y. (2003). *Tomato yellow leaf curl virus*, the intracellular dynamics of a plant DNA virus. *Molec. Plant Pathol.* **4**, 9–15.

AFRICAN CASSAVA MOSAIC VIRUS (ACMV)

Classification

A species in the genus *Begomovirus* in the family *Geminiviridae*. ACMV is a two component begomovirus. (See TYLCV previously.)

Symptoms and Host Range

Infection causes a severe mosaic and a 60 to 80% loss of yield. In nature the virus is only found in cassava; the experimental host range is narrow and includes several solanaceous species, such as *Nicotiana* spp.

Strains

There are several strains of ACMV differentiated on severity of symptoms.

Geographic Distribution

ACMV is found in many sub-Saharan African countries.

Transmission

Transmitted in a circulative nonpropagative manner by the whitefly *Bemisia tabaci* (see Chapter 12).

Particles

The particles are geminate, 20 × 30 nm, and are similar to those shown above for TYLCU.

Genome and Genome Organisation

The genome is divided between two molecules of circular ssDNA of about 2.78 kb (DNA A) and 2.72 kb (DNA B). The diagram shows the genome organisation; the circles represent the ssDNAs of components A and B; the black boxes, the common regions. The arced boxes within the circles are the ORFs, the functions being: AV1, the coat protein; AV2, cell-to-cell movement protein; AC1, the Rep protein; AC2, the transcriptional activator protein; AC3, the replication enhancer protein; AC4, unknown function; BV1, the nuclear shuttle protein; BC1, cell-to-cell movement.

Notes

ACMV is a major constraint to cassava cultivation in Africa. Interactions between it and closely related viruses have caused very severe epidemics (see Box 11.3).

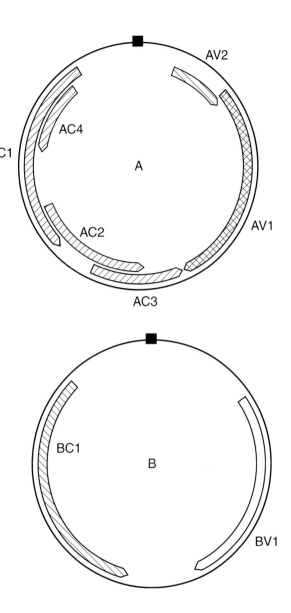

Further Information

Bock, K.R. and Harrison, B.D. (1985). African cassava mosaic virus. *Association of Applied Biologists Descriptions of Plant Viruses*, No. 297.

Fargette, D., Konate, G., Fauquet, C., Muller, E., Peterschmitt, M., and Thresh, J.M. (2006). Molecular ecology and emergence of tropical plant viruses. *Annu. Rev. Phytopathol.* **44,** 235–260.

ABUTILON MOSAIC VIRUS (ABMV)

Some Features of Other Begomoviruses

A two-component begomovirus with a genome organisation similar to that of ACMV. It has characteristic symptoms as shown by those caused by a West Indian isolate in *Abutilon sellovianum* var *marmorata*. [Figure courtesy of *Descriptions of plant viruses*, No. 373.] Infected abutilons are important as an ornamental plant.

Further Information

Jeske, H. (2000). Abutilon mosaic virus. *Association of Applied Biologists Descriptions of Plant Viruses*, No. 373.

BEAN GOLDEN MOSAIC VIRUS (BGMV)

Type species of the genus *Begomovirus*. An important virus of beans (*Phaseolus vulgaris*) in South and Central America causing a severe yellow or golden mottle of the leaves. It is a two-component begomovirus with a genome organisation similar to that of ACMV. It is transmitted by the whitefly, *Bemisia tabaci*.

Further Information

Goodman, R.M. and Bird, J. Bean golden mosaic virus. *Association of Applied Biologists Descriptions of Plant Viruses*, No. 192.
Morales, F.J. (2008). Bean golden mosaic virus. *Encyclopedia of Virology*, Vol. 1, 295–300.

COTTON LEAF CURL DISEASE

An important disease of cotton caused by a monopartite begomovirus and a satellite DNA ß and satellite DNA-1 (see Chapter 3 for DNA satellites). The disease causes severe losses in cotton in North Africa and the Indian subcontinent. The complex is transmitted by the whitefly, *Bemisia tabaci*. The genome organisation of the begomovirus component resembles that of TYLCV. On the Indian subcontinent, seven distinct begomovirus species have been associated with the disease.

Further Information

Mansoor, S., Amin, I. and Briddon, R.W. (2008). Cotton leaf curl disease. *Encyclopedia of Virology*, Vol. 1, 563–568.

GENUS *MASTREVIRUS*

Maize streak virus (MSV)

Classification

Type species of the genus *Mastrevirus* in the family *Geminiviridae*.

Symptoms and Host Range

The first symptoms on a susceptible maize cultivar are circular pale spots. These develop into chlorotic veinal streaks, 0.5 to 1.0 mm wide and a few mm to several cm long. The host range is restricted to the *Graminae* but it can infect >80 grass species.

Strains

There are nine major strains (MSV-A to MSV-I) that vary in severity of symptoms in maize; only MSV-A produces a severe disease in maize.

Geographic Distribution

MSV is found in sub-Saharan Africa and in the Indian Ocean islands of Madagascar, Mauritius, and La Réunion.

Transmission

The virus is transmitted by cicadellid leafhoppers in a persistent nonpropagative manner (see Chapter 12). The most important vector is *Cicadulina mbila*. Not transmitted mechanically or by seed.

Particles

Geminate particles 22 × 38 nm resembling those of begomoviruses.

Genome and Genome Organisation

The genome is a covalently closed circle of ssDNA, 2.6–2.8 kb.

The circle represents the ssDNA; the black box the common region. The arced boxes within the circle show the ORFs, the functions being: V1, cell-to-cell movement; V2, coat protein; C1 protein regulates the host DNA replication ability; C2 and C1: C2 (fusion protein) are replication associated proteins (see Chapter 8).

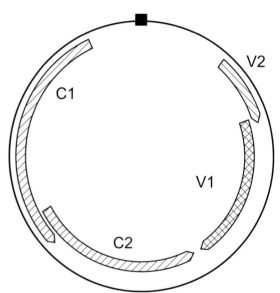

Notes

MSV is a major constraint to maize production in many African countries.

Further Information

Gafni, Y. (2003). *Tomato yellow leaf curl virus,* the intracellular dynamics of a plant DNA virus. *Molec. Plant Pathol.* **4**, 9–15, Wiley-Blackwell.

Martin, D.P. (2007). Maize streak virus. *Association of Applied Biologists Descriptions of Plant Viruses,* No. 416.

Martin, D.P., Shepherd, D.N., and Rybicki, E.P. (2008). Maize streak virus. *Encyclopedia of Virology,* Vol. 3, 263–271.

Profile No. 8

FAMILY *LUTEOVIRIDAE*

GENUS *LUTEOVIRUS*

Barley yellow dwarf virus—PAV (BYDV-PAV)

Classification

Type species of the genus *Luteovirus* in the family *Luteoviridae*.

Symptoms and Host Range

Stunting of plants and chlorosis or red/purple colour of leaves. Moderate natural host range limited to the family *Poaceae* causing yield losses of up to 30% in some cereal crops; also will infect some solanaceous species on artificial inoculation. (Figure from www. ag.nsdu.edu.)

Strains

Several variants of the virus that are classified as distinct species based mainly on their vector—for example, BYDV-PAV transmitted mainly by *Rhopalosiphon padi* and *Macrosiphon avenae* and BYDV-MAV, mainly by *M. avenae*.

Geographic Distribution

Worldwide

Transmission

Transmitted by aphids in the persistent, circulative, nonpropagative manner (see Chapter 12). Not mechanically or seed transmitted.

Particles

Isometric particles 25–30 nm in diameter with T = 3 symmetry (Figure; Bar = 20 nm). [Courtesy of *Association of Applied Biologists Descriptions of Plant Viruses*, No. 32.] Two proteins involved in the capsid (see Chapter 12).

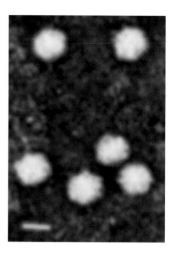

Genome and Genome Organisation

The genome is a single molecule of positive-sense ssRNA, 5.68 kb, the 3' end of which has a hydroxyl group, (the genomes of other genera in the family *Luteoviridae* have a 5' VPg; see Chapter 6).

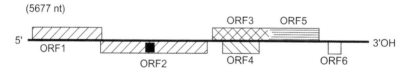

■ = RdRp motif, ❭ = read-through, 3'OH = 3' hydroxyl group. The genome encodes 6 ORFs the functions being: ORFs 1 and 2 viral replicase via a ribosomal frameshift (see Chapter 7); ORF3, major coat protein that forms a minor coat protein by read-through into ORF5 (see Chapter 7); ORF4, cell-to-cell movement; ORF6, function unknown; there are some minor ORFs within the major ones that may regulate transcription late in replication.

Notes

The luteoviruses cause serious losses to small grain cereal crops.

Further Information

Domier, L.L. (2008). Barley yellow dwarf viruses. *Encyclopedia of Virology*, Vol. 1, 279–285.
Domier, L.L. and D'Arcy, C.J.D. (2008). Luteoviruses. *Encyclopedia of Virology*, Vol. 3, 231–237.
Miller, W.A., Liu, S. and Beckett, R. (2002). Barley yellow dwarf virus: *Luteoviridae* or *Tombusviridae*. *Molec. Plant Pathol.* **3**, 177–184.
Rochow, W.F. (1970). Barley yellow dwarf virus. *Association of Applied Biologists Description of Plant Viruses*, No. 32.

GENUS *POLEROVIRUS*

Potato leafroll virus (PLRV)

Classification

Type species of the genus *Polerovirus* in the family *Luteoviridae*.

Symptoms and Host Range

Primary infections of potato (*Solanum tuberosum*) cause pallor of tip leaves, which may become rolled and erect. Secondary symptoms in plants grown from infected tubers are stunting and upward rolling of leaflets, especially those on the lower leaves. Left-hand panel of Figure shows infected plant surrounded by symptomless plants; right-hand panel is enlargement of infected plant showing leaf rolling. (Courtesy of S.A. Slack. Reprinted by permission from Compendium of Potato Diseases, 2nd ed., American Phytopathology Society, St. Paul, MN.) Causes considerable crop losses. Limited host range of about 20 species.

Strains

Several strains have been distinguished on symptom severity in potato and ease of aphid transmission.

Geographic Distribution

Worldwide where potatoes are grown.

Transmission

Transmitted by aphids in the persistent, circulative, nonpropagative manner (see Chapter 12). Not mechanically or seed transmitted.

Particles

Isometric particles 25–30 nm in diameter with T = 3 symmetry, similar to those for BYDV-PAV. Two proteins involved in the capsid (see Chapter 12).

Genome and Genome Organisation

The genome is a single molecule of positive-sense ssRNA, 5.68 kb, the 3′ end of which has a VPg.

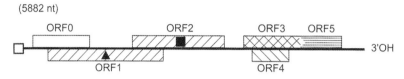

(5882 nt)

□ represents the 5′ VPg, ▲ the helicase motif, ■ = RdRp motif, ❭ = read-through, 3′OH = 3′ hydroxyl group. The genome encodes 6 major ORFs the functions being: ORF0, suppressor of RNA silencing; ORFs 1 and 2, viral replicase via a ribosomal frameshift (see Chapter 7); ORF3, major coat protein that forms a minor coat protein by read-through into ORF5 (see Chapter 7); ORF4, cell-to-cell movement; there are some minor ORFs within the major ones that may regulate transcription late in replication.

Notes

An important disease of potato.

Further Information

Harrison, B.D. (1984). Potato leafroll virus. *Association of Applied Biologists Descriptions of Plant Viruses*, No. 291.
Taliansky, M., Mayo, M.A., and Barker, H. (2003). Potato leafroll virus: A classic pathogen shows some new tricks. *Molec. Plant Pathol.* **4**, 81–89.

Profile No. 9

POTATO VIRUS X (PVX)

Classification

Type member of the genus *Potexvirus* in the family *Flexiviridae*.

Symptoms and Host Range

Natural infections of PVX cause a mild mosaic in potato (*Solanum tuberosum*), a mottle in tobacco (*Nicotiana tabacum*), and mosaic in tomato (*Lycopersicum esculentum*). It has a wide experimental host range of more than 240 species, mainly in the Solanaceae.

Strains

There are many minor variants of PVX.

Geographic Distribution

Worldwide in potato-growing areas.

Transmission

Most field transmission is by mechanical contact. It is not seed transmitted.

Particles

PVX has flexuous rod-shaped particles 515 nm long and 13 nm wide [Figure; bar = 100 nm]. [Courtesy of Brunt *et al.* (2000), in *Virus Taxonomy. Seventh Report of the International Committee on the Taxonomy of Viruses*, M.H.V. van Regenmortel *et al.*, Eds., pp. 975–981, Academic Press, San Diego.] The genomic RNA is encapsidated in a single species of coat protein.

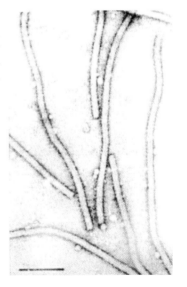

Genome and Genome Organisation

The PVX genome is a single molecule of positive-sense ssRNA, 5.84 kb, the 5′ end of which has a cap and the 3′ end is polyadenylated. The genome contains 5 ORFs.

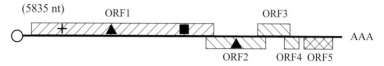

○ represents the 5′ cap, AAA shows a 3′ polyA sequence, + methyl transferase domain, ▲ helicase domain, ■ RdRp domain. ORF1 encodes the replicase; ORFs 2, 3, and 4 form the triple gene block for cell-to-cell movement; ORF5 encodes the coat protein. ORF1 is translated from the genomic RNA; ORFs 2–4 from subgenomic RNA 1 and ORF5 from subgenomic RNA2.

Notes

PVX is moderately important as a disease agent in potato and together with PVY is important in causing tomato streak disease.

Further Information

Koenig, R. and Lesemann, D.-E. (1989). Potato virus X. *Association of Applied Biologists Descriptions of Plant Viruses*, No. 354.
Ryu, K.H. and Hong, J.S. (2008). Potexvirus. *Encyclopedia of Virology*, Vol. 4, 310–313.

Profile No. 10

GENUS *POTYVIRUS*

POTATO VIRUS Y (PVY)

Classification

Type species of the genus *Potyvirus* in the family *Potyviridae*.

Symptoms and Host Range

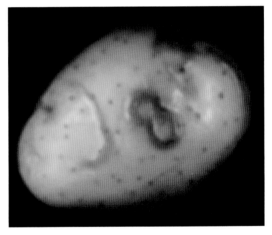

Symptoms vary according to host and strain of virus. Some strains are mild and cause a rugose mosaic (left-hand Figure); others cause a severe mosaic in potato with necrosis of tubers (right-hand Figure). [Figures courtesy of *Association of Applied Biologists Descriptions of Plant Viruses*, No. 242.] Moderate natural host range mainly in the family *Solanaceae*; experimental host range nearly 500 spp. In >70 genera of >30 plant families.

Strains

Three main strain groups PVY^O, PVY^N, and PVY^C, based on symptoms in *Nicotiana tabacum* cv Samsun, *Solanum tuberosum* ssp. *tuberosum*, and *Physalis floridana*.

Geographic Distribution

Worldwide in potato-growing areas.

Transmission

Transmitted by aphids in the nonpersistent manner; requires a virus-coded transmission factor (see Chapter 12). Can be transmitted mechanically; there is no evidence for seed transmission.

Particles

Long flexible particles about 750 nm long and 11 nm wide. Figure shows negatively stained preparation of PVY. [Courtesy of *Association of Applied Biologists Descriptions of Plant Viruses*, No. 242.]

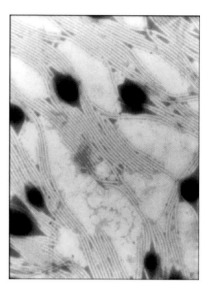

Genome and Genome Organisation

The genome is a single molecule of positive-sense ssRNA, about 9.7 kb, the 5′ end of which has a VPg and the 3′ end is polyadenylated (see Chapter 6). Potyviral genomes have very similar genome organisations—that of *Tobacco etch virus* is shown here.

(9704 nt)

AAA

| 35K | 52K | 50K | 6K | 71K | 6K | 21K | 27K | | 58K | 30K |
| P1 | HC-Pro | P3 | | | | Vpg | | | NIb | |

□ = VPg, ₐₐₐ = Poly-A. ▲ = helicase motif, ■ = RdRp motif. The potyvirus genome is expressed as a polyprotein that is processed to the gene products by proteinases (see Figure 7.4). The functions of the gene products are: P1, serine proteinase; HC-Pro a cysteine proteinase, helper component for aphid transmission, suppressor of RNA silencing; P3 + 6K, pathogenicity and host range determinant, 71K, cytoplasmic inclusion (CI) body protein RNA helicase, cell-to-cell movement, RNA replication; 6K, membrane anchoring; VPg, 5′ virus-linked genome protein; 27K, cysteine proteinase, priming RNA synthesis; (VPg + 27K proteins give the nuclear inclusion a - NIa); NIb, nuclear inclusion protein, RNA replicase; 30K, coat protein.

Notes

The potyvirus genus is the biggest genus of plant viruses with >110 species.

Further Information

Kerlan, C. (2006). Potato virus Y. *Association of Applied Biologists Description of Plant Viruses*, No. 242.
Kerlan, C. and Moury, B. (2008). Potato virus Y. *Encyclopedia of Virology*, Vol. 4, 288–301.
López-Moya, J.J. and Garcia, J.A. (2008). Potyviruses. *Encyclopedia of Virology*, Vol. 4, 314–324.

PAPAYA RINGSPOT VIRUS (PRSV)

Other Potyviruses

PRSV infection causes significant losses in papaya (*Carica papaya*). Leaves show mottling and distortion with ringspots on the fruit. The virus is found in most papaya growing countries.

There are several strains of PRSV that differ in severity of symptoms and in their genomic nucleic acid sequences.

Further Information

Gonsalves, D., Suzuki, J.Y., Tripathi, S., and Ferreira, S.A. (2008). Papaya ringspot virus. *Encyclopedia of Virology*, Vol. 4, 1–7.
Purcifull, D., Edwardson, J., Hiebert, E., and Gonsalvez, D. (1984). Papaya ringspot virus. *Association of Applied Biologists Descriptions of Plant Viruses*, No. 292.

PLUM POX VIRUS (PPV)

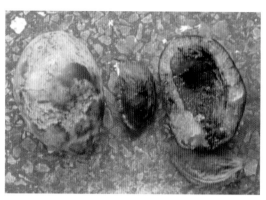

PPV causes Sharka disease or "plum pox" of *Prunus* spp (plum, apricot, peach, sour cherry, sweet cherry). Leaf symptoms are chlorotic ring spots and oak-leaf patterns; fruit symptoms are shallow rings and discolouring. [Figure courtesy of *Association of Applied Biologists Descriptions of Plant Viruses*, No. 410.] Premature fruit drop. The virus is found in Europe, North Africa, Asia Minor, China, and South and North America.

There are several strains of this important virus differing in symptoms in various hosts.

Further Information

Glasa, M. and Candresse, T. (2005). Plum pox virus. *Association of Applied Biologists Descriptions of Plant Viruses*, No. 410.
Glasa, M. and Candresse, T. (2008). Plum pox virus. *Encyclopedia of Virology*, Vol. 4, 238–242.

TOBACCO ETCH VIRUS (TEV)

TEV causes mottling and necrotic etching of some varieties of tobacco (*Nicotiana tabacum*); in other varieties it is less severe. It also causes diseases of peppers and tomatoes. It has a wide experimental host range of more than 120 species. It is found in North, Central, and South America. The study of TEV has contributed much to the understanding of the functioning of potyviruses.

Further Information

Purcifull, D.E. and Hiebert, E. (1982). Tobacco etch virus. *Association of Applied Biologists Descriptions of Plant Viruses*, No. 258.

Profile No. 11

RICE RAGGED STUNT VIRUS (RRSV)

Classification

Type species of the genus *Oryzavirus* in the family *Reoviridae.*

Symptoms and Host Range

Infected plants are stunted with white spindle-shaped enations on the back of twisted and ragged leaves. [Figure courtesy of P. Waterhouse.] Delayed flowering and unfilled grains lead to yield losses; generally losses are 10 to 20%, but severe infections can result in 100% loss. The host range is restricted to the genus *Oryza* and some other members of the *Poaceae.*

Strains

There are three strains of RRSV, -India, -Philippines, and -Thailand.

Geographic Distribution

RRSV has been reported from China, India, Indonesia, Japan, Malaysia, the Philippines, Sri Lanka, Taiwan, and Thailand.

Transmission

The virus is transmitted by the rice brown planthopper, *Nilopavarta lugens*, in a circulative propagative manner (see Chapter 12). Not mechanically or seed transmitted.

Particles

Isometric, 75–80 nm in diameter (Fig. bar = 50 nm); (courtesy of *Descriptions of plant viruses*, No. 248) with 12 A-type surface spikes; the inner "subviral" core is 57–65 nm in diameter and has 12 B-type spikes (see Chapter 5). Each particle contains a full complement of genome segments.

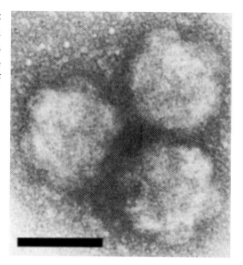

Genome and Genome Organisation

Ten segments of dsRNA ranging in size from 3.85 kbp to 1.16 kbp and totalling about 26 kbp. Each of the genomic segments is monocistronic, except segment 4, which is bicistronic. The segments code for: 1, B spike; 2, inner core capsid; 3, major core capsid; 4, RNA-dependent replicase and unknown; 5, capping enzyme/guanyltransferase; 6, unknown; 7, nonstructural protein; 8, precursor protease (major capsid); 9, vector transmission (spike); 10, nonstructural protein.

Notes

RRSV replicates in both plants and insects. It resembles many of the animal reoviruses.

Further Information

Geijskes, R.J. and Harding, R.M. (2008). Plant reoviruses (Phytoreoviruses and Fijiviruses). *Encyclopedia of Virology*, Vol. 4, 149–155.
Milne, R.G., Boccardo, G., and Ling, K.C. (1982). Rice ragged stunt virus. *Association of Applied Biologists Descriptions of Plant Viruses*, No. 248.

Profile No. 12

RICE STRIPE VIRUS (RSV)

Classification

Type species of the genus *Tenuivirus*, not assigned to any family.

Symptoms and Host Range

General stunting of the plant with chlorotic stripes and general chlorosis on leaves (Figure); early infection gives significant loss of yield of rice due to few or no flower panicles being formed. Host range limited to monocotyledons, mainly *Graminae*.

Strains

Some isolates differ in severity of symptoms.

Geographic Distribution

RSV causes important diseases in China, Japan, Korea, and Taiwan.

Transmission

Transmitted by planthoppers in the circulative propagative manner. The main field vector is the small brown planthopper, *Laodelphax striatellus*.

Particles

Filamentous nucleoprotein particles 500–2,000 nm long and 8 nm wide (Figure). Some particles may appear branched or circular.

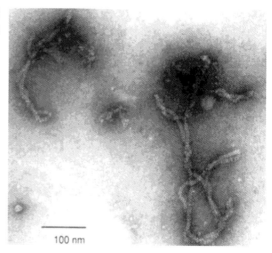

100 nm

Genome and Genome Organisation

The ssRNA genome of RSV is divided into 4 segments, the largest (RNA1) being negative sense, and the other three (RNA 2–4) being ambisense (see Chapter 7 for ambisense).

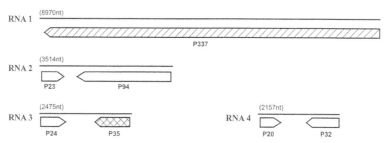

The product from the L segment is the RNA-dependent RNA polymerase (337 kDa). The functions of the two products from RNA2 are unknown. The P35 from RNA3 is the nucleocapsid protein; P24 is possibly a suppressor of RNA silencing. The P20 from RNA4 is a major nonstructural protein; the function of P32 is unknown.

Notes

RSV replicates both in plants and insects. Tenuiviruses have some similarities with members of the *Bunyaviridae*, particularly those in the genus *Phlebovirus*.

Further Information

Ramirez, B.C. (2008). Tenuivirus. *Encyclopedia of Virology*, Vol. 5, 24–26.
Toryama, S. (2000). Rice stripe virus. *Association of Applied Biologists Descriptions of Plant Viruses*, No. 375.

Profile No. 13

SONCHUS YELLOW NET VIRUS (SYNV)

Classification

Species in genus *Nucleorhabdovirus* in the family *Rhabdoviridae*.

Symptoms and Host Range

Infected plants show stunting, mosaic symptoms, veinal necrosis, yellowing and crinkling of leaves. Natural host range restricted to the Compositae; experimental host range includes *Nicotiana benthamiana*, [Figure courtesy of A. Jackson.] *N. clevelandii*, and *Chenopodium quinoa*, which is a local lesion host.

Strains

No strains known.

Geographic Distribution

Central and South Florida.

Transmission

Transmitted in a persistent manner by the aphid, *Aphis coreopsidis* in which it replicates.

Particles

Bacilliform particles 95 nm in diameter and 250 nm in length with surface spikes. The Figure of an electron micrograph of a negative stained particle shows the striated inner core, envelope, and glycoprotein spikes. [This article was published in *Encyclopedia of Virology*, Vol. 4, A.O. Jackson, R.G. Dietzgen, R.-X. Fang, M.M. Goodin, S.A. Hogenhout, M. Deng, and J.N. Bragg (B.W.J. Mahy and M.H.V. van Regenmortel, Eds.), Plant rhabdoviruses, pp. 187–196, Copyright Elsevier (2008).] For details of structure, see Chapter 5.

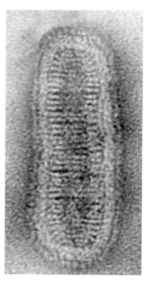

Genome and Genome Organisation

One component of negative-sense ssRNA, about 13.7 kb.

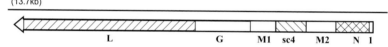

SYNV encodes 6 proteins that are translated from separate transcripts. The functions of the gene products are: L, RNA-dependent RNA replicase; G, glycoprotein that forms the spikes; M1 and M2, structural matrix proteins; sc4, cell-to-cell movement protein; N, nucleocapsid protein. For transcription and more on gene functions, see Chapter 7.

Notes

SYNV replicates both in plants and insect and resembles animal retroviruses; the gene product sc4 is not found in animal rhabdoviruses and demonstrates the adaptation of the genome to plants.

Further Information

Jackson, A.O. and Christie, S.R. (1979). Sowthistle yellow net virus. *Association of Applied Biologists Descriptions of Plant Viruses*, No. 205.

Jackson, A.O., Dietzgen, R., Goodin, M.M., Bragg, J.N., and Deng, M. (2005). Biology of plant rhabdoviruses. *Annu. Rev. Phytopathol.* **43**, 623–660.

Jackson, A.O., Dietzgen, R.G., Fang, R.-X-., Goodin, M.M., Hogenhout, S.A., Deng, M., and Bragg, J.N. (2008). Plant rhabdoviruses. *Enclopedia of Virology*, Vol. 4, 187–196.

Profile No. 14

TOBACCO MOSAIC VIRUS (TMV)

Classification

Type species of the genus *Tobamovirus*.

Symptoms and Host Range

 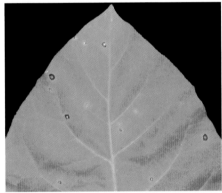

Symptoms of TMV in *N. tabacum*   Local lesions in *N. tabacum* NN
cv Samsun 3 weeks after infection   genotype 1 week after infection

TMV has a moderate host range including many solanaceous species. In *Nicotiana tabacum* cv Samsun the type strain causes initial vein clearing followed by a light and dark green mosaic, distortion, and blistering of systemically infected leaves (left-hand figure). Causes severe crop losses. In *N. tabacum* cv Samsun NN (right-hand figure) and *N. glutinosa* causes local lesions. [Figures courtesy of *Association of Applied Biologists Descriptions of Plant Viruses*, No. 370.]

Strains

Many strains differing in symptoms and host range; the Ob strain overcomes the N gene resistance.

Geographic Distribution

Worldwide

Transmission

Very infectious, transmitted by contact; see Chapter 12.

Particles

Rigid rod-shaped particles 18 × 300 nm; see Figure 4.1. The particles are very stable.

Genome and Genome Organisation

Linear (+)-sense ssRNA, type strain 6395nt.

○ 5′ cap; + methyl transferase; ▲ helicase; ❭ readthrough; ■ RdRp; ⚥ tRNA-like 3′ end. ORFs 1 and 2 Replicase; ORF3 movement protein; ORF4 coat protein.

Notes

TMV was the first pathogen to be recognised as a virus (see Chapter 1).

Further Information

Knapp, E. and Lewandowski, D.J. (2001). Tobacco mosaic virus, not just a single component virus anymore. *Molec. Plant Pathol.* **2**, 117–123.

Lewarndowski, D.J. (2008). Tobamovirus. *Encyclopedia of Virology*, Vol. 5, 68–71.

Van Regenmortel, M.H.V. (2008). Tobacco mosaic virus. *Encyclopedia of Virology*, Vol. 5, 54–59.

Zaitlin, M. (2000). Tobacco mosaic virus. *Association of Applied Biologists Descriptions of Plant Viruses*, No. 370.

Profile No. 15

TOBACCO RATTLE VIRUS (TRV)

Classification

Type species of the genus *Tobravirus*.

Symptoms and Host Range

TRV causes systemic necrotic flecks on tobacco (*Nicotiana tabacum*) leaves, stunting of the shoots, and sometimes death. In potato (*Solanum tuberosum*) it causes an important disease, spraign or corky ringspot, with necrotic rings in the tuber flesh. [Figure courtesy of *Association of Applied Biologists Descriptions of Plant Viruses*, No. 398.] The virus has a wide experimental host range of more than 400 species in more than 50 dicot and monocot families.

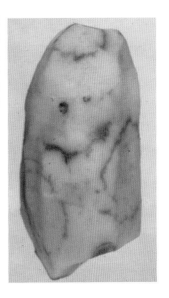

Strains

There are many strains of TRV distinguished on symptoms in test plants.

Geographic Distribution

The virus causes diseases in Europe, Japan, New Zealand, and North America.

Transmission

TRV is transmitted by nematodes (*Trichodorus* and *Paratrichodorus* spp.; see Chapter 12 for nematode transmission). It is also mechanically transmissible and seed transmitted.

Particles

The two genomic segments are each encapsidated in rigid rod (tubular) particles of length 190 nm and 50–115 nm (depending on isolate) and diameter 23 nm. Figure, bar = 100 nm. [Courtesy of *Association of Applied Biologists Descriptions of Plant Viruses*, No. 398.]

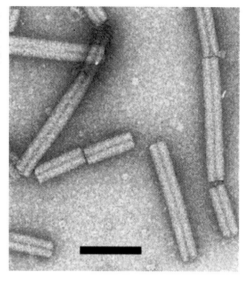

Genome and Genome Organisation

The genome is divided between two positive-sense ssRNA species (RNA1 and RNA2) each having a 5′ cap and the 3′ hydroxyl group. The size of RNA2 differs between isolates.

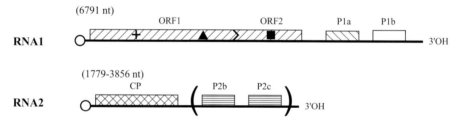

○ 5′ cap; + methyl transferase; ▲ helicase; ❭ readthrough; ■ RdRp; 3′OH = 3′ hydroxyl group. RNA1 has 4 ORFs: ORF1 reads through into ORF2 to give the replicase protein (194 kDa); the third ORF (P1a) is the 30 kDa cell-to-cell movement protein; the fourth ORF (P1b) expresses a 16 kDa protein that suppresses RNA silencing. The 5′ ORF of RNA2 of all strains encodes the coat protein; some strains have additional ORFs (P2b and P2c) that are involved in nematode transmission of the virus. P1 a and b and P2 b and c are expressed from subgenomic RNAs.

Notes

TRV causes many natural diseases in a range of dicot and monocot plants.

Further Information

MacFarlane, S.A. (2008). Tobravirus. *Encyclopedia of Virology*, Vol. 5, 72–75.
Robinson, D.J. (2003). Tobacco rattle virus. *Association of Applied Biologists Descriptions of Plant Viruses*, No. 398.

Profile No. 16

FAMILY TOMBUSVIRIDAE
GENUS TOMBUSVIRUS

Tomato bushy stunt virus (TBSV)

Classification

Type member of the genus *Tombusvirus* in the family *Tombusviridae*.

Symptoms and Host Range

 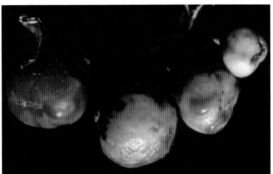

Depending on host and virus strain range from mild to severe stunting, bushy growth, necrosis, chlorotic spots, and crinkling of leaves. The left-hand panel of the Figure shows distortion and necrosis of tomato leaves; the right-hand panel shows deformation and chlorotic blotches on tomato fruit. [Courtesy of *Association of Applied Biologists Descriptions of Plant Viruses*, No. 382.] Restricted host range.

Strains

Three major strains differentiated on host range and symptoms.

Geographic Distribution

Europe, North and South America, and North Africa.

Transmission

No vector known but evidence for soil transmission. Mechanically transmissible; seed transmitted in some hosts.

Particles

Isometric 30 nm in diameter with T = 3 symmetry (see Chapter 15). The Figure shows a preparation of TBSV negatively stained in uranyl acetate; bar = 50 nm. [Courtesy of *Association of Applied Biologists Descriptions of Plant Viruses*, No. 382.]

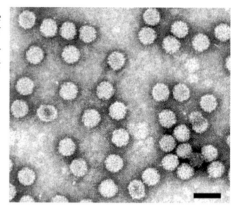

Genome and Genome Organisation

The genome is a single molecule of positive-sense ssRNA, 4.78 kb, the 3′ end of which has a hydroxyl group (see Chapter 6).

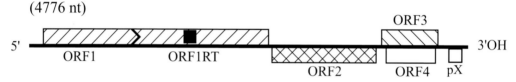

$\rangle$ = read-through of stop codon, $\blacksquare$ = RdRp motif, 3′OH = 3′ hydroxyl group. The genome encodes 5 ORFs the functions being: ORF1 viral replicase with read-through of a weak stop codon (see Chapter 7); ORF2, coat protein; ORF3, cell-to-cell movement; ORF4, suppressor of RNA silencing; pX, function unknown.

Notes

TBSV is important in studies on virus structure and virus-host interactions.

Further Information

Lommel, S.A. and Sit, T.L. (2008). Overview [Tombusviridae]. *Encyclopedia of Virology*, Vol. 5, 145–150.

Martelli, G.P., Russo, M., and Rubino, L. (2001). Tomato bushy stunt virus. *Association of Applied Biologists Descriptions of Plant Viruses*, No. 382.

Yamamura, Y. and Scholthof, H.B. (2005). *Tomato bushy stunt virus*: A resilient model system to study virus-plant interactions. *Molec. Plant Pathol.* **6**, 491–502.

GENUS *CARMOVIRUS*

Carnation mottle virus (CarMV)

Classification

CarMV is the type member of the genus *Carmovirus* in the family *Tombusviridae*.

Symptoms and Host Range

The virus causes a mild mottle and loss of vigour in carnations. Its natural host range is restricted to *Caryophyllaceae*. The experimental host range is of more than 30 species in 15 dicot families.

Strains

There are a few minor variants of CarMV distinguished on symptoms in test plants.

Geographic Distribution

Worldwide wherever carnations are grown.

Transmission

No natural vectors known. It is distributed by vegetative propagation. The virus is mechanically transmissible but not seed transmitted.

Particles

CarMV particles are icosahedral, 30 nm in diameter with T = 3 symmetry. The Figure shows a preparation of CarMV negatively stained in phosphotungstate; bar = 100 nm. [Courtesy of *Association of Applied Biologists Descriptions of Plant Viruses*, No. 7.]

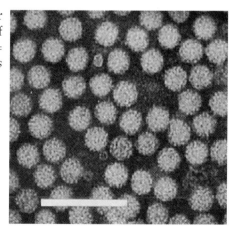

Genome and Genome Organisation

The 4 kb genome of CarMV is a single species of positive-sense ssRNA, capped at its 5′ end and with a 3′ hydroxyl group.

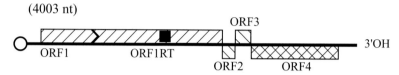

(4003 nt)

○ 5′ cap; ⟩ read-through; ▪ RdRp; 3′OH = 3′ hydroxyl group. The genome has four ORFs. ORF1 consists of two parts; the N-terminal part reading through to the C-terminal part to give the replicase. ORFs 2 and 3 are involved in cell-to-cell movement and ORF4 encodes the coat protein. ORFs 2 + 3 and ORF4 are expressed from subgenomic RNAs.

Notes

Carmoviruses such as CarMV and *Turnip crinkle virus* are important in molecular studies on simple plant viruses and also on associated satellite RNAs.

Further Information

Hollings, M. and Stone, O.M. (1970). Carnation mottle virus. *Association of Applied Biologists Descriptions of Plant Viruses*, No. 7.

Qu, F. and Morris, T.J. (2008). Carmovirus. *Encyclopedia of Virology*, Vol. 1, 453–456.

Profile No. 17

TOMATO SPOTTED WILT VIRUS (TSWV)

Classification

Type species of genus *Tospovirus* in the family *Bunyaviridae*.

Symptoms and Host Range

TSWV has a very wide host range of >800 species from >80 botanical families, both dicotyledons and monocotyledons. It also infects about 10 thrip species. The symptoms vary between and within host species and range from chlorosis, mottling, stunting, and wilting to severe necrosis of leaves, stems, and fruits. The figure shows necrotic blotching and ring-spots on tomato fruit. [Courtesy of *Association of Applied Biologists Description of Plant Viruses*, No. 412.]

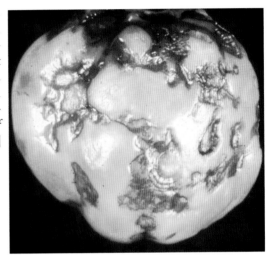

Strains

There are several strains of TSWV. Other tospoviruses have wide host ranges and can cause similar symptoms.

Geographic Distribution

TSWV is found in temperate regions worldwide.

Transmission

Transmitted by thrips in the circulative propagative manner (see Chapter 12). *Frankinella occidentalis* is an important vector especially in glasshouses. The virus is mechanically transmissible with difficulty and is not seed transmitted.

Particles

Membrane-bound quasi spherical particles 80–120 nm in diameter with surface projections 5–10 nm in length (see Chapter 5). The Figure shows a negatively stained preparation of TSWV; bar = 100 nm. [Courtesy of *Association of Applied Biologists Descriptions of Plant Viruses*, No. 412.]

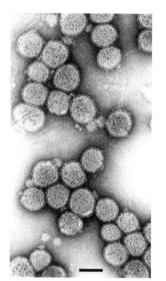

Genome and Genome Organisation

The genome is divided between three segments of ssRNA; the L segment is negative-sense 8.90 kb, the M and S segments are ambisense of 4.82 and 2.92 kb, respectively (see Chapter 7 for ambisense).

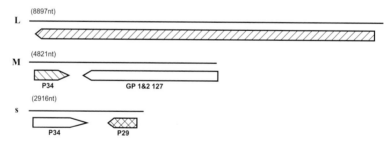

The product from the L segment is the RNA-dependent RNA polymerase (330 kDa). The P34 (NSm) from the M segment is involved in cell-to-cell movement and the GP 1 & 2 form the glycoprotein projection on the particle; the P34 (NSs) from the S segment is the suppressor of RNA silencing and the N protein forms the nucleocapsid.

Notes

TSWV is an important plant virus economically. It replicates both in plants and insect and resembles animal bunyaviruses.

Further Information

Adkins, S. (2000). *Tomato spotted wilt virus*—positive steps toward negative success. *Molec. Plant Pathol* **1**, 151–157.

Kormelink, R. (2005). Tomato spotted wilt virus. *Association of Applied Biologists Descriptions of Plant Viruses*, No. 412.

Pappu, H.R. (2008). Tospoviruses. *Encyclopedia of Virology*, Vol. 5, 157–162.

Profile No. 18

TURNIP YELLOW MOSAIC VIRUS (TYMV)

Classification

Type species of the genus *Tymovirus* in the family *Tymoviridae*.

Symptoms and Host Range

Causes a bright yellow to white or yellow green mosaic of leaves of Chinese cabbage (*Brassica pekinensis*). Limited host range mainly of *Cruciferae*. The Figure shows a white mosaic in turnip.

Strains

There are several strains of TYMV that differ in symptom severity and serologically.

Geographic Distribution

Europe

Transmission

TYMV is transmitted in a noncirculative manner by beetles. The virus is mechanically transmissible but not seed transmitted.

Particles

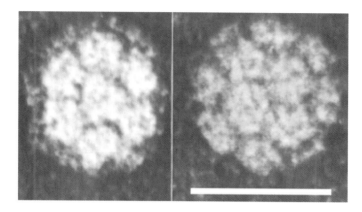

TYMV has icosahedral particles, about 30 nm in diameter with T = 3 symmetry. The Figure of particles negatively stained in uranyl acetate shows morphological subunits at the 5-fold axis (left-hand panel) and 2-fold axis (right-hand panel); bar = 30 nm. [Courtesy of *Association of Applied Biologists Descriptions of Plant Viruses*, No. 230.] The particles are made up of one coat protein species and are very stable, some particles containing the single genome species and others being "empty."

Genome and Genome Organisation

The genome is a single molecule of positive-sense ssRNA, 6.32 kb with a 5′ cap and 3′ tRNA-like structure.

○ 5′ cap; + methyl transferase; ◯ protease; ▲ helicase; ■ RdRp; ⚲ tRNA-like 3′ end. The genome has three ORFs, ORF0 encoding a protein of 69 kDa that is involved in cell-to-cell movement and silencing suppression; ORF1, a 206 kDa replicase protein that is processed between the helicase and RdRP domains by the protease domain to give products of 141 and 66 kDa; ORF2 encodes the 20 kDa coat protein.

Notes

Study of the full and empty particles of TYMV led to the recognition of an infectious RNA genome. Its replication involves the chloroplast membrane.

Further Information

Dreher, T.W. (2004). *Turnip yellow mosaic virus*: Transfer RNA mimicry, chloroplasts and a C-rich genome. *Molec. Plant Pathol.* **5**, 367–376.
Haenni, A.-L. and Dreher, T.W. (2008). Tymoviruses. *Encyclopedia of Virology*, Vol. 5, 199–208.
Matthews, R.E.F. (1980). Turnip yellow mosaic virus. *Association of Applied Biologists Descriptions of Plant Viruses*, No. 230.

Index

Note: Page numbers followed by *f* refer to figures; page number followed by *t* refer to tables; and *b* refer to boxes.